P9-AFF-178

THE FOUNDATIONS OF EDUCATIONAL EFFECTIVENESS

Titles of Related Interest

REYNOLDS *et al.*
Advances in School Effectiveness: Research and Practice

ELLEY
The IEA Study of Reading Literacy

NEVO
School-Based Evaluation: A Dialogue for School Improvement

Journals of Related Interest

International Journal of Educational Research
Teaching and Teacher Education

THE FOUNDATIONS OF EDUCATIONAL EFFECTIVENESS

by

Jaap Scheerens & Roel J. Bosker

Pergamon

U.K.	Elsevier Science Ltd, The Boulevard, Langford Lane, Kidlington, Oxford OX5 1GB, U.K.
U.S.A.	Elsevier Science Inc., 665 Avenue of the Americas, New York, NY 10010, U.S.A.
JAPAN	Elsevier Science Japan, Higashi Azabu 1-chome Building 4F, 1-9-15, Higashi Azabu, Minato-ku, Tokyo 106, Japan

First edition 1997

Library of Congress Cataloging in Publication Data
Scheerens, J. (Jaap)
The foundations of educational effectiveness/by Jaap Scheerens & Roel J. Bosker.—1st ed.
p. cm
Includes bibliographical references (p.) and index.
ISBN 0 08 042769 3
1. Educational evaluation. 2. Educational productivity
I. Bosker, (Roel J.) II. Title.
LB2882.75.S293 1997
379.1'58—dc21

British Library Cataloguing in Publication Data
A catalogue record for this book is available from the British Library

ISBN 0 08 042769 3

Printed and bound in Great Britain by Redwood Books Ltd

Contents

Preface		**vii**
PART ONE: FOUNDATIONS		**1**
Chapter One	The Conceptual Map of School Effectiveness	3
Chapter Two	Modeling Educational Effectiveness	35
Chapter Three	Size, Stability, Consistency and Scope of School Effects	69
PART TWO: THE KNOWLEDGE BASE		**97**
Chapter Four	The Meaning of the Factors that are Considered to Work in Education	99
Chapter Five	The Knowledge Base on Effectiveness-Enhancing Conditions, Part 1: Qualitative Reviews	139
Chapter Six	The Knowledge Base on Effectiveness-Enhancing Conditions, Part 2: Quantitative Research Syntheses	211
Chapter Seven	An International Comparative School Effectiveness Study Using Reading Literacy Data	239
PART THREE: REDIRECTION		**263**
Chapter Eight	Theories on School Effectiveness	265
Chapter Nine	Redirection of School Effectiveness Inquiry	299
References		323
Author Index		341
Subject Index		343

PREFACE

Do schools really differ with respect to their impact on pupils? How big is this effect? Is it the same for all subject domains? How can it be assessed? What are the factors that cause the effect? And can this effect be found in all types of educational contexts, such as both in primary and in secondary schools, or in various countries around the world? What theoretical explanations are behind research findings in this area? These were just some of the questions that puzzled us when starting the research program on school effectiveness at the University of Twente. Almost a decade later we felt that we had to make up the balance, if only because every now and then we asked ourselves the question: is the idea of school effectiveness not a myth? Is this "much ado about nothing", to quote Shakespeare? Then should we not be at the front to demystify it? This was actually the driving force when we started to write, summarizing the work that was done so far. At least we satisfied our needs, and came up with an answer. Since we are continuing our work in this area the reader may be able to guess what the answer might be.

Many have contributed to this book in one way or another, and of course we are very grateful to these people. We especially would like to thank Hennie Brandsma, Aletta Grisay, Peter Hill, Ken Rowe, Pamela Sammons and Sally Thomas, for their valuable comments on Chapter Five of this book, where we describe their, as we view it, exemplary sound research into school effectiveness.

Furthermore, as stated before, the content of this book is the result of a joint effort made by all those working in the research program. In that respect we should mention the contributions made by Hans Luyten's work on consistency and stability of school effects, which helped us enormously in writing Chapter Three. Moreover, without the joint work with Bob Witziers on meta-analyses in the field of school effectiveness, viable parts of this book (especially parts of Chapters Two and Six) could never have been written. Maria Hendriks reanalyzed questionnaires and other survey instruments, and this analysis served as part of the input to Chapter Four. We would also like to thank Lisenka Alblas and Carola Groeneweg for the administrative support in writing the book.

The book is written by the two of us, and is thus a joint effort. Nevertheless, as is suggested by the order of our names, different parts of the book were more the responsibility of one author. Jaap Scheerens initiated the scope and outline of chapters of the book, and is mainly responsible for the text of Chapters One, Four, Five and Eight. Roel Bosker is mainly responsible for Chapters Three and Seven, Chapters Two, Six and Nine are jointly written.

PART ONE: FOUNDATIONS

CHAPTER ONE

THE CONCEPTUAL MAP OF SCHOOL EFFECTIVENESS[1]

Introduction

What is the meaning of the statement that a school is "effective"? In educational discussion the term effective is often associated with the quality of education. Some authors (Corcoran, 1985) give an even broader meaning of the word by speaking of the general "goodness" of a school. Other concepts which, rightly or wrongly, are used as synonyms for effectiveness include efficiency, productivity and the survival power of an organization.

It is clear that a more precise definition is required. Moreover, we also run up against the problem that effectiveness is defined differently according to various disciplines. In this chapter economic, organization-theoretical and pedagogical definitions will be considered. The economic and organization-theoretical definition of school effectiveness will be considered in consecutive sections of this chapter. Particular emphasis will be placed on the organization-theoretical approach since this book focusses on the functioning of schools as organizations. The pedagogical definition of school effectiveness will be discussed under the heading of the underlying model of school effectiveness research.

The organization-theoretical analysis points out that effectiveness may be defined in terms of various criteria and that a range of elements and aspects of schooling may be chosen as point of impact of action aimed at increased performance. These elements and aspects of schooling are labeled "modes of schooling". Control-theoretical and contingency-theoretical perspectives add to the overall pluriformity and contextual dependence of effectiveness

[1]Parts of this chapter are an updated version of Chapter 1 of Scheerens, J. (1992). *Effective schooling. Research theory and practice*. London: Cassell, plc. This material is used with permission of the publisher.

phenomena. The broad conceptual framework on school organizational effectiveness that is developed by considering diverging criteria, a range of modes of schooling and contextual constraints is compared to the implicit model of current school effectiveness research. Apart from providing clarity of definitions of school effectiveness the comparison of the broad conceptual framework with the implicit model of school effectiveness has the following objectives, also with regard to the interpretation of the material that is presented in subsequent chapters of this book:

1. to indicate implicit choices in the predominant approach in empirical school effectiveness research;
2. to indicate possible "white spots" and underdeveloped areas that may be used in redirecting future empirical school effectiveness research;
3. to provide an evaluative framework to assess the relevance of the school effectiveness research results for school improvement.

Economic Definitions of Effectiveness

In economics, concepts such as effectiveness and efficiency are related to the production process of an organization. Put in a rather stylized form, a production process can be summed up as a turnover or transformation of inputs to outputs. *Inputs* of a school or school system include pupils with certain given characteristics and financial and material aids. *Outputs* include pupil attainment at the end of schooling. The transformation *process* or *throughput* within a school can be understood as all the instruction methods, curriculum choices and organizational preconditions which make it possible for pupils to acquire knowledge. Longer term *outputs* are denoted with the term *outcomes* (see Table 1.1).

Table 1.1 Analysis of factors in the education production process

Input	Process	Output	Outcome
Funding	Instruction methods	Final primary school test scores	Dispersal on the labor market

Effectiveness can now be described as the extent to which the desired level of output is achieved. *Efficiency* may then be defined as the desired level of output against the lowest possible cost. In other words, efficiency is effectiveness with the additional requirement that this is achieved in the cheapest possible manner. (For variations on the meaning of efficiency refer to Boorsma & Nijzink, 1984, pp. 17–21; Windham, 1988.) Cheng (1993) offers a further elaboration of the effectiveness and efficiency definitions, by incorporating the dimension of short-term output versus long-term outcomes. In his terms: *technical* effectiveness and technical efficiency refer to "school outputs limited to those in school or just after schooling (e.g. learning behavior, skills obtained, attitude change, etc.)". *Social* effectiveness and efficiency are associated with "effects on the society level or the life-long effects on individuals (e.g. social mobility, earnings, work productivity)" (Cheng, 1993, p. 2). When crossing these two dimensions four types of school output are discerned (see Table 1.2).

It is vitally important for the economic analysis of efficiency and effectiveness that the value of inputs and outputs can be expressed in terms of money. For determining efficiency it is necessary that input costs such as teaching materials and teachers' salaries are known. When the outputs can also be expressed in financial terms efficiency determination is more like a cost–benefit analysis (Lockheed, 1988, p. 4). It has to be noted, however, that a strict implementation of the above-mentioned economic characterization of school effectiveness runs up against many problems.

These problems begin with the question of how the desired output of a school should be defined, even if one concentrates on the short-term effects. For instance, the production or returns of a secondary school can be measured by the number of pupils who successfully pass their school-leaving diploma. The unit in which production is measured in this way is thus the pupil having passed his or her final examination. Often, however, one will want to establish the units of production in a finer way and will want to look, for instance, at the grades achieved by pupils for various examination subjects. In addition,

Table 1.2 Distinction between school effectiveness and school efficiency (from Cheng, 1993, p. 4)

	Nature of school output	
Nature of school input	In school/Just after schooling Short-term effects Internal (e.g. learning behavior, skills obtained)	On the society level Long-term effects External (e.g. social mobility, earnings, productivity)
Non-monetary (e.g. teachers, teaching methods, books)	School's technical effectiveness	School's societal effectiveness
Monetary (e.g. cost of books, salary, opportunity costs)	School's technical efficiency (internal economic effectiveness)	School's societal efficiency (external economic effectiveness)

there are all types of choices to be made with regard to the scope of effectiveness measures. Should only performance in basic skills be studied; or should the concern also be with higher cognitive processes, and should social and/or affective returns on education be established? Other problems related to economic analysis of schools are the difficulty in placing a monetary value on inputs and processes and the prevailing lack of clarity on how the production process operates (precisely what procedural and technical measures are necessary to achieve maximum output).

Relevant to the question of how useful one regards the characterizion of effectiveness to be in economic terms is the acceptability of the school as a metaphor for a production unit.

Organization-Theoretical Views on Effectiveness

Organizational theorists often adhere to the thesis that the effectiveness of organizations cannot be described in a straightforward manner. Instead, a pluralistic attitude is taken with respect to the interpretation of the concept in question. By that it is assumed that which interpretation will be chosen depends on the organization theory and the specific interests of the group posing the question of effectiveness (Cameron & Whetten, 1983, 1985; Faerman & Quinn, 1985). The main perceptions on organization which are used as background for a wide range of definitions on effectiveness will be briefly reviewed.

Economic rationality

The economic description of effectiveness mentioned earlier is seen as deriving from the idea that organizations function rationally, that is to say, purposefully. Goals that can be operationalized as pursued outputs are the basis for choosing effect criteria, which are the variables by which effects are measured (i.e. student achievement, well-being of the pupils, etc.). There is evidence of economic rationality whenever the goals are formulated in the sense of outputs of the primary production process of the school. In the entire functioning of a school other goals can also play a part, for instance, having a clear-cut policy with regard to increasing the number of enrollments. Also, with regard to this type of objective, a school can operate rationally; however, this falls outside the specific interpretation given to economic rationality. Effectiveness as defined in terms of economic rationality can also be identified as the productivity of an organization. In education the rational or goal-oriented model is mainly propagated via Tyler's model, which can be used for both curriculum development and educational evaluation (Tyler, 1950). From the remaining perceptions on organization, to be discussed shortly, the economic rationality model is dismissed as being both simplistic and out of reach. It is well known in the teaching field how difficult it is to reach a consensus on goals and to operationalize and quantify these goals. From the position that other values besides productivity are just as important for organizations to function, the rational model is regarded as simplistic.

The organic system model

According to the organic system model, organizations can be compared to biological systems which adapt to their environment. The main characteristic of this approach is that organizations openly interact with their surroundings. Thus, they need in no way be passive objects of environmental manipulation but can actively exert influence on the environment themselves. Nevertheless, this viewpoint is mainly preoccupied with an organization's survival in a sometimes hostile environment. For this reason, organizations must be flexible, namely to assure themselves of essential resources and other inputs. According to this viewpoint flexibility and adaptability are the most important conditions for effectiveness in the sense of survival. A result of this could be that the effectiveness of a school is measured according to its yearly intake, which could partly be attributed to intensive canvassing or marketing of the school.

No matter how remarkable this view on effectiveness may seem at first glance, it is nevertheless supported by an entirely different scientific sphere — microeconomics of the public sector. Niskanen (1971) demonstrated that public sector organizations are primarily targeted at maximizing budgets and that there are insufficient external incentives for these organizations, including schools, to encourage effectiveness and efficiency. In this context it is interesting to examine whether canvassing activities of schools mainly comprise displaying acquired facilities (inputs) or presenting output data such as previous years' examination results.

Finally, it should also be mentioned that it is conceivable that the inclination towards inputs of the organic system model coincides with a concern for satisfying outputs, namely in those situations where the environment makes the availability of inputs dependent on quantity and/or quality of earlier realized achievements (output).

The human relations approach of organizations

If in the open system perception of organizations there is an inclination towards the environment, with the human relations approach the eye of the organization analyst is explicitly focussed inward. This fairly classical school of organizational thought has partly remained intact even in more recent organizational characterization. In Mintzberg's concept of the professional bureaucracy, aspects of the human relations approach reoccur, namely in emphasizing the importance of the well-being of the individuals in an organization, the importance of consensus and collegial relationships as well as motivation and human resource development (Mintzberg, 1979). From this perception, job satisfaction of workers and their involvement with the organization are likely criteria for measuring the most desired characteristics of the organization. The organizational theorists who share this view regard these criteria as effectiveness criteria.

The bureaucracy

The essential problem with regard to the administration and structure of organizations, in particular those that, like schools, have many relatively autonomous subunits, is how to create a harmonious whole. Appropriate social interaction and opportunities for personal and professional development (see *The human relations approach*) provide a means for this. A second means is provided by organizing, clearly defining and formalizing these social relations. The prototype of an organization in which positions and duties are formally organized is the "bureaucracy". According to this perspective, certainty and continuity concerning the existing organization structure form the effectiveness criterion. Bureaucratic organizations are well known for their tendency to produce more bureaucracy. The underlying motive behind this is to ensure the continuation, or better still, the growth of one's own department. This continuation can start operating as an effect criterium in itself.

The political model of organizations

Certain organizational theorists see organizations as political battlefields (Pfeffer & Salancik, 1978). Departments, individual workers and management staff use the official duties and goals in order to achieve their own hidden, or less hidden, agendas. Good contacts with powerful outside bodies are regarded as very important for the standing of either the department or the individual. From a political perspective it is difficult to determine the effectiveness of the organization as a whole. The interest is more in the extent to which internal groups succeed in complying with the demands of certain external interested parties. In the case of schools these bodies could be school governing bodies, parents of pupils and the local business community.

It has already been mentioned that organizational concepts on effectiveness depend not only on theoretical answers to the question of how organizations are pieced together but also on the position of the factions posing the effectiveness question. On this point, there are differences among these five views on

Table 1.3 Organizational effectiveness models

Theoretical background	Effectiveness criterion	Level at which the effectiveness question is asked	Main areas of attention
(Business) economic rationality	Productivity	Organization	Output and its determinants
Organic system theory	Adaptability	Organization	Acquiring essential inputs
Human relations approach	Involvement	Individual members of the organization	Motivation
Bureaucratic theory; system members theory; social psychological homeostatic theories	Continuity	Organization + individual	Formal structure
Political theory on how organizations work	Responsiveness to external stakeholders	Subgroups and individuals	Interdependence, power

organizational effectiveness. With regard to the economic rationality and the organic system model, the management of the organization is the main actor posing the effectiveness question. As far as the other models are concerned, department heads and individual workers are the actors that want to achieve certain effects. The chief characteristics of the organization-theoretical perceptions on effectiveness are summarized in Table 1.3.

The diversity of views on effectiveness taken by organizational theory leads to the question of which position should be taken: should one operate from a position of there being several forms of effectiveness, should a certain choice be made, or is it possible to develop from several views, one all-embracing concept on effectiveness?

Options for dealing with multiple criteria of organizational effectiveness

There are various possibilities for dealing with the reality of multiple criteria of organizational effectiveness:

1. to make a preference ordering, and designate one of the criteria as essential or ultimate (Scheerens, 1992);
2. to treat the criteria as competing values (Faerman & Quinn, 1985);

3. to use a hierarchical ordering of the criteria and place them in a dynamic framework of application (Cheng, 1993).

Means–goal ordering of criteria

Scheerens (1992) proposes a means–goal ordering of criteria of organizational effectiveness, whereby productivity is seen as the ultimate criterion and the other criteria are seen as supportive conditions. His formulation raises the possibility of assessing investment in one of the supportive criteria as inefficient or counterproductive when, for instance, far too many resources are required than is strictly needed for efficient production. He sees this as just another instance of "goal displacement" (Etzioni, 1964).

This means–goal character of the various effectiveness perceptions is illustrated in Fig. 1.1, in which the arrows pointing in both directions express the fact that satisfaction, solidarity and the oneness of structures can be seen as both the cause and the effect of high productivity. It is not implausible that staff of a productive organization are more satisfied than the personnel of one which is hardly productive.

Placing the relationship pattern of Fig. 1.1 within an educational context, one can see as an example of the criterium adaptability specific measures, instruments and organizational forms to make the school curriculum more relevant to the needs of the labor market. Another example is to have an organizational structure that can withstand drastic changes in the direct environment such as externally imposed increases in scale (Gooren, 1989). An example of acquiring vital resources could be machines that are needed for technical education and which can sometimes be obtained through contacts with local industry. Examples of process-support criteria like solidarity, motivation and continuity are: working on a shared vision of

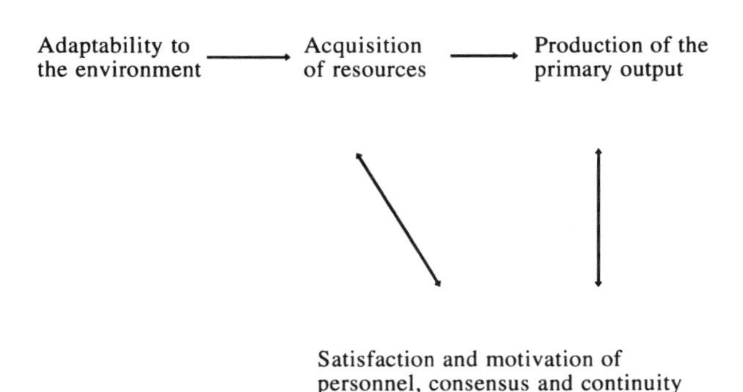

Figure 1.1 Means–goal relationship between effectiveness criteria.

education within a teaching team, working with specialist departments and their accompanying consultative bodies and a certain amount of decision making delegated from the school management to the teaching staff.

The criteria as competitive values

An integral harmonious model of alternative effectiveness criteria was discussed in the previous section. It has already been stated that the whole thing becomes unstuck when the means become goals in themselves. (For further information on this phenomenon of "goal displacement" see Etzioni, 1964, pp. 10–12.) One can also see the various alternative criteria in competition with one another, particularly when one does not want to choose between effectiveness criteria and support conditions. This viewpoint has been elaborated upon by Faerman and Quinn (1985). They emphasize the fact that different organizations vary in the extent to which they focus on productivity, adaptability to the environment, continuity and solidarity. The first two categories of criteria express an external orientation, while the second two have an internal inclination. The more energy spent on one or two of these criteria, the less time can be automatically spent on the others. An interesting elaboration of this competing values framework is the assumption that the pattern of priorities among these criteria changes during the course of an organization's development. Young organizations would be largely externally oriented, to obtain vital resources, while more established organizations would be more concerned with consensus and internal stability. Ultimately, more of a balance could exist, whereby productivity and acquiring input again receive slightly more emphasis than the internal support values.

The competing values perspective can be linked with the means–goal perspective by the assumption that the support conditions (see Fig. 1.1) could reach certain optimum values, whereby too much emphasis on consensus, for instance, would be literally counterproductive.

Dynamic application of the criteria

Cheng (1993), following Cameron (1984), uses a related but more elaborate set of "models of organizational effectiveness" than the four basic models chosen herein. The essence of these models is summarized in Table 1.4. Cheng emphasizes that schools have to deal with multiple constituencies and unbalanced pressures from environmental constraints in which there may also be competition for giving priority to a certain effectiveness criterion. He takes the capacity of schools, in the long run, to deal with these conflicting pressures and "unbalances" as the key characteristic of school effectiveness: "... school effectiveness is the extent to which a school can adapt to the internal and external constraints and achieve the multiple goals of its multiple constituencies in the long run" (Cheng, 1993, p. 17).

From this position Cheng recognizes a hierarchy in the various models of organizational effectiveness, with the organizational learning model as the one that comes closest to his definition of school effectiveness. His approach can be seen as a further elaboration of Faerman and Quinn's idea to gear different effectiveness criteria to developmental stages of the organization.

Despite the diversity of criteria of organizational effectiveness, and the various ways in which one could see the priority of some over others, in this book, whenever we speak of school effectiveness, we will mean the productivity criterion of effectiveness.

Table 1.4 Characteristics of evaluation models of school effectiveness (from Cheng, 1993, p. 11)

Model	Definition of school effectiveness	When the model is useful
	A school is effective if …	The model is useful when …
Goal model	it can achieve its stated goals	goals are clear, consensual, time-bound and measurable
System-resource model	it can acquire needed resources and inputs	there is a clear relationship between inputs and outputs
Internal process model	the schooling process can be smooth and healthy	there is a clear relationship between school process and outcomes
Strategic-constituencies model	all of the powerful constituencies are at least minimally satisfied	the demands of the powerful constituencies are compatible and cannot be ignored
Legitimacy model	it can survive as a result of engaging in legitimate activities	the survival and demise of schools must be assessed
Organizational learning model	it can learn to deal with environmental changes and internal barriers	the schools are new and developing, or the environmental change cannot be ignored
Ineffectiveness model	there is an absence of characteristics of ineffectiveness	there are no consensual criteria of effectiveness, or strategies for school improvement are needed

Modes of Schooling as Points of Impact for Attaining Effectiveness

In the previous section it was established that the overall concept of school effectiveness can be differentiated according to normative criteria related to various schools of thought in organizational science.

The concept of school effectiveness may be divided, according to a very general framework, into a domain of *effects* and a domain of *causes* or *means*, and these domains may be differentiated further. In doing so the question that is dealt with concerns the distinction of all possible features of the functioning of schools that are malleable in order to reach the effects that are aimed for. Such a broad perspective is needed to obtain as complete a picture as possible on elements and aspects of schooling and school functioning that are potentially useable in improving effectiveness. In this way it will also be possible, in subsequent chapters, to assess the coverage of the bulk of school effectiveness research with respect to the complete set of potential points of impact of effectiveness-enhancing measures.

According to well-known distinctions in organizational science (e.g. Mintzberg, 1979; de Leeuw, 1986), the following categories can be used as a core framework to distinguish further the elements and aspects of school functioning (referred to here as modes of schooling):

- goals
- the structure of positions or subunits (*Aufbau*)
- the structure of procedures (*Ablauf*)
- culture
- the organization's environment
- the organization's primary process.

Goals

In organizational effectiveness thinking goals can be seen as the major defining characteristic of the effectiveness concept itself. In the previous section it was established that different goal areas, or effectiveness criteria, can be used to apply effectiveness assessment operationally.

When goals are not taken as given in effectiveness assessment, but as options or directions that the organization can choose, this further emphasizes the relativity of the organizational effectiveness concept. In fact, the question of whether an organization chooses the "right" goals or objectives can be seen as a fundamental question that precedes the question of instrumental rationality, concerning the attainment of given objectives. In this respect the well-known distinction between "doing the right things" and "doing things right" is at stake. In its turn the question of the "rightness" of a particular choice of organizational goals can be seen as instrumental to meeting the demands of stakeholders in the external environment of the organization. In the case of schools, for instance, these may be demands from the local community or from parents' associations.

Further options of choice with respect to goals are:

- various priorities in further specification of the overall goals (in the case of schools, for instance, the relative priority of cognitive versus non-cognitive objectives and the relative emphasis on basics versus other subjects);
- the levels or standards of goal attainment that are strived for: to the degree that schools are relatively autonomous they may set absolute standards that every pupil should achieve or they may adapt achievement standards to the initial achievement level of pupils;
- whether or not attainment levels are differentiated for different subgroups of pupils.

Finally, an assignment of organizations is to ensure that goals or attainment targets are shared among the members of the organizations. This is particularly relevant for organizations like schools, where teachers traditionally have a high degree of autonomy. In control theory the phenomenon of unifying the goals of organizational subunits (i.e. departments and individual teachers, in the case of schools) is known as goal coordination.

The structure of positions (Aufbau)

The position structure of the organization is defined as the complete set of static relationships of functions and units within the different levels of the organization (van Kesteren, 1996, p. 97).

Parts of the position structure are:

- the management structure (whether or not a school has a middle-management structure, e.g. heads of departments);
- the support structure (whether or not the school has a professional, next to a domestic support staff);
- the division of tasks and formal positions; for instance, the distinction between classroom teacher (mostly the case in primary schools) and subject-specific teacher (mostly the case in secondary schools);
- the grouping of teachers and students: the school can opt for a particular strategy in assigning teachers to subgroups of pupils (e.g. deliberately matching "strong" teachers with "weak" students, Monk, 1989); students can be in year grades, or non-graded classrooms and in ability groups within classes.

The structure of procedures (Ablauf)

The structure of procedures comprises the dynamic whole of organizational arrangements which concern the functioning of the organization over time: decision-making processes, planning procedures, policy formation and coordination. In fact, everything contained under the general label of "management" is part of the Ablauf structure.
Relevant functional management areas are:

- general management
- production management
- marketing management
- personnel management
- financial management.

General management concerns the overall control, the final responsibility and the coordination over the other functional management areas.

Production management takes care of the control of the primary production process of the organization and its technological requirements.

In the most general sense, marketing management is aimed at gearing the organization's functioning and output to the demands of customers and stakeholders. Aspects of marketing are the identification of potential customers, researching the needs of potential customers, determining the organization's position *vis-à-vis* competitors, and using all of this information to make choices with respect to client groups, products, price and presentation (advertising, information-provision).

Personnel management concerns the control of all tasks that are related to the recruitment, selection, appraisal, advancement, training and discharge of staff. More fashionable terms for this functional management area are human resources management (HRM) and human resources development (HRD).

Financial and administrative management in the most general sense concerns the control over the financial and material assets and resources of the organization.

Although some of these functional management areas are more prominent than others, all of them are relevant when considering the management of schools. Particularly when schools become more autonomous, several functional management areas gain in importance: marketing management, financial management and personnel management. General management, in the sense of coordination, is of particular interest in professional bureaucracies such as schools, where the professional workers in the operating core of the organization have a high degree of autonomy. When, at the same time, the school seeks to accomplish a general mission, it takes particular management skills to align these independent professionals, while at the same time respecting their professional autonomy.

The concept of educational or instructional leadership, which will be covered further in subsequent chapters, is to be seen as an example of production management.

It should be noted that in all of these functional management areas there are always four essential management activities present: planning, coordinating, assessing and controlling.

To the degree that the position and procedural structure of the organization are highly regulated and formalized versus more fluid and informal, the organization is called mechanistic versus organic. In its turn, the degree to which a structure is mechanistic rather than organic is considered relevant with respect to environmental circumstances. A more dynamic and complex environment requires a more flexible organic structure than a more stable and predictable environment.

Communication and cooperation among subunits and members of the organization can be seen as the result of orchestration by the management of the organization, but also as an autonomous aspect of the structure of procedures. With respect to the functioning of schools the distinction between formal and informal cooperation and communication is relevant. Traditionally, teachers were seen to operate as kings in their own classrooms, who needed little or no consultation and task-oriented communication with their colleagues. According to more recent ideas about the quality of schools, integration, communication and cooperation are emphasized. An example is the image of the "learning organization", where cooperative effort and mutual consultation form one of the key characteristics.

The organizational structure of schools in terms of both *Aufbau* and *Ablauf* clearly provides a broad area of conditions that are malleable, in principle, in order to improve effectiveness. A key question when considering practical instances of school restructuring, is to what extent such efforts are indeed targeted at effectiveness enhancement and, then, at which particular effectiveness criterion (e.g. production improvement, or improvement of the resource acquisition function) are they aimed. By examining school effectiveness research traditions and the knowledge base that has been built by means of this research, the relative success of organizational restructuring in comparison to other modes of school functioning can be assessed (cf. Wang *et al.*, 1993).

The cultural dimension

The concept of organizational culture is somewhat related to both goals and structure. It is related to goals in the sense that it is also about normative positions, not just with respect to the desired output of the organization, but also concerning the processes and interactions that are associated with being part of the organization. It is related to structure in the sense that it is also concerned with interrelationships between units and members of the organization, on the more informal next to the formal level.

Despite these connections the concept of organizational culture is usually considered to have sufficient integrity to be treated as a separate major area of organizational functioning. It is usually defined as the set of shared meanings, collective norms and views on interaction and collaboration. As such, culture is considered of great importance in providing the normative glue that holds the organization together.

Maslowski (1995), in summarizing the literature on organization-cultural aspects, distinguishes three aspects:

1. The *substance or direction* of a culture. In the school effectiveness research literature, substantive dimensions that have been emphasized are an achievement-oriented ethos, and a safe and orderly climate.
2. The *homogeneity* of the culture, that is the degree to which the organizational culture is shared among the members of the organizations. This aspect of organizational culture is of particular relevance to educational organizations, which have been described as "loosely coupled", "professional bureaucracies" and even "organized anarchies". Although there are some important general factors such as common training and a relatively stable tool base of skills inherent in the concept of the professional bureaucracy, strengthening the consistency of practice and the cohesion and consensus among teachers is usually seen as an important factor of increasing school effectiveness.
3. The *strength* of the culture, i.e. the degree to which cultural elements more or less coercively influence the attitudes and behavior of the members of the organization. The strength of the organization's culture could be seen as a relative phenomenon. Given the supposedly loosely coupled nature of most schools some degree of strengthening of the culture will generally be seen as conducive to increased effectiveness.

It is a matter of debate whether culture is a mode of schooling that is directly or indirectly malleable. On the one hand, cultural change can be seen as the by-product of common structural arrangements or structural modification. For example, inducing more participative management as an example of restructuring may appeal to the different ways in which teachers interact, both formally and informally, with each other and with the head teacher.

Schein (1985) distinguishes several other indirect mechanisms to change organizational culture, apart from structural (re)design and common procedures: the design of buildings and interial decoration, stories and myths about the organizations, and formal agreements on principles. In addition, Schein mentions five direct mechanisms:

1 Priorities set by the organization's leader(s).
2 The leader's reaction to critical events.
3 The enactment of desired behavior.
4 The setting of norms and standards for delivering rewards and providing status.
5 Criteria with respect to hiring and firing.

Ahead of chapters in which the school effectiveness knowledge base is reviewed, it should be mentioned that a considerable number of factors that are hypothesized to be enhancing effectiveness refers to the cultural dimension, for example: a shared achievement-oriented ethos, high expectations of pupils' achievement and a safe and orderly climate. The usefulness of the cultural mode of school functioning as a malleable condition for school improvement will be revisited in Chapter Eight.

The environment

The school's environment should not just be seen in terms of a set of external conditions which have an influence on the organization's functioning, but also as a set of conditions that the school, to some extent, can and must manipulate. Even in routine functioning the school has to acquire a relatively stable stream of pupils entering the school and, at the end of the school career, generate pupils that have attained skills and knowledge that sufficiently meet the needs of parents, the society or higher forms of education. To the degree that there is competition among schools for acquiring a sufficiently large intake of pupils and to the degree that the "consumers" of education place high demands on the "products" that schools deliver, an active school policy is required to manage these aspects of the environment.

Buffering against external pressure, or protecting the school's primary processes against all kinds of disturbances is another kind of managerial activity that deals with the environment. In the Dutch educational system, for example, the head teacher has been seen as the one who filters all kinds of missives from the Ministry of Education.

Finally, the school can attempt to influence relevant constituencies, in order to enhance resource acquisition, for example, the use of technical equipment from local industries for educational purposes.

The primary process

The core transformation process of an organization, to be abstractly represented in terms of transforming inputs into outputs, is usually indicated as the primary process. An organization may have more than one primary process. Universities, for instance, have two: education and research.

In primary and secondary schools there is just one primary process, namely the process whereby cohorts of enrolling pupils are receiving instruction and tuition, in order to be able to leave the school at the end of the period of schooling endowed with sufficient additional knowledge and skills to meet the demands of the next level of schooling, the labor market or the society at large.

In education the *curriculum* can be seen as the major planning document or blueprint of the desired primary process. The curriculum, in addition to laying out choices with respect to the contents or subject matter areas to be taught, may contain guidelines for the deliverance and presentation of these contents. As such, in a general sense, curriculum documents may contain guidelines with respect to *educational technology*. Using this term in the broad sense of the knowledge base required for the transformation of inputs into outputs (not just the use of information and communication hardware and software), the technology used in schools may be further qualified.

Dimensions that are used to qualify technology in general are: uncertainty, interdependence and complexity.

Uncertainty depends on variations in inputs, the frequency of exceptions as opposed to routine procedures in the transformation process, and diversity in the output that must be produced. When these criteria are applied to schooling, a first observation must be that a school has certain options to "make things easier or more difficult". The key dimension here is the level of aspiration of schools to deliver individualized instruction, that is the degree to which schools make more or less of individual differences between pupils on intake and of the degree to which instruction is differentiated. In practice there will be strong impulses to limit uncertainty in educational technology with respect to all of these areas (inputs, transformation process and outputs). The usual practice is that groups of pupils are united according to age level and that differentiation is limited to three ability levels in a classroom at most (Slavin, 1996). Output variation is strongly limited by external regulations such as examination requirements and the demands of higher educational levels.

Different levels of interdependence are pooled, sequential and reciprocal interdependence (Thompson, 1967). Pooled interdependence is the least demanding type and merely requires a shared use of resources and facilities. Sequential interdependence implies that there is a fixed sequence in the production process. Sequential interdependence is clearly evident in schools where pupils pass from one grade to the next. Curriculum alignment, in the sense that the teacher in a particular grade level takes into account the content covered in the previous grade, is therefore an important requirement in schools.

Reciprocal interdependence is the most demanding type of interdependence, where there is a cycle of interdependence, B depends on A, but in the next phase A's intervention depends on B's reaction. An example in the field of education would be Reading Recovery programs, where pupils are regularly tutored on a one-to-one basis if they lag behind.

The complexity of teaching is determined by the number of tasks and role specifications and the number of locations where the work is to be carried out. The complexity of teaching in terms of task and role specification can be considered as rather low. In secondary education where teachers teach a limited number of subjects to different grades there is a fair degree of complexity in terms of locations.

In summary, it is hard to classify unilaterally the primary process of learning and instruction in schools in terms of uncertainty, interdependence and complexity. The average situation would give rise to qualifying the process as fairly predictable and stable, with only moderate to low interdependence levels; however, much would depend on the aspiration level of the school and the educational philosophy. For example, some instructional strategies, such as direct teaching, reduce uncertainty by providing a detailed structure, whereas others, inspired by constructivism, focus on giving leeway to independent learning and in doing so allow for a larger degree of uncertainty.

It is clear that the primary process offers the most direct approach to improving productivity. The implication for school effectiveness is to seek effectiveness-enhancing conditions primarily as part of the classroom-level instructional process. As will be elaborated in subsequent chapters, instructional effectiveness forms a research tradition in itself that can be more or less integrated within the broader field of school effectiveness.

Apart from focussing at this transformation process, the primary process can also be controlled through the selection of inputs, in this case students.

School effectiveness

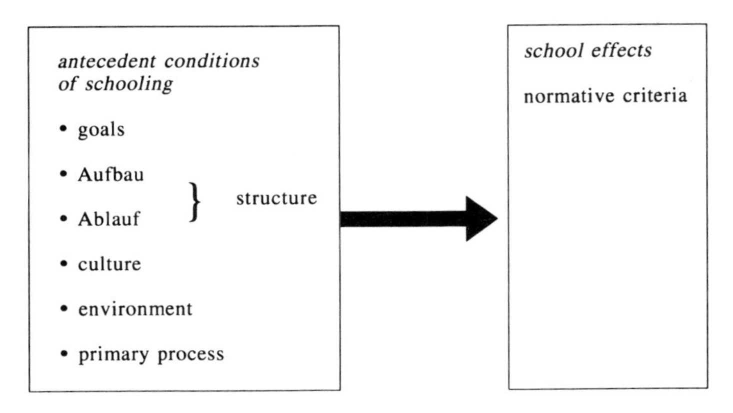

Figure 1.2 Schematic representation of school effectiveness.

Summary: the anatomy of schools as organizations

The aim of this section was to elaborate points of impact of potentially causal factors in the overall effectiveness equation, depicted schematically in Fig. 1.2. For this, a dissection of the anatomy of schools as organizations was

Table 1.5 Modes of schooling

Goals
- goals in terms of various effectiveness criteria
- priorities in goal specifications (cognitive–non-cognitive)
- aspirations in terms of attainment level and distribution of attainment
- goal coordination

Aufbau (position structure)
- management structure
- support structure
- division of tasks and positions
- grouping of teachers and students

Ablauf (structure of procedures)

• general management • production management • marketing management • personnel management (including HRM, HRD) • financial and administrative management • cooperation	planning coordinating controlling assessing

Culture
- indirect measures
- direct measures

Environment
- routine exchange (influx of resources, delivery of products)
- buffering
- active manipulation

Primary process
- curricular choices
- curriculum alignment
- curriculum in terms of prestructuring instructional process
- pupil selection
- levels of individualization and differentiation
- instructional arrangements in terms of teaching strategies and classroom organization

required. When the general domains of antecedent conditions in Fig. 1.2 are specified further, as above, the result is a list of elements and aspects of schooling that can, in principle, be actively manipulated to enhance effectiveness. (As mentioned earlier, for our purposes effectiveness is considered as productivity, unless explicitly stated otherwise.) For these elements and aspects we use the term *modes of schooling*.

In the process of school improvement choices can be made as to which mode, or combination of modes, will be selected for active manipulation. With regard to school effectiveness research and the review of the knowledge base this type of research has yielded, it becomes possible to check which modes have been investigated most often, which modes have proven to be promising areas with respect to enhancing effectiveness and which modes, until now, have been "white spots" in the field of school effectiveness research. These modes are summarized in Table 1.5.

Completion of the Conceptual Framework: Control Theoretical and Contingency Perspectives

In the previous section it was shown that there are multiple effect criteria that can be strived for by a school and multiple modes of school organizational functioning that may serve as points of impact for enhancing effectiveness.

The actual potential of a school or an external constituency to evoke certain manipulations on one or several of the modes, in order to attain certain effects, depends on three additional desiderata:

1. the possibilities for a school manager or an external change agent to actively manipulate or control internal and external conditions;
2. situational characteristics, or contingency factors;
3. instrumental knowledge about means to end relationships: which changes in the state of a certain mode of schooling lead to which (levels of) a certain type of effect?

Control theory, a specific strand within axiomatic systems theory, can be used to put the first and third points into perspective, by distinguishing the potential to control by a certain change agent and the controllability of the system. The general paradigm of contingency theory is suited to deal with the second point.

The third point also refers to the practical use that the knowledge base created by school effectiveness research might potentially have in the area of school improvement.

The control paradigm

Control is described by de Leeuw (1986) as "every form of directed influence". In order to specify schematically the situations in which control takes place, such situations are modeled as consisting of a controller (C), a system to be controlled (CS) and an environment (E).

The flow of control actions and information feedback between these components of a control situation are shown in Fig. 1.3.

The overall complexity of the control task depends on the level of ambition of the goal(s), the certainty or uncertainty of the environment and the controllability of the system to be controlled.

The *controllability* is seen as a characteristic of the system to be controlled. When a controlled system is not well integrated, for instance when teachers have divergent ideas about the pedagogic mission of their school, this can be seen as detrimental to the controllability.

The *potential for control*, in contrast, is solely a characteristic of the controller. It depends on the controller having an appropriate model of the production process in CS, (y = f(x,u)), C disposing of sufficient control measures (i) and sufficient information processing capacity in C in order to employ incoming information about the environment for an appropriate choice of intervention, also considering the goal(s).

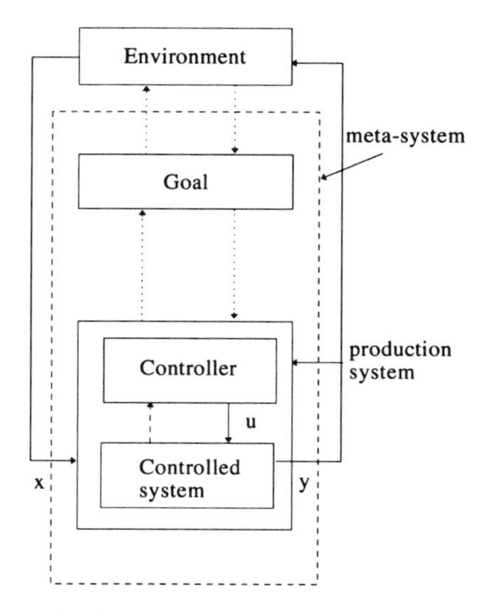

Figure 1.3 Control of a production system (after van Kesteren, 1996, p. 92). Solid arrow u: control measures; solid arrow x: influx of inputs; solid arrow y: output; striped arrow: feedback from CS to C; dotted arrow: interaction about goals.

When a controller uses more control actions than are efficient, considering limitations in the controllability of CS, de Leeuw speaks of "control fuzz". An example of control fuzz would be putting a formula 1 racing pilot in a VW Beetle or Morris Mini.

This rather sketchy introduction to some of the concepts of control theory is used to underline the following points.

1. To stress the centrality of the goal concept in organizational effectiveness thinking, both as a bridge function between environmental demands and the organization and as an indicator of the complexity of effectiveness-oriented action (e.g. with respect to ambition levels).
2. To make the point that some interventions will be more feasible than others, and that certain modes of schooling are more controllable than others. For example, ambitions towards a more flexible personal policy in schools may run up against stifling governmental regulations and/or rigid protective assurances that are the result of collective bargaining with trade unions.
3. To emphasize the open interaction of organizations with their environment and thus underline the fact that in some situations particular effectiveness criteria will be more likely to be chosen than in others. For example, when the number of enrollments drops, schools may need to invest more in resource acquisition (where pupils are considered as resources) than in a more stable situation.

When this last point is considered, it leads automatically to the contingency perspective, which goes further in seeing the effectiveness of organizational arrangements as dependent on situational factors, of which environmental complexity/predictability is just one of several contingency factors.

The contingency perspective

Contingency thinking was developed towards the end of the 1950s. Woodward's (1965) study of British companies set the stage for a new perspective on organizational functioning, shattering the established assumption that there was "one best way to organize". Situational conditions appeared to determine which method of organizing (in terms of position and procedures structure, Aufbau and Ablauf) had the best results.

The main theses of contingency theory may be briefly summarized as follows.

1. The organizational structure should fit the organization's main goal, e.g. Chandler's (1962) well-known dictum: "structure follows strategy".

2. The organizational structure should fit characteristics of the environment. Relevant environmental characteristics are complexity, predictability and (in)stability.
3. The organizational structure should fit characteristics of the technology.
4. Structural characteristics of an organization should fit among themselves. Mintzberg (1979, pp. 217–220) uses the term "configuration" for the thesis that structural characteristics should form a consistent pattern.
5. Contingency factors and structural characteristics should be consistent. Mintzberg calls this last thesis the "configuration plus" or "extended configuration" thesis (Mintzberg, 1979, p. 220). It departs from the implicit causal assumption in the first three theses in which contingency factors are the independent variables and structural characteristics the dependent variables. "Configuration plus" leaves room for the organization's shaping and modifying of contingency factors.

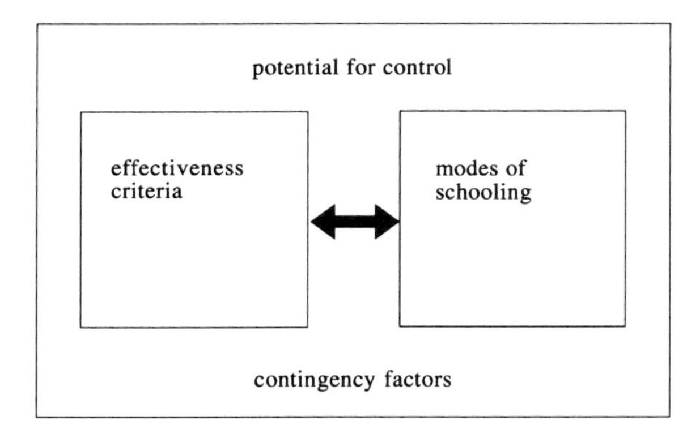

Figure 1.4 Choice of effectiveness criteria and modes of schooling (seen as points of impact of improvement-oriented action) depending on contingency factors and potential for control.

A more extensive treatment of contingency theory will be given in Chapter Eight, which investigates relevant theories to explain school effectiveness phenomena. In the present context the point to be made is that contingency thinking indicates yet another source of complexity in dealing with school effectiveness.

Contingency thinking can be used as a meta-principle to make choices with respect to effectiveness criteria and modes of schooling more or less plausible. This is schematically depicted in Fig. 1.4.

Examples of relevant contingency factors embedded in current developments in educational systems in most industrialized countries are:

1. the use of information technology in the primary process of education and training, offering new opportunities for more standardized instruction and coordination among teachers and classrooms;
2. changing patterns of governance and school autonomy, usually increasing autonomy in the financial domain, sometimes geared to increased centralization on outcome control (national assessments, accountability requirements; van Amelsvoort *et al.*, 1995); these developments require adaptations in school management, requiring more investment either in financial management or in production management;
3. reorientation of educational philosophy, with new trends emphasizing embedded cognition, general cognitive skills and independent learning inspired by a constructivist orientation to learning and instruction; these reorientations require different instructional strategies as well as managerial adaptations (Scheerens, 1994);
4. enlarging scale (school size) in some countries, which pushes effectiveness criteria associated with smooth organizational functioning more to the fore;
5. demographic tendencies, sometimes paired with market-inspired policies which lead to more intense competition between schools for pupil enrollments; such developments put a certain premium on adaptation and responsiveness as likely effectiveness criteria and emphasize marketing management as an important mode.

Comparison of the Conceptual Framework with the Underlying Model of School Effectiveness Research

Having presented a broad conceptual framework in which the pluriformity and contingency of school effectiveness orientations on situational characteristics was underlined, the predominant model used in empirical school effectiveness research will be generally introduced, to allow comparison with the conceptual framework in the final section of this chapter.

A broad outline of the underlying model of school effectiveness research

In educational effectiveness-oriented research the term "educational effectiveness" designates causal models of educational outcomes that may or may not contain school-level variables. The term "school effectiveness" is used in the more restricted sense of outcome-oriented models that explicitly contain school-level variables (Bosker & Scheerens, 1994, p. 160). An important defining characteristic of school effectiveness is that it uses an outcome measure as its criterion that is adjusted for prior achievement and/or other relevant student background characteristics. In this way the added value of schooling can be separated from overall development or innate growth of students. The supposed causal relationship between malleable conditions of schooling and (adjusted) outcomes (also to be seen as attained objectives) is another distinguishing characteristic in the operational definition of school effectiveness. Effectiveness refers to means-to-ends relationships (where ends are considered in terms of desired levels of output) that have a logical structure similar to cause–effect relationships.

Several loose ends to this definition deserve further attention. First, it should be noted that the concept of school effectiveness is "empty": in principle, any type of outcome could be used as the criterion to assess effectiveness. In practice, the most widely employed output data in school effectiveness research are test results in basic skills, such as language and arithmetic in primary schools, and in the mother tongue, foreign languages and mathematics in secondary schools. A related question is whether judgments about a school's effectiveness should be based on all possible outcomes, or on a subset, or even whether effectiveness should be considered as an outcome-specific feature of schooling. This implies that a school could be effective in one subject-matter domain, and not in the other.

Second, similar questions about the comprehensiveness versus subject or context specificity could be asked in the sense of the stability of school effectiveness over time, the consistency of effectiveness across grade levels and teachers, subgroups of students (differential effectiveness), contexts (e.g. urban or rural) and the sustained effect of primary schooling at the secondary level. Most of these issues are amenable to empirical investigations which together have been termed "foundational effectiveness studies" (Scheerens, 1993). Important progress has been made in this area (e.g. Luyten, 1994; Thomas, 1995; Goldstein & Sammons, 1995; Bosker & Witziers, 1995; Gray *et al.*, 1995). Thomas *et al.* (1995), for example, collated several of these issues in a study aimed at building "a definition of school effectiveness that encompasses a range of different outcomes and takes into account the contribution of different regional, socioeconomic and educational policy contexts". Chapter Three of this book continues to establish the state of the art in this domain.

Another peculiarity concerning the way that school effectiveness is operationally defined in empirical research in most cases is that it is a relative concept depending on naturally occurring variance in attainment levels among schools. This implies, for example, that (unadjusted) attainment levels of effective schools in one country may be lower than attainment levels in ineffective schools in another country. Effectiveness in terms of value added is defined irrespective of absolute attainment level.

Finally, it should be noted that in most cases the establishment of process–outcome relationships in school effectiveness research depends on correlational studies. Correlational relationships can be distinguished from associations between variables that can be established in experimental research. The main difference is that, strictly speaking, with correlations it is impossible to obtain an unequivocal insight into the question of whether one variable can be seen as the cause of the other, even though a certain impression can be gained. Experimental research gives more certainty on this score. The fact that most school effectiveness research is correlational creates problems in interpreting results. While it is largely assumed that school characteristics are the causes and levels of achievement the results (outputs), in some cases the reverse is true. Several school effectiveness studies have established that in schools where teachers have high expectations of pupils, the level of achievement is higher. One interpretation here could be that pupils are motivated by the positive attitude of the teachers. Another interpretation, just as plausible, is that the teachers, on the basis of knowing the levels of achievement of their pupils, have realistic expectations, whereby high expectations go hand-in-hand with high levels of achievement. In short, the high expectations characteristic can be both a cause and a result of high levels of achievement.

Position with regard to economic and organization-theoretical definitions of effectiveness

The above general outline of school effectiveness research will be compared to the economic and organization-theoretical definitions of effectiveness discussed in previous sections. Cameron and Whetten (1983) have developed a checklist to determine organization-effectiveness models. This will be applied to the general outline on school effectiveness research.

Question 1: *From whose perspective is effectiveness to be judged?*
This question draws attention to the practical implication of school effectiveness research, which is not always self-evident. This research can also be seen as an area of scientific research in which theory development is uppermost and practical application of results takes second place. Notwithstanding, most areas of school effectiveness research are narrowly linked to specific application. Political questions on education on unequal opportunity, developing compensatory education and encouraging school improvement programs are probably more important sources of inspiration for school effectiveness research than pure scientific questions. According to Ralph and Fennessey (1983), school effectiveness research is dominated by both the school improvement perspective and educational support institutions. They

mention the effective schools movement and by that imply that in adopting results of school effectiveness research there is a discernable "ideology of reform". According to them this ideological movement explains the important practical significance given to the results of school effectiveness research.

It is still difficult to answer the question of which area of the education service should make the most use of school effectiveness research. There are examples of application both within local school guidance services and for Department of Education initiatives, for instance, concerning in-depth studies attached to national assessment programs and evaluation studies.

A third category of potential users comprises the consumers of education, parents and pupils. Last but not least, schools could be inspired to use these results to improve their own practices.

Question 2: *Which area of activity within an organization determines effectiveness?*
In some of the models of organizational effectiveness covered in the preceding section the emphasis was either on the activities of the management to acquire, for instance, essential resources, or on activities directed at increasing staff motivation. In school effectiveness research the results of the primary production process — pupils' attainment levels — are mainly studied. In addition, one looks for predictors or determinants of these results (outputs). In effective schools research these determinants are defined as characteristics of

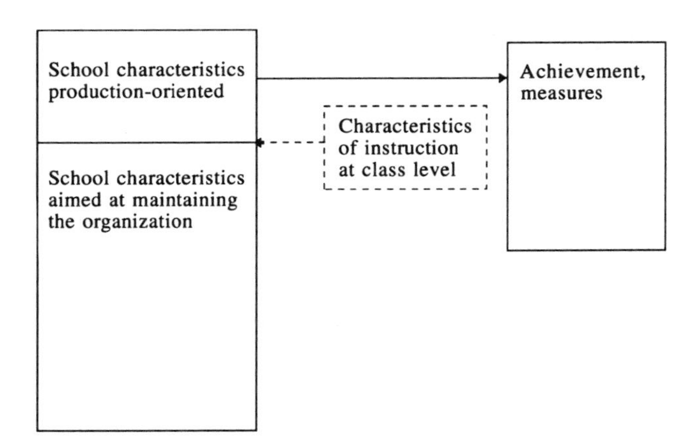

Figure 1.5 Area of activity (reproduced in terms of effectiveness determinants and criteria) of school effectiveness research. The variables in the dotted box are seen as classroom characteristics aggregated to school level. The broken arrow indicates the aggregation of instruction characteristics, measured at class level, to school level.

the school. Sometimes these characteristics, e.g. management, organization and curriculum, can be defined purely at school level. Other characteristics can be seen as the sum total of education in the various classes, i.e. classroom characteristics aggregated to school characteristics.

The main areas on which school effectiveness research focusses are shown in Fig. 1.5.

Question 3: *At which level of the organization is effectiveness analyzed?*

This question has already been answered. As a rule, school effectiveness is analyzed at school level. More, however, can be said on this question on the grounds of research technicality. The output variable in school effectiveness research, e.g. the scores of final primary school year tests, are determined for each individual pupil. School characteristics are only defined at school level. Thus the concern here is to relate variables defined at school level to those at pupil level. This has the advantage of keeping the information at the lowest aggregate level (the pupil) in the analysis, especially when one is establishing these sorts of links and wishes to correct for the background characteristics of pupils. When it is recognized that the classroom is to be seen as a crucial intermediary level between the school and individual pupils' achievement, it makes sense to include variables at this level when determining the impact of school characteristics on pupils' performance. Specific technical problems may occur, linked to relating data measured at different levels to one another, but multilevel analysis techniques provide a solution.

Question 4: *How is effectiveness defined in terms of time?*

Subsidiary factors which play a role here are (a) the frequency with which effectiveness is determined, and (b) the specific measurement moments in management and/or production processes by which effectiveness is gauged.

With regard to the first factor it is conceivable that instruments exist which keep a more-or-less permanent check on whether an organization functions well and produces sufficiently. At national level, monitoring by means of educational indicators makes such permanent quality control possible. At individual school level a school management information system fulfills this role (Essink & Visscher, 1987). With regard to the prototype of school effectiveness research, the effectiveness is determined once only.

Examples of research whereby the output of schools is determined at several points in time (longitudinal study) are scarce. Examples of studies where this is the case are given in a subsequent chapter.

The second factor, (b), concerning the point at which effectiveness is gauged, generally occurs at the end of training. According to some critics of school effectiveness research it is unreasonable to call a school effective just because the average level of achievement appears to be relatively high at the end of the last school year. Ralph and Fennessey (1983) state, for instance, that an effective school should be able to demonstrate relatively high levels of attainment in every school year. Moreover, they feel that schools only deserve the label "effective" when they have performed well over several years, in other words when school effectiveness appears as a stable factor. As stated earlier, this has given rise to a new strand of school effectiveness studies, described as "foundational studies" by Scheerens (1993), in which the scope and stability of school effectiveness comprise the main research focus.

Question 5: *What sort of data is used to form an opinion on effectiveness?*

Cameron and Whetten presented objective data as opposed to subjective data. In effective schools research objective data are mainly used to measure effect criteria, whereas in establishing determinants of effectiveness subjective opinions from those directly involved can often be decisive. In some studies use is made of expert assessments (thus essentially subjective data; e.g. Hoy & Ferguson, 1985).

Question 6: *What standards or measures are used in order to make effectiveness judgments?*

Organizational effectiveness assessments can be made by comparing similar sort of organizations, like schools, to one another (cross-sectional comparative measurement), by assessing the same organization at various points in time (longitudinal comparative measurement) or by comparing the output achieved to an absolute standard such as a prefixed target figure.
In school effectiveness research the cross-sectional comparative measurement is used most frequently. More specifically, comparisons between schools can be in the nature of (a) placing "extremely good" schools against "extremely bad" schools; (b) comparing a particular school with standard data that could be established for each country or province; or (c) comparing schools that have taken part in an improvement program with schools that have not.

An important decision to be made when using comparative measurements is the size of the difference between schools which is to be considered meaningful.

In summary, it can be established that the underlying model for school effectiveness research compared to other models for organization effectiveness can be described as a multilevel, process-product model of learning achievement propelled by the quest for knowledge of school reformers and national policy makers in which as much use as possible is made of objective data, a short-term perspective is discernible and assessment standards are largely comparative.

The problem of defining school effectiveness could have been approached more directly by simply pointing out the obvious common ground that it shares with the economic typification of effectiveness (within the broader perspective of efficiency and productivity) and with the related organizational model of economic rationality. However, a conscious choice has been made for also discussing alternative effectiveness views within the field of organization theory. This broader conceptual framework is regarded as necessary in order to reach a more considered determination of position. When considering the coverage of the modes of schooling a general observation, which runs ahead of a basis for this type of conclusion when the empirical research evidence is reviewed in subsequent chapters, is that school management, school culture and instruction are modes that have received most of the attention in empirical school effectiveness research.

Contingency thinking is not strongly represented in most empirical school effectiveness studies; however, exceptions include:

1. studies on contextual effectiveness (cf. Stringfield & Teddlie, 1990);
2. studies dealing with configuration of school management and instructional conditions (Bosker & Scheerens, 1994; Bosker, 1990b);
3. studies on differential effectiveness (Sammons *et al.*, 1995b).

Considerations about the potential for control within the context of enhancing school organizational effectiveness have not been part of empirical school effectiveness studies.

This line of thinking about organizational effectiveness is mostly relevant for considering the dynamics of school improvement and school effectiveness, e.g. in analyzing processes over time of schools becoming either more or less effective.

Summary and Conclusions

In this chapter, the process of delineation of the conceptual map of school effectiveness began by referring to economic definitions of effectiveness. Comparisons with economic definitions of effectiveness and efficiency indicated that the bulk of current empirical school effectiveness research considers the relationship between non-monetary inputs and short-term outputs: in Cheng's (1993) terminology, technical effectiveness.

Organization theoretical approaches to organizational effectiveness indicated a range of models, each emphasizing a different type of criterion to judge effectiveness, with productivity, adaptability, involvement, continuity and responsiveness to external stakeholders as the major categories. Comparison of this range of effectiveness criteria to the implicit model used in most empirical school effectiveness studies showed that the productivity criterion is the predominant criterion in actual research practice. This position can be legitimized from the point of view of a means-to-end ordering of the criteria, with productivity taken as the ultimate criterion (Scheerens, 1992). Such a position is contested, however, by other authors who see the criteria as "competing values" (Faerman & Quinn, 1985), or who opt for a more dynamic interpretation where the predominance of any single criterion would depend on the organization's stage of development (Cheng, 1993).

In recognizing that effectiveness is essentially a causal concept, in which means-to-end relationships have a meaning similar to cause–effect relationships, three major components exist in the study of organizational effectiveness:

1. the range of effects;
2. the points of impact of actions to attain particular effects (indicated as modes of schooling);
3. functions and underlying mechanisms that explain why actions impinged on certain modes lead to effect attainment.

In this chapter modes of schooling were described using the following main categories of an organization's anatomy as a basic framework:

- goals
- organization structure, with respect to both the structure of positions and the structure of procedures (including management functions)
- culture
- environment
- primary process/technology.

Each of these main categories was treated as an area that, in principle, can be manipulated or influenced by the school or an external change agent. In comparing the list of modes to the current practice of empirical school effectiveness research, it appeared that the structure of procedures (particularly school management), culture, and instructional conditions have received most of the attention.

While the discussion on effectiveness-enhancing mechanisms is postponed to a subsequent chapter, two additional perspectives were used to complete the conceptual map of school effectiveness.

Control theory draws attention to the limits of management and control, by introducing concepts such as controllability, potential for control and control fuzz (indicating that control measures may be ineffective or even counterproductive, owing to low controllability).

Contingency theory is based on the notion that there is no one best way to organize and that the effectiveness of procedural and structural arrangements depends on situational constraints, related to the goals, the environment and the technology of the organization.

Both perspectives point at additional desiderata in enhancing organizational effectiveness and also show the limitations of the practical application of research outcomes that, to some extent by necessity, abstract from intricacies associated with actual potential to take improvement-oriented action and with contextual constraints.

Van Kesteren (1996) includes most of the pluriformity of perspectives discussed in this chapter in his definition of organizational effectiveness:

> Organizational effectiveness is the degree to which an organization, on the basis of competent management, while avoiding unnecessary exertion, in the more or less complex environment in which it operates, manages to control internal organizational and environmental conditions, in order to provide, by means of its own characteristic transformation process, the outputs expected by external constituencies (translated from van Kesteren, 1996, p. 94).

Conclusions about the implications of the broad conceptual framework for the conduct and possible redirection of school effectiveness research, and the relevance of this research for educational practice, will be postponed until a closer look is taken at the state of the art of school effectiveness research, the knowledge base that it has yielded and further theoretical interpretation of the research findings. All of this will be covered in the following chapters.

CHAPTER TWO

MODELING EDUCATIONAL EFFECTIVENESS[1]

Introduction

Returning to the modes of schooling, as discussed in Chapter One, it is possible to define more specific approaches to educational effectiveness in terms of each of the modes, for example, instructional effectiveness, school cultural effectiveness, resource effectiveness. Following Scheerens and Creemers (1989), the term "educational effectiveness" is used to refer to the effectiveness of the educational system in general (comprising all modes of schooling). They use the term "instructional effectiveness" to refer to the effectiveness of education at the classroom level and describe "school effectiveness" as the effectiveness of the school as an organization and as an educational system.

In this chapter contributions to the conceptualization of effectiveness in education will be used that have focussed on different modes and school organizational levels of schooling. This is why the title of this chapter includes the term "educational effectiveness" rather than "school effectiveness".

A model specifies or visualizes complex phenomena in a simplified or reduced manner. In more abstract terms it is described in terms of a set of units (facts, concepts, variables) and a system of relationships among these units.

The modeling approach to educational effectiveness is therefore aimed at further clarification of the core elements and relationships of the more general concepts that were used in the first chapter. A distinction will be made

---

[1]The third section of this chapter is based on a previously published article: Bosker, R. J. & Scheerens, J. (1994). *International Journal of Educational Research*, 2, 159–180.

between conceptual and formal models. In the case of conceptual models only verbal descriptions and diagrams will be used as the "modeling language". Formal models consist of mathematical equations.

Advances in the modeling of educational effectiveness take place in each of the distinct research traditions in the field. Scheerens (1992, p. 31) distinguishes five of these research traditions, which will be discussed further in Chapter Three. For the purposes of this chapter it suffices to distinguish three disciplinary backgrounds to educational effectiveness modeling:

1. the economic approach, focussed on "education production functions" (Monk, 1992);
2. the educational psychological approach to effective instruction and learning conditions (Creemers, 1994);
3. the generalist–educationalist approach to integrated, multilevel school effectiveness[2] modeling (Bosker & Scheerens, 1994).

A preliminary component in modeling school effectiveness is the modeling of school effects. In the most general terms (see Chapter One) educational effectiveness models consist of a set of effect measures (A), a set of antecedent conditions that are expected to raise effects (B), and a function specifying the relationship between A and B.

As stated in Chapter One, when one is interested in the effects of schooling, seen as a set of malleable conditions, one should somehow partial out the impact of innate abilities, prior achievement and other variables describing the background of pupils. In the section on *modeling school effects*, different types of adjustment to be made in order to obtain value-added effect measures will be discussed.

In the third and final section of this chapter the elaboration and formalization of integral multilevel school effectiveness models will be dealt with.

[2]It should be noted that in the case of such integrated, multilevel models the distinction between school effectiveness and educational effectiveness models becomes rather arbitrary. In this book the term "school effectiveness" will mostly be used, since the primary interest of the authors is the school organization and management perspective.

Conceptual Models

Education production functions

Economically oriented education production function research initially focussed on estimating "the relationships between the supply of selected purchased schooling inputs and educational outcomes, controlling for the influence of various background features" (Monk, 1992, p. 308). Resource input variables such as pupil/teacher ratio, teacher salary and overall measures of per pupil expenditure were of primary interest in the earlier studies (Hanushek, 1986).

The most common equation in education production function research is as follows (Elberts & Stone, 1988, p. 293, citing Hanushek, 1979):

$$A_{it} = \mathrm{f}(B_{it}, P_{it}, S_{it}, I_i)$$

where A_{it} is the student outcome of ith student at time t, B_{it} is the vector of family background influences of the ith student cumulative to time t, P_{it} is the vector of influence of peers of the ith student cumulative to time t, S_{it} is the vector of school inputs of the ith student cumulative to time t, and I_i is the vector of innate abilities of the ith student.

The function may be linear, consisting of main effects and interaction terms (e.g. allowing for the possibility that resource inputs are differentially effective for subgroups of pupils) or non-linear. Brown and Saks (1986), for instance, use a log-linear model to assess the impact of instructional time.

The basic education production function model has the following characteristics:

1. selection of resource inputs as the major type of antecedent conditions;
2. measurement of direct, rather than causally mediated effects (in this respect, Monk speaks of "fundamentally primitive black-box formulations": 1992, p. 309);
3. data at one level of aggregation, either micro (pupil) level data or aggregated school, or even district level data;
4. it is a static, rather than a dynamic model.

The overall rationale behind these models is the assumption that increased input will lead to increments in outcomes although, as stated earlier, functions other than linear, e.g. the notion of diminishing returns, may be used to qualify this basic assumption further.

Monk (1992) discusses elaborations of the basic production function equation, which encompass the other two research traditions referred to in this section. This development shows that there is a blending of approaches in the field of educational effectiveness.

Under the heading of *Relationships between configurations of inputs and outcomes*, he explicitly refers to the effective schools research tradition and discusses studies that included school organizational conditions and teachers' perceptions.

Other approaches, which he sees as extensions of the basic education production function model, are evaluations of school improvement programs such as "Success for All" (Slavin *et al.*, 1990) and "process studies" which concentrate on economic models of decision making within the educational system (cf. Correa, 1995).

As the most promising future orientation, Monk proposes studies which focus at the classroom level and draw on the economic process studies. He discusses various hypothetical configurations which are based on either "engaged" or "accommodating" behavior of teachers. Engagement means an active teacher who enjoys the challenges of teaching a particular group of pupils. Accommodation occurs when a teacher protects himself or herself in order to be able to cope with a difficult class.

Instructional effectiveness

Instructional effectiveness or "educational productivity" models (Walberg, 1984) were initially similar to the basic education production function model discussed in the previous section. The most important differences are:

1. a different choice of antecedent conditions: in addition to "time" (a variable that is also encountered in studies of economists working in the education production function tradition), variables such as "content covered" and "quality of instruction", as well as "psychological" variables such as learning aptitudes and motivation to learn are included;
2. use of microlevel data only, i.e. data collected at the classroom level.

Other characteristics (monolevel, static, direct effects) correspond to the basic education production function model.

The Carroll model (Carroll, 1963) is usually considered as the starting point of modeling instructional effectiveness. It consists of five classes of variables that are expected to explain variations in educational achievement. All classes of variables are related to the time required to achieve a particular learning task. The first three factors are directly expressed in terms of amounts of time, while the two remaining factors are expected to have direct consequences for the amount of time that a student actually needs to achieve a certain learning task. The five classes of variables are:

1. *aptitude*: variables that determine the amount of time a student needs in order to learn a given task under optimal conditions of instruction and student motivation;

2. *opportunity to learn*: the amount of time allowed for learning;
3. *perseverance*: the amount of time a student is willing to spend on learning the task or unit of instruction.
4. *quality of instruction*: when the quality of instruction is suboptimal, the time needed for learning is increased;
5. *ability to understand instruction*, e.g. language comprehension, the learners' ability to figure out independently what the learning task is and how to go about learning it (Carroll, 1963, 1989).

The model can be seen as a general, encompassing causal model of educational achievement. In a more recent attempt to formulate an encompassing model of educational productivity (Walberg, 1984) the basic factors of the Carroll model remained intact, while an additional category of environmental variables was included. The Walberg model is shown in Fig. 2.1.

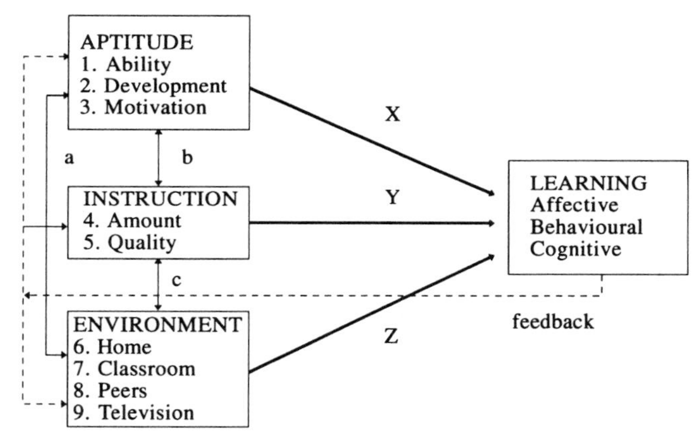

Figure 2.1 Causal influences on student learning. Aptitude, instruction, and the psychological environment are major direct causes of learning (shown as bold arrows X, Y and Z). They also influence one another (shown as arrows a, b and c), and are in turn influenced by feedback on the amount of learning that takes place (shown as broken arrows). (Source: Walberg, 1984.)

Numerous research studies and meta-analyses have confirmed the validity of the Carroll model (see Chapter Five). The Carroll model was the basis for Bloom's concept of mastery learning (Bloom, 1968) and is also related to "direct instruction", as described by Rosenshine (1983).

Characteristics of mastery learning are:

1. clearly defined educational objectives;
2. small discrete units of study;

3. demonstrated competence before progress to later hierarchically related units;
4. remedial activities keyed to student deficiencies;
5. criterion-referenced rather than norm-referenced tests (Block & Burns, 1970).

Direct instruction also emphasizes structuring the learning task, frequent monitoring and feedback and high levels of mastery (success rates of 90–100% for initial tasks) in order to boost the self-confidence of the students.

The one factor in the original Carroll model that needed further elaboration was "quality of instruction". As Carroll pointed out himself in a 25-year retrospective of his model, the original formulation was not very specific about the characteristic of high-quality instruction, "but it mentions that learners must be clearly told what they are to learn, that they must be put into adequate contact with learning materials, and that steps in learning must be carefully planned and ordered" (Carroll, 1989, p. 26).

The cited characteristics of mastery learning and direct instruction are to be seen as a further operationalization of this particular factor, which is of course one of the key factors (next to providing optimal learning time) for a prescriptive use of the model. It should be noted that Carroll's reference to the need for students to be placed in adequate contact with learning materials developed into a concept of "opportunity to learn" which differed from his own. In Carroll's original formulation, opportunity to learn is identical to allocated learning time, whereas opportunity to learn is now mostly defined in terms of the correspondence between learning tasks and the desired outcomes. Synonyms for this more common interpretation of opportunity to learn are: "content covered" or "curriculum alignment" (Berliner, 1985, p. 128). In more formal mathematical elaborations the variable "prior learning" has an important place (Aldridge, 1983; Johnston & Aldridge, 1985).

The factor "allocated learning time" has been further specified in later conceptual and empirical work. Karweit and Slavin (1982), for instance, divide allocated learning time (the clock time scheduled for a particular class) into procedural time (time spent on keeping order, for instance) and instructional time (subject matter-related instruction) and time on task (the proportion of instructional time during which behavior appropriate to the task at hand took place).

Ability to understand instruction can be seen as the basis for further elaboration in the direction of learning to learn, metacognition, etc. The comprehensiveness of the Carroll model is shown by this potential to unite two schools of instructional psychology, the behavioristically inclined structured teaching approaches and the cognitivist school (cf. Bruner, 1966; De Corte & Lowyck, 1983).

In a more recent formulation the Walberg model is tested as a structural equation model on science achievement, indicating more complex, indirect next to direct relationships (Reynolds & Walberg, 1990) (see Fig. 2.2).

The data used were from the IEA Second Science Study. Striking results are the high impact of prior achievement and the low coefficient for the direct effect of instructional quality on science achievement in grade 8.

As stated above, starting from the initial Carroll model, an important development has been to "fill in" further the black box of "quality of instruction". Making use of the principles of mastery learning and direct instruction, Creemers (1994) has proposed a more elaborate model in which three main aspects of "quality of instruction" are distinguished: curriculum, grouping procedures and teacher behavior. Each of these components contains a set of effectiveness-enhancing conditions, which is roughly similar across the three components. Creemers calls this the consistency principle: "... the same characteristics of effective teaching should be apparent in the different components. It is even more important that the actual goals, structuring, and

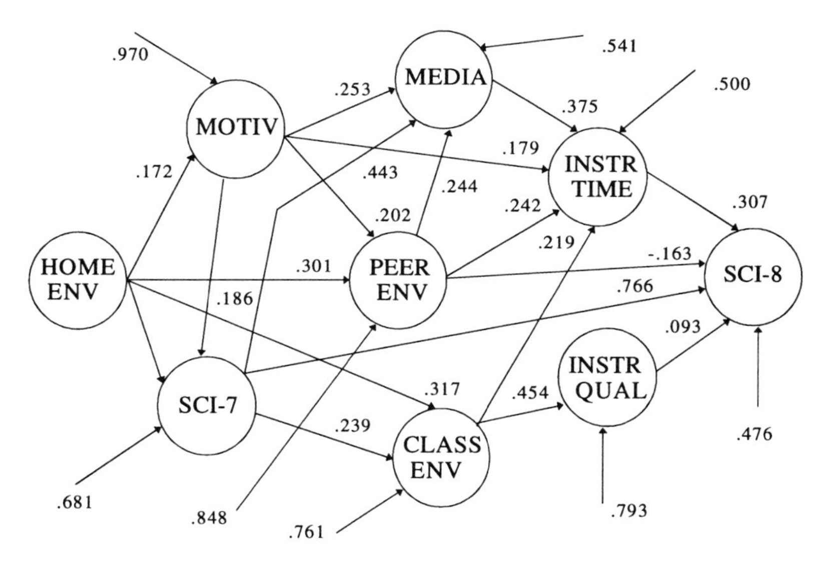

Figure 2.2 Significant (T > 2.0) standardized effects of revised model. Home env: home environment; motiv: motivation; SCI-7: grade 7 science achievement; media: mass media; peer env: peer environment; class env: class environment; instr time: instructional time; instr qual: instructional quality; SCI-8: grade 8 science achievement. Unattached arrows for each construct are unexplained variances. (Source: Reynolds & Walberg, 1990.)

evaluation in curricular materials, grouping procedures, and teacher behavior are in the same line. ... In this way a synergetic effect can be achieved" (Creemers, 1994, p. 11).

This modeling of the quality of instruction is visualized in Fig. 2.3. Structuring and the cybernetic cycle of evaluation, feedback and corrective action can be seen as the basic factors behind instructional quality in each of the three domains.

Another more recent development in modeling instructional effectiveness is the emerging new paradigm inspired by constructivism. Constructivism claims that reality is more in the mind of the knower, but does not go as far as denying external reality altogether (solipsism); however, some radical constructivists do come very close to a position of complete denial. The image of *student learning* that goes with constructivism underlines the active role of the learner. Students are to be confronted with "contextual" real-world environments, or "rich" artificial environments simulated by means of interactive media. Learning is described as self-regulated with ample opportunity for discovery and students' own interpretation of events.

Learning strategies, learning to learn and reflecting on these learning strategies (metacognition) are as important as mastering content. Different ways of finding a solution are as important as the actual solution itself. Terms such as "active learning" (Cohen, 1988), "situated cognition" (Resnick, 1987) and "cognitive apprenticeship" (Collins *et al.*, 1988) are used to describe student learning.

The other side of the constructivist coin includes approaches to *teaching* and *instructional technology* that enable students "to construct their own meaningful and conceptually functional representations of the external world" (Duffy & Jonassen, 1992, p. 11). The teacher becomes more of a coach, who assists students in "criss-crossing the landscape of contexts", looking at the concept from a different point of view each time the context is revisited (Spiro *et al.*, 1992, p. 8). Cohen (1988) uses the term "adventurous teaching" for this approach.

There is less emphasis on structuring goals, learning tasks and plans in advance; goals are supposed to emerge when situated learning takes place and plans are not so much supposed to be submitted to the learner as constructed in response to situational demands and opportunities.

Learning situations must be such that students are invited to engage in sustained exploration (real-life contents or simulated environments). Some authors writing from this perspective state that "transfer" is the most distinguishing feature (Tobias, 1991), whereas others mention argument, discussion and debate to arrive at "socially constructed meaning" (Cunningham, 1991).

The role of assessment and the evaluation of students' progress is hotly debated. Radical constructivists take the position that performance on an

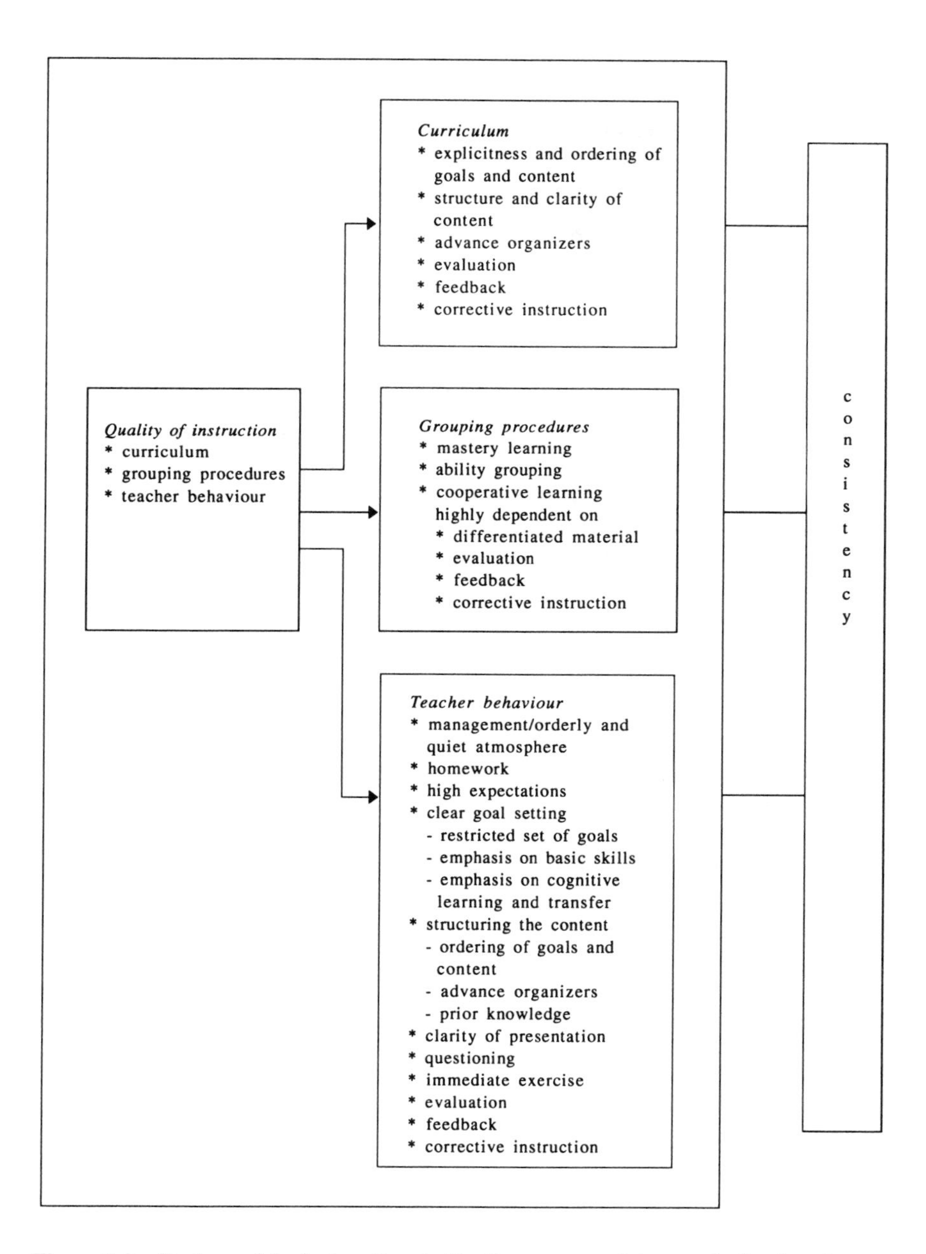

Figure 2.3 Basic model of educational effectiveness: consistency of effective characteristics and components. (Source: Creemers, 1994, p. 12.)

actual learning task is the only legitimate way of assessment, since distinct external evaluation procedures could not do justice to the specific meaning of a particular learning experience for the student.

Others (e.g. Jonassen, 1992) come to the conclusion that assessment procedures from a constructivist perspective should merely be different: goal-free, rather than fixed on particular objectives, formative rather than summative, and oriented to assessing learning processes rather than mastery of subject matter. Appraisals of samples of products, portfolios and panels of reviewers that examine authentic tasks are also mentioned as acceptable procedures.

In Table 2.1 some of the major distinguishing features of learning and instruction according to the constructivist position are contrasted with characteristics of more traditional instructional models such as direct instruction and mastery learning. Bipolar comparisons such as the one in Table 2.1 run the risk of oversimplification and polarization whilst also constructing men of straw. It should be emphasized that less extreme constructivist views can be very well reconciled with more objectivist approaches (cf. Merrill, 1991). Also, more eclectic approaches are feasible, as can be seen when more teacher-controlled and learner-controlled instructional situations are used alternately (cf. Boekaerts & Simons, 1993).

Creemers (1996) considers the changed perspective on the role of the student as the essential difference between the newer, constructivist views on

Table 2.1 Comparison of traditional and constructivistic instructional models (Source: Scheerens 1994)

Traditional instruction	Instruction inspired by constructivism
Emphasis on basic skills	Bias towards higher order skills
Subject matter orientation	Emphasis on learning process
Structured approach • prespecified objectives • small steps • frequent questioning/feedback • reinforcement through high percentage of mastery	Discovery learning Rich learning environment • intrinsic motivation • challenging problems
Abstract-generalizable knowledge	Situation-specific knowledge; learning from cases
Standardized achievement tests alternative procedures	Assessment; less circumscribed

learning and instruction, and the older models: passive in the models originating from the Carroll model, and active, picturing a student who conducts knowledge and skills through working with context, in the newer models.

Brophy (1996) also points at a way to integrate the established principles of structured classroom management and self-regulated learning strategies. Elements of effective classroom management, such as "preparation of the classroom as a physical environment suited to the nature of the planned academic activities, development and implementation of a workable set of housekeeping procedures and conduct rules, maintenance of student attention and participation in group lessons and activities, and monitoring of the quality of the students' engagement in assignments and of the progress they are making toward intended outcomes", are equally relevant when instruction is seen as helping students to become more autonomous and self-regulated learners (Brophy, 1996, pp. 3–4).

When it comes to implementing the new instructional principles, Brophy points to a guided, gradual approach where learning goals and expectations are clearly articulated, and students are helped by means of modeling and providing cues. He also stresses the fact that, initially, students may need a great deal of explanation, modeling and cuing of self-regulated learning strategies. As they develop expertise, this "scaffolding" can be reduced.

Integrated, multilevel educational effectiveness models

Since the mid-1980s a blending of approaches to educational effectiveness has taken place. This development has already been referred to in the section *Education production functions*. Another development in this direction is the work of authors who have attempted to integrate the findings of school effectiveness research, research on instructional effectiveness, and the early input–output studies (Stringfield & Slavin, 1992; Scheerens, 1992; Creemers, 1994).

The main characteristics of the resulting integral or comprehensive educational effectiveness models are as follows.

1. Antecedent conditions are classified in terms of inputs, processes and context of schooling, in other words according to a basic systems model of an organizationally and contextually embedded production process.
2. The model has a multilevel structure, where schools are nested in contexts, classrooms are nested in schools and pupils are nested in classrooms or teachers.
3. In recent formulations complex causal structure, multilevel nature, dynamic aspects and non-recursiveness are present as further elaborations.

The Scheerens' model

A first example of this approach is Scheerens' integrated model of school effectiveness (Scheerens, 1990) (Fig. 2.4). The model is based on a review of the instructional and school effectiveness research literature. The general assumption is that higher level conditions somehow facilitate lower level conditions (the nature of these cross-level relationships will be further elaborated later in this chapter).

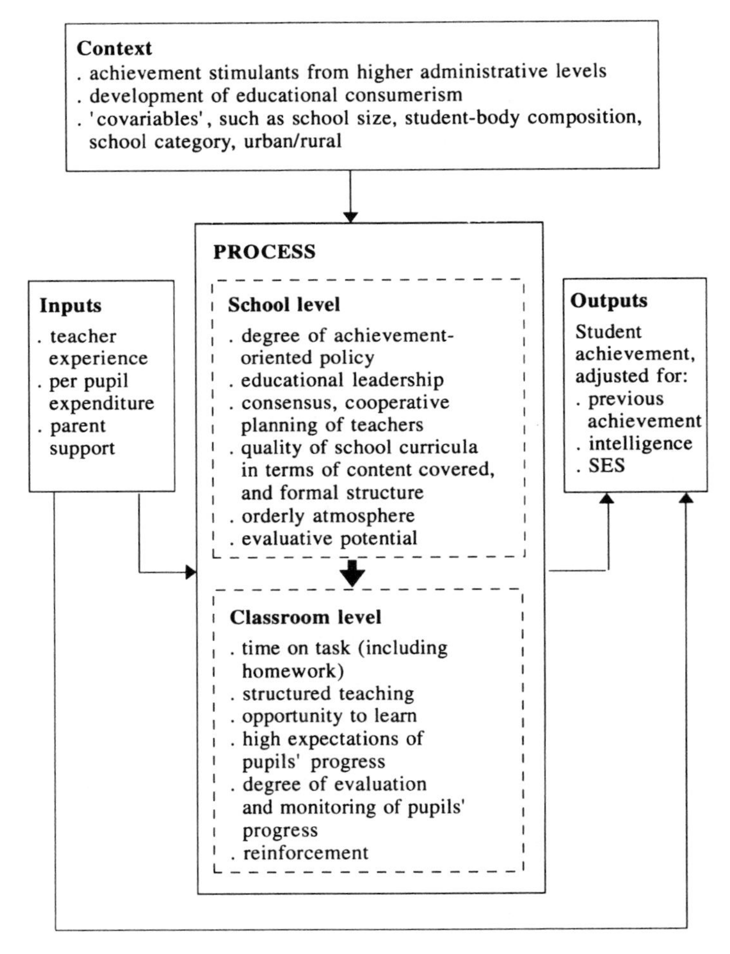

Figure 2.4 An integrated model of school effectiveness. (From Scheerens, 1990.)

The Slavin/Stringfield model

A second example is the QAIT/MACRO model developed by Stringfield and Slavin (1992). QAIT stands for quality, appropriateness, incentive and time; MACRO is the acronym for meaningful goals, attention to academic focus, coordination, recruitment and training, and organization.

This model has four levels:

1. the level of the individual student and learner;
2. (para)professionals who are in direct interaction with students;
3. schools, with head teachers, other school level personnel, and programs, "which affect student learning by affecting the ways in which students, teachers and parents act and interact" (Stringfield & Slavin, 1992, p. 36);
4. the above-school level, comprising the community, the school district, state and federal sources of programming, funding and assessment.

The student level part of the model is the Carroll model, discussed in a previous section.

The teaching and classroom level part of the model is Slavin's "theory of effective classroom organization", which focusses on those parts of the Carroll model that are potentially in control of the teacher, namely:

Quality: opportunity to learn, time on task, and principles such as: teachers make frequent presentations and demonstrations, are enthusiastic, ask clear and appropriate questions, provide clear feedback, provide guidance after students answer incorrectly, incorporate student comments and interests into lessons, prepare students for assignments, and circulate among students during academic work (Stringfield & Slavin, 1992, p. 39).

Appropriateness: the difficulty level of the subject matter, to which the problem of gearing instruction to individual prior attainment levels is related.

Incentive: stimulating motivation to learn. According to Brophy (1987), four preconditions for motivating students are: a supportive environment, appropriate levels of challenges and difficulty, meaningful learning objectives, and moderation in the use of any one incentive or motivational strategy (Brophy, 1987, cited in Stringfield & Slavin, 1992, p. 40).

Time on task, ultimately expressed in terms of actual learning time, is framed by actual teaching time and officially scheduled time.

According to Stringfield and Slavin, these characteristics of effective instruction are borne not only by the teacher, but also by parents and special programs.

At the school level clear statement of *goals* and the sharing of these goals by all staff are emphasized. Attention to academic functioning to a large extent depends on what other authors have described as "instructional leadership". "Recruitment of prospective teachers, development of all staff, and, when necessary, the movement of longitudinally unsuccessful teachers out of the school", emphasizes the human resource development mode of schooling, which is not so common in empirically oriented school effectiveness models.

School organization features, the last component of MACRO, emphasize coordination, structuring the school day and use of support staff.

Features at the above-school, contextual level that are mentioned are: relationships with parents and the local community, the school district, special programs (such as Chapter 1 in the U.S.A.) and school funding arrangements determined by federal and state governments.

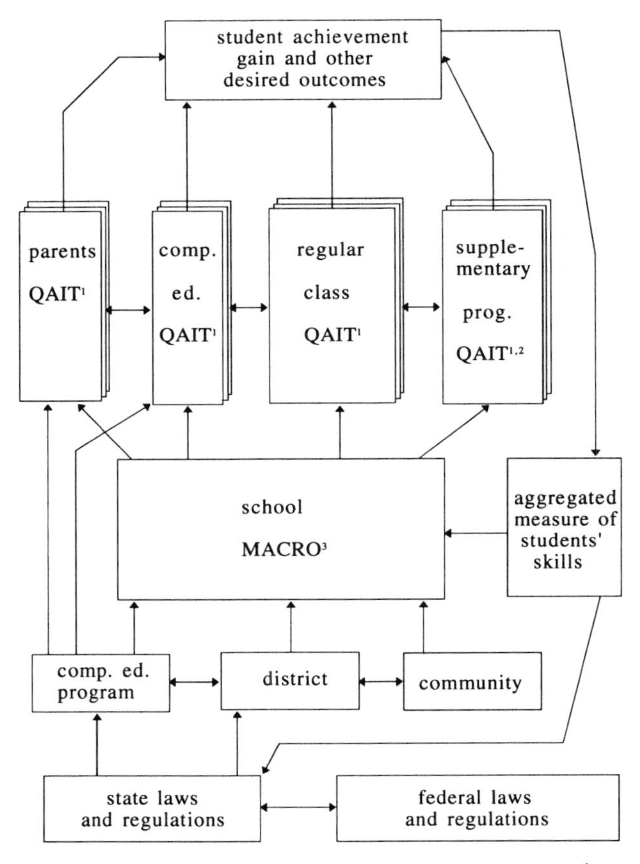

Figure 2.5 A hierarchical elementary education effects model. [1]QAIT: quality, appropriateness, incentive, time of instruction; [2]special education, bilingual education, etc. [3]MACRO: meaningful goals, attention to academic functions, coordination, recruitment and training, organization. (Source: Stringfield & Slavin, 1992.)

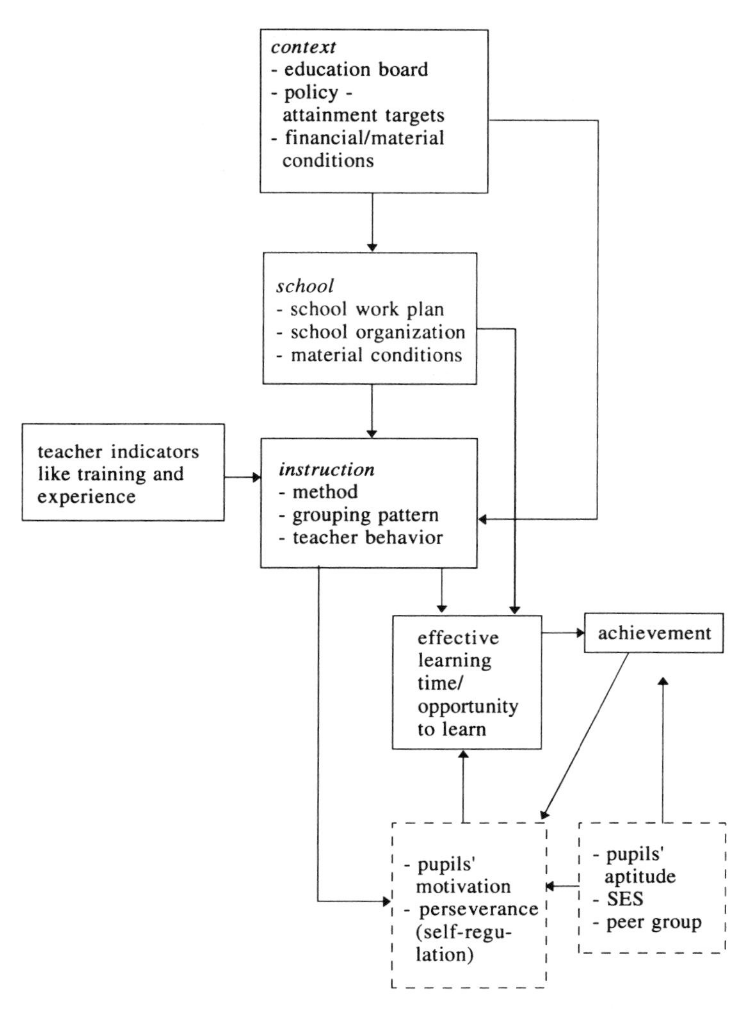

Figure 2.6 Creemers' model of school learning (Creemers, 1994).

The complete QAIT/MACRO model is depicted in Fig. 2.5. The authors also add a dynamic dimension to the model by showing some illustrative scenarios of how schools could develop *in*effectiveness over time, as the cumulative result of teachers in higher grades having to compensate for weak teachers in lower grades, an increased proportion of disadvantaged pupils and decreased teacher morale.

The QAIT/MACRO model is an encompassing model with reference to modes and levels of schooling. Another interesting feature is the idea that there are several sources or bearers of effective tuition (teachers, parents and programs). This idea is similar to Creemers' consistency principle in components of effective instruction. One final observation that should be made is that the QAIT modeling on the student and teacher level is more elaborated than the MACRO parts of the model.

The Creemers' model

A third and final example of a comprehensive educational effectiveness model is Creemers' model on school learning. This model is depicted in Fig. 2.6. The core of this model is again closely related to the Carroll model. The instruction block was further elaborated in the overview, cited earlier in Fig. 2.1. In this model too, higher level school organizational and contextual conditions are seen as facilitative with respect to lower level conditions.

In the subsequent sections of this chapter the modeling of school effects and the modeling of school effectiveness will be further elaborated and presented in a more formal way.

Operational Questions in the Modeling of School Effects

When trying to find school effects — to resolve questions regarding the relative importance of schools — the following three points need to be addressed.

1. Which criterion should be used?
2. Which definition should be applied, i.e. unadjusted gross or value-added measures?
3. What kind of statistical caveats should be avoided in estimating school effects, in other words how can the multilevel phenomenon (estimating the effects of institutions on individuals) be dealt with correctly?

Criterion definition

The major challenges to the choice of a particular effectiveness criterion are critical questions concerning the ultimacy, fairness, validity and economy of the actual output measures.

It is assumed that school effectiveness is primarily concerned with output, and therefore should not dwell upon views of organizational effectiveness in which input and process characteristics are also seen as effectiveness criteria (cf. Chapter One; Cameron & Whetten, 1983; see the earlier statement of our positon in Chapter One where the basic perspective on school effectiveness is considered to be the productivity perspective).

Achievement or attainment measures

The predominant criterion in school effectiveness studies from various disciplinary origins is achievement. Attainment measures depend on formal levels in the school careers of pupils. Roughly speaking, educational attainment scores express the level that individuals or groups of pupils have reached after a certain number of years of schooling. Examples of discrete attainment levels are the end of the primary school period and the end of the secondary school period. Attainment scales can become quite differentiated, particularly when an educational system consists of many school types, to which societal value is attributed in various degrees (e.g. Bosker & van der Velden, 1989).

When discussing the option of choosing either attainment measures or achievement measures, various underlying dimensions for this choice can be discerned. First, the choice may depend on different connotations of effectiveness (e.g. maximalization of output versus enhancing quality). Second, preferences concerning band width versus specifity of output measures could determine the choice, i.e. the question of whether an overall output measure or a more narrowly defined performance indicator is to be preferred. Third, the question of the predominance of a more practical versus a scientific interest in establishing effectiveness may lie at the background of this choice.

Attainment measures are close to the economic notion of effectiveness as maximization of output, where output is measured as the amount of product resulting from a particular production process. In education, "pupils who pass their exams" can be seen as the products of the process of schooling. Achievement, in contrast, fits more neatly into an interpretation of effectiveness in terms of "quality". Achievement tests as effectiveness criteria capitalize on more fine-grained quality differences of the units of output.

Attainment measures are cruder output measures than achievement tests, but at the same time they usually imply a broader coverage of the whole spectrum of educational objectives. The passing of a final examination (attainment indicator) depends on achievement in many subjects, whereas achievement tests in school effectiveness studies are often limited to arithmetic and language tests.

School effectiveness is both a subject of scientific inquiry and an "applied" field of interest in educational policy and management. When issues of consumer demands, monitoring of schools and accountability are at stake one cannot do without attainment indicators. In the case of inquiry into determinants of school effectiveness, i.e. input–output, or input–process–output studies, researchers will require output indicators that differentiate more strongly between qualities of the "units of output" and thus prefer achievement tests.

In summary, it is our contention that attainment measures are called for when purely economic and applied perspectives of effectiveness predominate or in case one wishes to explore, in the tradition of sociology of education, the contribution of schools to a person's attainment of status. Achievement measures are more likely to be chosen when quality of education is at stake and when a more psychological interest in cognitive development (or an educational interest in schools as organizations) predominates.

Finally, it should be mentioned that the choice between attainment and achievement can be avoided in two ways: (a) by using both measures, or (b) by a decision-oriented use of achievement tests (as when performance standards in the form of cutting scores on tests determine further career options).

Intermediate and "ultimate" effectiveness measure

Are attainment or achievement measures obtained at the end of a particular period of schooling to be considered as the ultimate productivity measures or would only longer term "civil effects" of schooling, such as employment or job level reached by graduates (e.g. Bosker *et al.*, 1996), qualify as such? Or, moving in the other direction on the scale of ultimacy of effectiveness measures, could we use intermediate effects such as attendance and drop-out rates as substitute effectiveness criteria? (e.g. Rutter *et al.*, 1979; Bosker & Hofman, 1994).

In our opinion, searching for ultimate school effects is like looking for the holy grail, since one can go on and on in stating even more ultimate effects. The most likely, albeit arbitrary, points in the school careers of pupils to measure school effects are indeed when a particular period of schooling is terminated and transition takes place to a higher school type or into the labor market. Postschool effect measures could be seen as important in macrolevel applications of educational indicators for purposes of monitoring national school systems. Also, postschool effect measures could be seen as important criteria for gaining insight into the predictive validity of effect measures at the end of the period of schooling. Attendance and drop-out rates are better treated as process measures in school effectiveness studies, because they generally function as means rather than as desired ends of schooling.

General versus curriculum-specific achievement tests

When the decision is taken to use achievement rather than attainment output data, there is a further option in the choice of the type of test. Madaus *et al.* (1979) have provided arguments in favor of curriculum-specific tests and against the use of general achievement tests (e.g. the Scholastic Aptitude Test). One of their arguments is that larger school (or class) effects are demonstrated when curriculum-specific tests (exams in their case) are used. Before offering a few lines of thiought in determining the choice of output measure along this particular dimension, we would like to state that "general versus curriculum-specific" achievement measures should be seen as a continuum with many discrete scale points rather than a dichotomous choice between two extremes.

Varying from curriculum-specific to general-aptitude measures, the following types of measures are discernible:

- authentic assessment by trained teachers (e.g. Rowe & Hill, 1996);
- trained test items;
- content specific measures;
- Rasch scales of narrow content areas;
- subject-specific tests;
- general scholastic aptitude tests;
- intelligence tests.

A general guideline to choose from these alternatives would be to use the more specific measures up to the degree that the application purpose is closer to the microsituation of classroom instruction. A line of thinking which perhaps offers a more fundamental solution to this problem of choice, within the content of school effectiveness research, would be to choose the type of outcome measure that has the greatest predictive validity with respect to the more ultimate educational effects. For example, when measuring achievement at the end of a specific stream of vocational education, content-specific measures may be preferable, assuming a close connection between the curriculum and the skills required in the job situation. Note, however, that this latter kind of criterion choice further depends on the theory one holds about the relationship between education and the labor market. Departing from a credentials or screening theory, certification itself would be the best criterion for school effectiveness, whereas achievement is more connected to the human capital philosophy.

Implicit definitions of school effectiveness

Although people generally use the school effect concept without any hesitation, there are many different underlying operationalizations. Some, of course, are concerned with the subject matter areas in which the performance or ability of the pupils was assessed (see also Chapter Three), while others are concerned with gross versus value-added indicators. More importantly, when using value-added measures different corrections are being made before the actual school effects are assessed, i.e. different covariates are used in different research studies. When looking at the type of covariates used, four different approaches can be distinguished, which will be called gross school effects, unpredicted achievement-based school effects, learning gain-based school effects, and unpredicted learning gain-based school effects.

Gross school effects

The first operational definition uses as the measure for school effect the mean (uncorrected) achievement score of pupils in a certain school. The value of this definition lies in its use within a criterion referenced framework: if a standard is set (or if a growth continuum is specified) a priori this gross school outcome measure provides the information to judge whether the school, on average, performs above, at or below the standard. It does not, however, imply that all pupils within that school meet the standard. This definition can be labeled the gross school effect. In operational terms it is the mean achievement, i.e. it is averaged over the pupils within a school with a correction for sampling error (this latter point will be elucidated further on).

School effects based on unpredicted student achievement

The second operational definition starts from unpredicted achievement. In this case a prediction equation is estimated from the data, where achievement is predicted from aptitude, socioeconomic status, age, gender, ethnicity-status, and other student variables. The reasoning behind this approach is that schools widely differ in their pupil populations, and that, since the aforementioned variables have a strong relationship with achievement, their effects on achievement should be partialled out. Most of these variables are static and not subject to much, if any, change, although the aptitude of a child may alter. For this reason the aptitude assessment should ideally take place before or at school entrance.

School effects based on learning gain

The third operational definition can be seen as a specific case of the second one: achievement is predicted from prior achievement or, more appropriately, the difference between the two is used. Once again the same argument applies as in the case of the aptitude assessment. If prior achievement is assessed at a later point in time than school entrance, the school effect transforms into the effect of a school on its pupils within a certain time-interval.

School effects based on unpredicted learning gain

The last and seemingly strictest definition combines the previous two defiitions. Using a posttest score corrected for a preassessment score, this score is in turn corrected for aptitude, socioeconomic status, age, gender, ethnicity status and other student variables, since these are related to pupils' learning progress. In this case, prior achievement as well as aptitude should ideally be assessed at school entrance.

In the literature on school effectiveness school effects are usually referred to as value-added measures, i.e. some kind of intake adjustment is applied. However, there are at least three different sets of value-added-based school effects. With the exception of the first definition of a school effect (gross school effect), where no adjustment for intake differences between schools is employed, the value-added-based definitions can be made stricter by correcting not only for the student level effect of the covariates but also for the potential extra effects of their aggregates. It is often observed that, for example, the average socioeconomic status of the student populations of schools has an effect over and above the individual socioeconomic status variable. This indicates that being a working-class child has a negative effect on achievement, but being in a school with a majority of working-class children has a substantial extra negative effect on achievement as well.

In all cases, however, causal attribution (showing that schools cause the effect) is difficult to achieve. Raudenbush and Willms (1996), following Rosenbaum and Rubin, argue that, since randomization cannot be achieved, assignment of students and schools to treatments should be "strongly ignorable". This qualification expresses that the different treatment outcomes for a student and a school "are conditionally independent of treatment assignment given a set of covariates" (Raudenbush & Willms, 1996, p. 312), implying that value-added measures should be preferred. However, it is possible that value-added-based school effects underestimate true school effects. "Privileged" students are more likely to choose "good practice" schools, so that controlling for a student's socioeconomic status, for instance, results in overadjustment. To complicate things further, it should be pointed out that in compensatory systems, school effects may appear only after adjustments for covariates are made. In compensatory systems unfavorable scores on the covariates are compensated by (and thus correlated with) good educational practices. Moreover, these school effects then can disappear again after context variables, which are negatively associated with good educational practices for the same reason, have been taken into account. This matter will be taken up again in Chapter Nine (see also the research by Grisay, described in Chapter Six).

Statistical issues

Having discussed different definitions of school effects, the next question is: what statistical models can be used to estimate these effects? The first definition given, the gross school effect, can be thought of as the mean achievement; however, it is actually averaged over the pupils within a school with a correction for sampling error. This latter correction is crucial, especially when there is only a restricted number of students per school for which achievement scores are available. The idea is that in estimating school effects a two-stage sample is employed. In the first stage, schools are selected at random from the population of schools, and then, within each selected school, students from a certain grade are selected at random. This may not look like a random sample when all students from a certain grade level are included in the sample. Therefore one could view this as a time sample: students are selected who are in grade 8 in the summer of 1996, and these can be seen as a random sample from all grade 8 students in this school in, say, the years 1990 to 2000. In a two-stage sample some variation will show up as between-schools variance by chance.

An example (taken from Snijders & Bosker, forthcoming) may clarify this point. Suppose we have a series of 100 observations as in the random digits table shown in Table 2.2. The core part of the table contains the random digits. Now suppose that each row in the table is a school, so that for each school there are observations on 10 students, and that the random digits are test scores. The averages of the scores for each school are in the last column. There seem to be vast differences between the randomly constructed schools if one looks at the variance in the school averages (106 to be precise). The total variance between the 100 students is 814. According to the outcomes we would conclude that the schools vary widely with respect to their average test scores, since they account for 13% (106/814 × 100%) of the variation in achievement. It is known, however, that in reality these schools only differ by chance.

Statistical theory tells us that the total between-schools variance is:

"true" variance BETWEEN =

observed variance BETWEEN – sampling error variance BETWEEN

The between-schools sampling error variance is:

sampling error variance BETWEEN = (variance WITHIN schools/n_j)

Table 2.2 True and sampling error variance between schools (random digits table from Glass & Stanley, 1970, p. 511)

School number	Scores for the students (random digits)	School average
01	60 36 59 46 53 35 07 53 39 49	44
02	83 79 94 24 02 56 62 33 44 42	52
03	32 96 00 74 05 36 40 98 32 32	45
04	19 32 25 38 45 57 62 05 26 06	32
05	11 22 09 47 47 07 39 93 74 08	36
06	31 75 15 72 60 68 98 00 53 39	51
07	88 49 29 93 82 14 45 40 45 04	49
08	30 93 44 77 44 07 48 18 38 28	43
09	22 88 84 88 93 27 49 99 87 48	69
10	78 21 21 69 93 35 90 29 12 86	53

The variance within schools is the sum over the schools of the sum of squares within the schools divided by $N \times n_j$, in which N is the number of schools in the sample, and n_j is the (harmonic) mean of the number of students per school (in this case n_j=10). For our random digits table the variance within the schools is 789. The estimate for the true BETWEEN-schools variance now is:
true variance BETWEEN = 106 – (789/10) = 27.

Using this example we see that in reality schools account only for 3% of the variation in achievement, a figure so low that statistically it does not differ significantly from 0.

The best estimate – in the long run – of the effect of a particular school on its students is not the average but a reliability-weighted average. The reliability weight for school j is defined as:

w_j = true between/[true between + (within/n_j)].

This weight is then: 27/[27+(789/10)]=0.25. The estimated school effect for school j is:

School effect$_j$ = observed average$_j \times w_j$ + (1-w_j)overall average,

which for the first school is (44 × 0.25) + ((1–0.25) × 47) = 46, and for the most outlying school with average 69 it is 53. Notice that the true variance between schools as well as the deviation from these new estimated school effects from the overall average are much smaller than the initial estimates.

This statistical sidestep was necessary to explain that it is not the simple school averages that are used in the gross school effect definition, but that these have to be weighted with a factor that causes something known as regression to the mean. The less reliable the information for a certain school the more its effect is estimated using information from the rest of the sample.

Having shown how the gross school effect can be estimated, we can now turn to the estimation of value-added-based school effects. Since the idea is that there are one or more covariates for which adjustment in the outcome measure is necessary, the principle to be employed is that of an analysis of covariance adjustment, in a multilevel context.

To give an idea of how this approach works a graphic impression is presented in Fig. 2.7. The horizontal axis of the graph depicts student IQ, ranging rather arbitrarily from 0 to 100. The vertical axis represents student achievement on an arbitrary scale from 0 to 100. The situation is depicted for five schools. The line on the graph represents the estimated regression of achievement on IQ; it can also be viewed as a prediction formula: for every point increase in IQ, achievement is predicted to increase by almost one point. From the graph it can be seen that the students from the school represented by × are all above the regression line (they all achieve above expectation, and therefore may be called overachievers), whereas the students from the school represented by ◇ are all under the regression line (underachievers). The average distance between the observed achievement scores for the students of a given school and their predicted achievement scores can now be used as a value-added-based school effect measure, although a correction for sampling error must also be made here.

Although the example deals with only one covariate, the extension to more covariates (SES, gender, ethnicity, etc.) is straightforward.

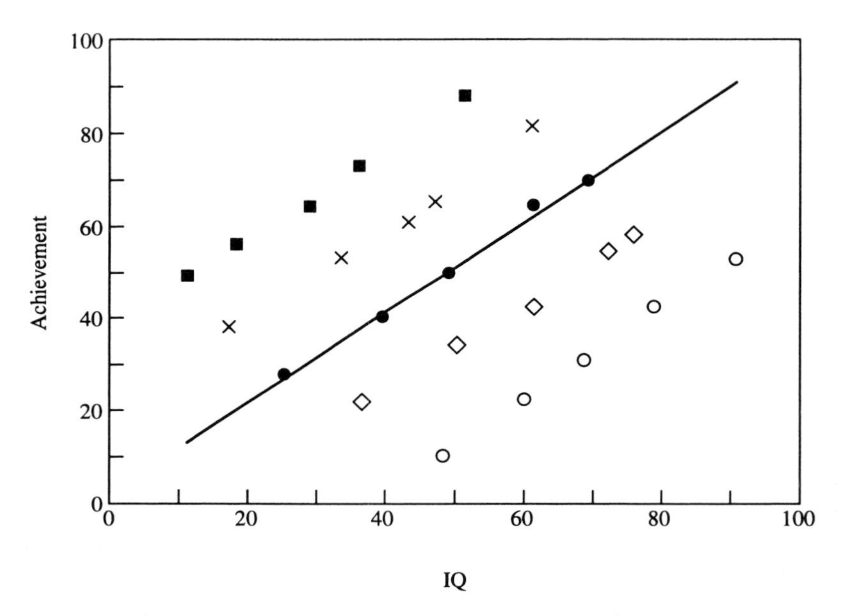

Figure 2.7 Relation between achievement and IQ for five schools.

Formalization of Alternative Interpretations of Cross-Level Facilitation

The general notion of higher level variables in some way facilitating lower level variables in the production of educational outcomes conceals rather divergent interpretations of these cross-level relationships.

First, relationships between conditions at higher and lower levels can take the shape of *contextual effects*. In this case, the aggregates of certain attributes defined at microlevel are seen as having an additional causal influence. For instance, Erbring and Young's (1979) model on endogenous feedback predicts that, should a school have a majority of effective teachers and sufficient feedback among the staff, the minority of less effective teachers will be stimulated to become more effective. Other contextual effects that are, as a matter of course, important in educational organizations, relate to classroom and peer group composition. To some extent contextual effects, acting as selection mechanisms, work against improving education by means of optimizing organization variables. However, it should be realized that selection and allocation decisions, subject to the distribution of authority within national educational systems, can be partly controlled by school management. Later, a formal multilevel model will be presented, from which the relative magnitude of contextual effects in comparison with genuine cross-level facilitating can be assessed.

Second, conditions at higher levels can act as mirrors to conditions at lower levels. As Berliner (1985, p. 143) observes: "The evidence on effective classrooms and effective schools is amazingly congruent". Features such as achievement pressure, high expectations of pupils' achievement, monitoring and an orderly atmosphere have been found to be relevant in studies of effective classrooms as well as effective schools. Congruence of factors that have meaning at different levels could be thought of as creating a consistent school culture. According to this interpretation higher level facilitation ismore likely to provide a general supportive background than a purposeful and direct manipulation of lower level conditions, and this interpretation is considered next.

Third, higher levels can be thought of as overt measures creating effectiveness-enhancing conditions at lower levels. The concept of instructional leadership comes under this heading, as do increasing the allocated learning time, the recruitment of effective teachers, the selection of teaching materials with effectiveness-enhancing characteristics, stimulating evaluations at classroom level, keeping records of pupils' progress, and many more.

Fourth, conditions at higher levels can serve as incentives to promote efficiency-enhancing conditions at lower levels. Higher level conditions include rewards for teachers from their superiors for effective teaching and monetary grants from their district if schools reach certain achievement standards. Educational consumer demand for effective schooling is another type of external incentive. This view of cross-level facilitation reflects a somewhat restricted input–output view, where school processes are treated as a black box and where standards or attainment targets are the main category of inputs.

Fifth, conditions at higher levels can serve as material facilities for conditions at lower levels (a more restricted case of the second "mirror" category). One example is a computerized, school-monitoring system implemented at school level to provide teachers with more sophisticated means of monitoring student progress.

Sixth, and finally, higher level conditions may serve as buffers to protect efficiency-enhancing conditions at lower levels. These high-level conditions include maintenance functions carried out by school directors, such as safeguarding student-enrollment figures, dealing with extramural pressures and governmental regulations, and representational activities. This view implies minimal expectations of the direct influence of school management on the actual education production process, which in this scenario would be left entirely to the autonomous teachers.

In the next section, some of these alternative interpretations will be formulated more precisely in terms of alternative specifications of multilevel or structural models. In a similar way, some further intricacies of the interrelationships of the nested-layers image of school functioning will be described in terms of alternative specifications of formal models.

Alternative Causal Specifications

When turning to the question of alternative causal specifications within a global framework of schools as nested layers, the following competing models are distinguished.

Additive versus interactive models

According to additive models, higher level conditions are seen as increments to variables operating at the lower level, e.g. achievement-oriented policy at school level adds to the effects of an achievement-oriented attitude of teachers. In the interactive models higher level conditions impinge on the (causally interpreted) relationship between lower level antecedent conditions and the criterion variable as, for instance, when the evaluative potential of a school (e.g. the availability of a pupil monitoring system) is thought of as determining the impact of direct instruction on individual learners (since direct instruction is facilitated by this administrative support system). In terms of multilevel modeling the comparison of these two interpretations involves an interest in comparing intercepts (additive model) versus an interest in comparing slopes (interactive model).

The additive model can be written as:

$$Y_{ijk} = \beta_{0jk} + \beta_1 P_{ijk} + R_{ijk} \qquad \text{pupil level} \qquad (2.1a)$$

$$\beta_{0jk} = \gamma_{00k} + \gamma_{001} T_{jk} + U_{0jk} \qquad \text{teacher level} \qquad (2.1b)$$

$$\gamma_{00k} = \delta_{000} + \delta_{001} S_k + V_{00k} \qquad \text{school level} \qquad (2.1c)$$

in which Y_{ijk} represents the achievement score of pupil i in class j in school k, β_{ojk} is the class-specific intercept, P_{ijk} represents, for example, the ability of pupil i in class j in school k, β_1 is the regression coefficient and R_{ijk} is the pupil level error term. In (2.1b) the class-specific intercept, which can be interpreted as the mean class achievement score corrected for ability when this latter variable is conveniently transformed (by subtracting the average $P_{...}$) to have zero mean, is modeled as a function of the school-specific intercept γ_{00k}, a teacher-level variable T_{jk}; γ_{001} is the regression coefficient, and U_{0jk} is the class-level error term. In (2.1c) the school-specific intercept is modeled as a function of the grand mean δ_{000}, a school-level variable S_k with accompanying regression coefficient δ_{001}; V_{00k} is the school-level residual. The interactive model is somewhat different:

$Y_{ijk} = \beta_{0jk} + \beta_1 P_{ijk} + R_{ijk}$	pupil level	(2.2a)
$\beta_{0jk} = \gamma_{00k} + \gamma_{01k} T_{jk} + U_{0jk}$	teacher level	(2.2b)
$\gamma_{00k} = \delta_{000} + \delta_{001} S_k + V_{00k}$	school level	(2.2c)
$\gamma_{01k} = \delta_{010} + \delta_{011} S_k + V_{01k}$	school level	(2.2d)

Equations (2.2a)–(2.2d) differ from the first set of equations in that they have school-specific regression coefficients γ_{01k} for the regression of class-mean achievement β_{0jk} on the teacher variable T_{jk} incorporated in the model. These school-specific regression coefficients are then modeled as a function of an overall regression coefficient δ_{010}, a school variable S_k, with regression coefficient δ_{011}, and a school-level error term V_{01k} which expresses the school-specific deviation from the overall regression of achievement on the teacher variable. Combining the equations by mere substitution shows that a term $\delta_{011}(S_k \times T_{jk})$ is included in the model: this is the cross-level interaction. A graphic representation of the models is given in Figs 2.8 and 2.9.

Examples of tests of additive versus interactive multilevel models of educational effectiveness are presented by Gamoran (1991) and Bosker *et al.* (1990). Gamoran found that in general the additive model might be preferred when studying educational effects, at least in models with only two levels. Content coverage and the quality of instructional discourse were instructional

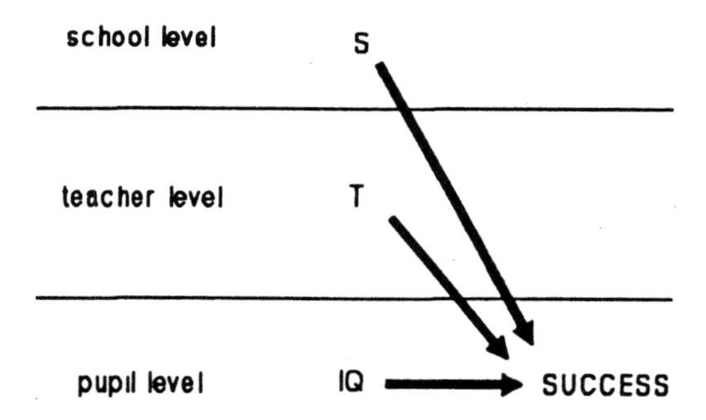

Figure 2.8 The additive model.

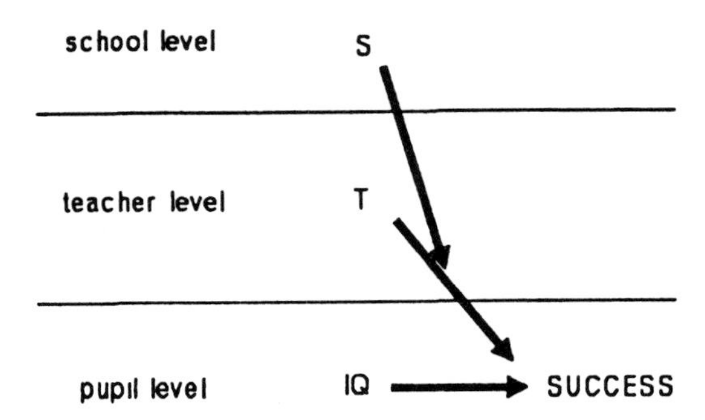

Figure 2.9 The interactive model.

factors that added to achievement. Nevertheless, he also found evidence that initially high-achieving pupils especially benefitted from the amount of curricular content covered. Bosker *et al.* undertook research on organizational-level conditions for effective instruction, thus working with the full three-level model as depicted before. They concluded that effective instruction (measured as actual instruction time) is related to achievement only in those cases where the degree of formalization of rules in school is below average. Instruction effects on achievement, although present in an additive sense, are also to some degree contingent on the organization of the school. Aitkin & Zuzovsky (1994), pleaing for a new paradigm in educational research, argue that all educational models should be interactive until proven not to be empirically valid. In that case cross-level interaction effects between the teacher and pupil levels might be included, as well as third-order interaction terms (pupil × teacher × school interaction effects). Their case is strong, since the additive model is a special case of the interactive model (namely when $\delta_{011} = 0$). This observation also leads to the conclusion that the additive model is scientifically more powerful, since it is more parsimonious than the interactive model.

Contextual versus "genuine" multilevel effects

A basic challenge of the nested-layers perspective on school functioning is the thesis that school effectiveness is largely determined by selection mechanisms (effective schools are schools that attract good pupils, good teachers and good administrators). According to this competing perspective, "high-level causation" would be largely determined by the contextual effects of aggregates (e.g. weak pupils do better in classes where average achievement is higher). Issues of contextual versus genuine multilevel effects can be settled by including both types of variables in multilevel models and examining the relative magnitude of regression coefficients.

The formalization of these models can be achieved by simply including $P_{..k}$ (the school average ability of pupils in school k) in model (2.1c). Next to equations (2.1a) and (2.1b), we then have:

$$\gamma_{00k} = \delta_{000} + \delta_{001}S_k + \delta_{002}P_{..k} + V_{00\delta} \qquad \text{school level} \qquad (2.3c)$$

and the test of genuine school effects is then concerned with the test that $\delta_{001} \neq 0$, not withstanding the inclusion of the effect of $P_{..k}$. It is important to note that contextual effects are only present when it can be proven that $\delta_{002} \neq \beta_1$.

A graphic representation of the contextual effects model is presented in Fig. 2.10.

Examples of these kind of contextual versus "genuine" effects are presented in Bryk and Raudenbush (1992). They compared Catholic and public schools, and demonstrated that the contextual effect of socioeconomic status was twice as strong as the individual effect. More importantly, however, it was shown that the sector effect (Catholic versus public) although still significant, was far more smaller once the contextual effects were taken into account.

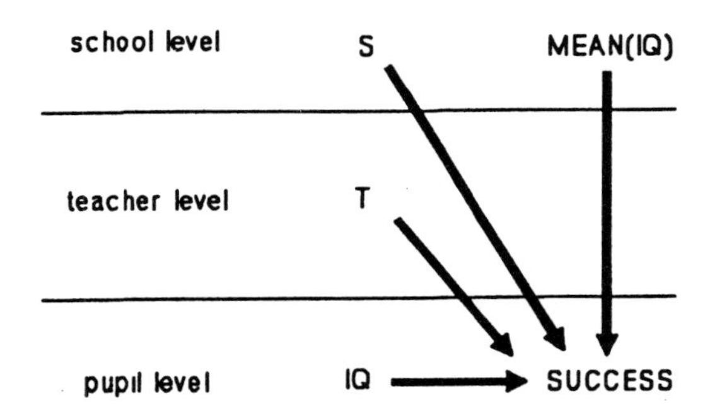

Figure 2.10 Contextual or genuine school effects.

Indirect versus direct causal effects

Conditions that are "more than one level up" with respect to educational achievement can be seen either as direct causes of achievement or as indirectly influencing achievement via intermediate levels. It should be noted that this sort of competing causal model cannot simply be settled by comparing different specifications of the usual LISREL-type or path-analytic models. Instead multilevel path-analytic techniques are required, and these are presently being developed. In the absence of these models one could assess direct and indirect effects as follows:

$$Y_{ijk} = \beta_{0jk} + \beta_1 P_{ijk} + R_{ijk} \quad \text{pupil level} \quad (2.4a)$$
$$\beta_{0jk} = \gamma_{00k} + \gamma_{001} T_{jk} + U_{0jk} \quad \text{teacher level} \quad (2.4b)$$
$$\gamma_{00k} = \delta_{000} + \delta_{001} S_k + V_{00k} \quad \text{school level} \quad (2.4c)$$
$$T_{jk} = \varepsilon_{000} + \varepsilon_{001} S_k + W_{00k} \quad \text{school level} \quad (2.4d)$$

The model differs from the former ones by (2.4d), in which the teacher-variable T_{jk} serves as the criterion that is predicted from the school variable S_k.

An indication of the existence of indirect effects can be found by assessing that δ_{001} is zero, while it differs from zero when T_{jk} is deleted as a predictor from the model (2.4b). S_k should have an effect on T_{jk}, i.e. ε_{001} should differ significantly from 0.

Models (2.4a) – (2.4d) form a set of multilevel structural equations which can be estimated either as in ordinary path models, or by applying the BIRAM model (McDonald, 1994) or BUGS (Spiegelhalter *et al.*, 1994), or by using a technique suggested by Muthén (1994) that is used in the preprocessor STREAMS (Gustafsson & Stahl, 1996) for LISREL or EQS or AMOS. A graphic representation of the indirect effects model is presented in Fig. 2.11.

One might view the earlier presented example of Bryk and Raudenbush (1992) as a means of demonstrating that the Sector effects in American high schools are mediated by contextual effects. A clear example of the path model approach is given by Kreft and Aschbacher (1994). These authors assessed the effects of a new curriculum on student achievement, and found empirical evidence for the hypothesis that the effects are mediated by teacher satisfaction. Hill *et al* (1995a) demonstrated that educational leadership (measured as support of teachers by their principal) affects teacher practices and attitudes (i.e. teacher–student interaction and professional culture), but no direct effects nor indirect effects on student achievement could be demonstrated. This latter example will be dealt with in more detail in Chapter Five.

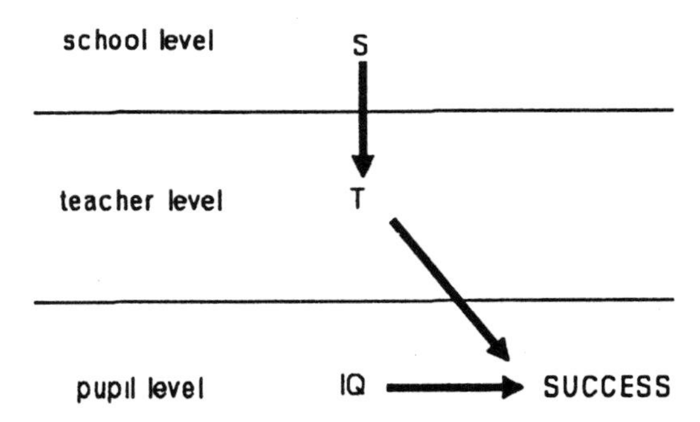

Figure 2.11 The indirect model.

Additive versus synergetic interpretations

School effectiveness researchers, confronted with very low correlations between their antecedent conditions and achievement, have sometimes sought refuge in the thought that joint effects of several variables, which individually appear to be of marginal influence, would "do the trick". The question is whether this magic of the whole being more than the sum of its parts is amenable to more precise and formal specification. When confronted with a set of school predictor variables one might investigate the synergetic interpretation by allowing for higher order interactions in the model. This results in a complex interactive model of educational effectiveness. However, the number of interactions potentially of interest grows exponentially with the number of predictors available. In research practice one therefore constructs ideal types by means of cluster analysis on the school-level predictor variables. The cluster analysis then searches for groups of schools that are as different from each other as possible, while within each group the schools are as much alike as possible. The theoretical reason for constructing such a typology is found in the configuration hypothesis of Mintzberg (1979): organizations are effective only if they succeed in finding a consistent pattern of structuring.

A graphic representation of the synergetic effect is presented in Fig. 2.12. Empirical evidence to support this approach is presented by Bosker (1990b): whereas no single organizational variable was linked to educational attainment of pupils, a configuration of cohesive, transparent, and goal-oriented characteristics proved to be the most successful organizational structure.

Reezigt *et al.* (1994) constructed typologies to find consistency within schools, but the consistent configurations did not affect student achievement more than the inconsistent configurations.

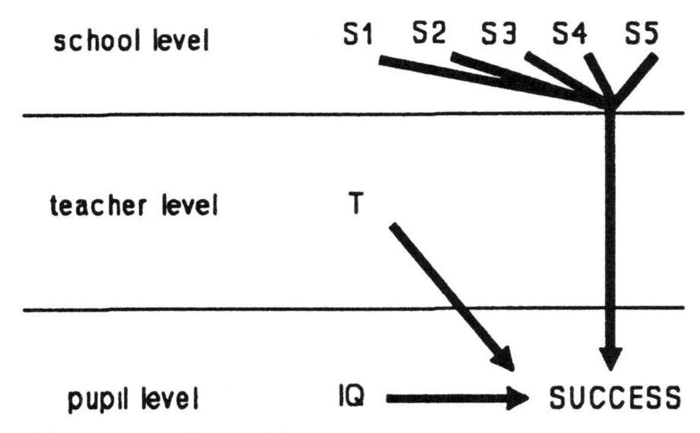

Figure 2.12 The synergetic model.

Recursive versus non-recursive models

Negative correlations between variables that are thought to enhance effectiveness and achievement are no exception in school effectiveness research. The inherent ambiguity in correlational research allows for the interpretation that, for instance, instructional processes are adapted to achievement levels. It is not at all implausible that several interrelationships among key variables of school effectiveness models are in fact recursive. The idea of recursive systems is depicted in Fig. 2.13.

Questions about the recursiveness or non-recursiveness of certain interrelationships within school effectiveness models can be tackled in three ways:

1. by means of experimental research;
2. by means of alternative path-analytical models;
3. by means of system-dynamic models.

De Vos (1989) presented a theoretical model that shows some recursive features: individual achievement contributes to the mean group achievement, which in its turn affects the standard to be set by the teacher for the class. The discrepancy between the individual achievement and the standard set by the teacher affects the learning gain to be made. This process is repeated in the next cycle, and so on. Empirical evidence of recursive relations in the field of school organizations is virtually non-existent, whereas the recursive interpretation seems all too plausible. Bosker and Guldemond (1994) simulated de Vos' model in a hierarchical system-dynamics model for secondary schools, and explained why efficiency, equality and effectiveness are to some degree hardly reconcilable goals in education. What is needed, however, to give some empirical basis to the recursive models, is longitudinal research at school level: following school changes and their potential causes over time.

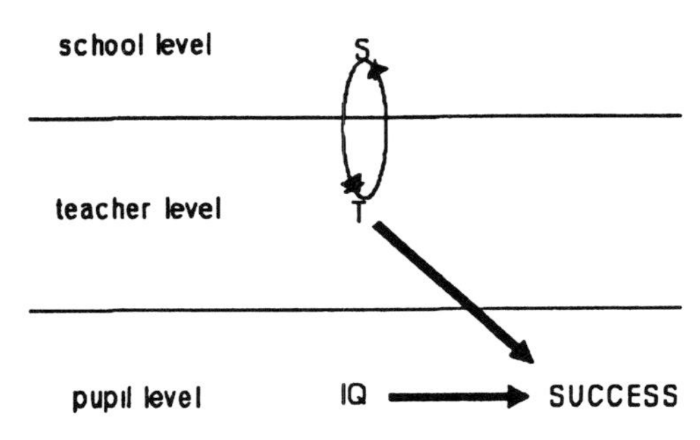

Figure 2.13 The recursive model

System-dynamic approaches also highlight the important question of which exogenous factors can break repetitive cycles or feedback loops. Conceptually this issue has to do with the primary levers of school effectiveness. The idea of meta-feedback which originates from the image of the learning organization is also to be tackled by means of this methodology.

CHAPTER THREE

SIZE, STABILITY, CONSISTENCY AND SCOPE OF SCHOOL EFFECTS

Introduction

Having outlined the concept and the modeling of school effectiveness in the two previous chapters, this chapter will focus on two critical questions:

1. What is the magnitude of school effects?
2. Can one think of the school effect concept as being unidimensional?

To answer the first question the results of a meta-analysis will be reported in the first section of this chapter, after which the unidimensionality question will be addressed.

The Magnitude of School Effects

In *Inequality of educational opportunity* Coleman *et al.* (1966) gave estimates for the relative importance of the factor "school" in accounting for differences between pupils' achievement. Differences between secondary schools in the U.S.A. accounted for 10.32% and 13.89% of the variation in mathematics achievement of 14-year-old white and black pupils, respectively. However, half of these differences could be accounted for by intake differences between the schools, as the schools served different student populations. Correcting for socioeconomic status differences the revised estimates for white and black students were 4.95% and 8.73%, respectively. As such, the Coleman study led to a rather pessimistic view on the possibilities of schools and education.

This pessimistic view was reinforced by other studies conducted at the time. A very influential study in this respect is the study by Jencks *et al.* (1972). Jencks' study differed from the Coleman study by refraining from using percentage of variance accounted for by school membership. Instead, he introduced the effect size measure normally used in presenting results of (quasi-)experiments, i.e. the difference between the experimental condition and the control group relative to the standard deviation of the criterion variable in the control group condition. He then used the square root of the variance accounted for by schools. The effect sizes for school differences (without intake corrections) were then 0.32 and 0.37 for white and black students, respectively. With intake corrections these values reduce to 0.23 and 0.28. Jencks then contended that, if previous achievement had been taken into account, the effect sizes would have been not larger than 0.17 and 0.20, respectively. Given these effect sizes it is no wonder that Jencks *et al.* (1972) contended that schools do not matter much with respect to student achievement, thus reinforcing the pessimistic view on the possibilities of schools and education.

The question regarding the magnitude of school effects was readdressed during the 1980s by school effectiveness research (Brookover *et al.*, 1979; Edmonds, 1979; Rutter *et al.*, 1979).

Although there are many books and articles that try to summarize the results of school effectiveness research (e.g. Purkey & Smith, 1983; Levine, 1992; Levine & Lezotte, 1990; Scheerens, 1992; Reynolds, 1992), of which many touch upon the existing differences in achievement between schools, there has been no systematic review of the question regarding the magnitude of school effects until now. In this section we try to fill this gap by presenting the results of a meta-analysis conducted on studies addressing the magnitude of school effects (Bosker & Witziers, 1996). More specifically, we will deal with questions regarding the importance of the school without taking account of student characteristics (the so-called gross school effects) and the importance of schools after controlling for these characteristics (the so-called value-added-based or net school effects). In particular, this last question is very relevant for those involved in the educational enterprise. From the results presented policy makers and others involved in educational improvement can assess to what extent policies and efforts aimed at improving school achievement are likely to be successful. If, for instance, the differences in achievement between schools were very small after correcting for relevant student characteristics, the effects of these policies and efforts would be rather limited. For educational researchers, particularly in the field of school effectiveness, our results might provide a framework regarding the question of what to expect from their work. Again, if differences in achievement between schools are small after taking account of student characteristics, the relationship between school characteristics and outcome measures will, by definition, also be very small. Finally, for parents and students the results might be helpful in the sense that the search for a "good" school is particularly relevant in a situation where differences between schools are significant.

Methodological guidelines regarding meta-analyses stress the importance of an exhaustive sample of studies for answering the research question at hand (White, 1994). In order to obtain an exhaustive sample several methods were employed. The first method consisted of thorough and systematic research of documentary databases containing abstracts of empirical studies. Of particular importance in this respect are ERIC, School Organization and Management Abstracts, Educational Administration Abstracts and the Sociology of Education Abstracts. Although these abstracts cover the most important scientific journals, they do not cover all scientific journals. The second method therefore consisted of paging through volumes of relevant educational scientific journals not covered by these abstracts (e.g. *Journal of School Effectiveness and School Improvement*) to search for articles relevant for the research questions. The third method consisted of a systematic examination of books and articles reviewing the state of the art of school effectiveness research. These three methods resulted in a database of several hundreds of potentially relevant articles, books and research reports. The last method consisted of reading all of the available material and reviewing the references used in this material to determine whether any important research material was missing. Given this rather exhaustive and systematic search, one would be tempted to conclude that this search has yielded a representative sample of the universe of relevant studies. However, it should be underlined that there is reason for doubt. For instance, our sample contained a rather limited set of studies into the magnitude of school effects conducted in the U.S.A. This could mean that studies relevant for this meta-analysis are not conducted on a large scale in that country, in contrast to the situation for countries such as England and The Netherlands. However, given the amount of educational research being conducted in the U.S.A., a more plausible reason for this limited size could be that although studies may have been conducted, they are not reported in forums accessible to the reviewers. This raises the wider issue of whether publication bias plays an important role in this study. The term publication bias refers to the fact that studies with significant effects have more chance of being published in scientific journals than studies without these effects. In the case of the meta-analysis on the magnitude of school effects, however, where the focus is on an implicit non-directional hypothesis (i.e. that schools differ), publication bias does not seem likely. Bias is especially unlikely to occur as many studies on the magnitude of school effects are a by-product of studies into the effects of educational practices, policy-evaluation studies or international assessment projects. It is clear that in the meta-analysis not all of the material collected was used. Although exhaustiveness is an important issue, other criteria are also important. First, only studies on either horizontal (i.e. comprehensive) school systems or schools with the same curricular track in vertical, selective school systems were used because, in categorical school systems, the size of school effects mainly reflects differences between curricular tracks due to (self-) selection of students.

Another important aspect in meta-analysis is the quality of the research studies at hand; the validity of the results of a meta-analysis is largely determined by the quality of the individual studies in the analysis. In this study quality was largely assessed from a statistical point of view. Studies before 1986 had to rely on (in hindsight) unsound statistical procedures (Aitkin & Longford, 1986), mainly because of the lack of appropriate statistical procedures available to deal with the hierarchical nature of the data. This situation has improved dramatically over the last decade, implying that the only studies used in this meta-analysis, given the research problem at hand, used sound statistical procedures. This implies that only studies using multilevel procedures were involved in the analysis, although there were some exceptions to this rule, for example, the study by Fitz-Gibbon (1992), which uses a random effects analysis of variance (ANOVA).

In total, 168 studies were selected for the meta-analysis. The studies on net school effects were rather heterogeneous with respect to the choice of covariates, and thus with respect to the (implicit) definition of school effect. For this reason only gross and net (i.e. value-added-based) school effects were distinguished. Since some of these studies contained replications (by studying either different cohorts or different subject matter areas), the number of replications for the meta-analysis on gross and net school effects was 153 and 89, respectively.

Bearing in mind that the term school effect suggests causation, but that this is actually difficult to demonstrate, the following questions should be asked.

1. What is the empirical magnitude of school effects, both gross, i.e. uncorrected, and net, i.e. corrected?
2. Does the magnitude of the school effect vary across subject domains, sectors and/or countries?

In order to answer these questions, the results of studies into the magnitude of school effects will be synthesized by means of a statistical meta-analysis.

Table 3.1 contains some descriptive statistics on the studies selected. Very few studies are focussed solely on net school effects because in most studies researchers start from an unconditional model, i.e. a model without any predictors, thus producing gross school effects estimates to start with. Also, many research studies into effective educational practices produce an unconditional model (and thus the gross effects), and then proceed to the models in which all predictors are included, thus skipping the intermediate phase of estimating models of net school effects.

From Table 3.1 it can be deduced that the studies vary neatly over sectors and countries, which were grouped into six main categories. Furthermore, most studies focussed on language and/or mathematics, while only a few used composite or science achievement scores.

Table 3.1 Characteristics of the 168 replications used in the meta-analysis

Measure	Gross	79	(47%)
	Nett	15	(9%)
	Both	74	(44%)
Sector	Primary	84	(53%)
	Secondary	74	(47%)
Subject	Language	81	(48%)
	Mathematics	72	(43%)
	Composite	11	(7%)
	Science	4	(2%)
Country	The Netherlands	55	(33%)
	United Kingdom	35	(21%)
	Europe—others	20	(13%)
	North America	25	(15%)
	Other industrialized	19	(11%)
	Third World	6	(3.6%)

The multilevel model for meta-analysis

The research design for the meta-analysis is determined by the following assumptions:

1. the studies to be reviewed are assumed to be a sample from a population of studies on school effectiveness;
2. studies with unreliable instruments and smaller sample sizes produce less reliable results, which manifests itself in larger standard errors accompanying the estimated effects;
3. for each study improved estimates of the effects can be obtained by using information from the other studies (since all studies are considered to be a sample from the same population);
4. differences in estimated effects between studies may be caused by the research designs employed.

The statistical model suited for this situation is the well-known random effects model (Raudenbush, 1994). The hierarchical structure of the data is such that originally studied units are thought to be nested within the particular study to be used in the meta-analysis. Using this approach an estimate of the average effect size is obtained, where the mean is corrected for the reliability of the information to be used. Moreover, a prediction model can be formulated to explain why certain studies may come up with larger effect sizes than others. This approach is characterized by using the variables on the research design of each study, thus reducing the between-study variance in effect sizes, and eventually resulting in a more reliable estimate of the mean average effect size.

For the meta-analysis the multilevel model approach as suggested by Raudenbush and Bryk (1985) is applied. The selected studies are considered to be a sample from the population of studies on school effects. Many studies, however, produce more than one result. Some studies, for example, used the same sample of schools for different cohorts of students, while still other studies focussed on two or more subject matter areas. The results per cohort and the results per subject matter area were treated as nested within studies. For this reason one can distinguish between the level of studies and the level of replications within studies. Nested under each replication are the tertiary units: the schools.

The standardized effect size d for comparing a set of population means (Cohen, 1969, p. 284)will be considered as the magnitude of the school effect:

$$d = \sqrt{(\rho/(1-\rho))} \tag{3.1}$$

in which ρ, the intraclass correlation, is the ratio of the variance between schools to the total variance. Although in equation (3.1) the intraclass correlation is used, the effect size d can also be expressed as the ratio of the standard deviation of the school means to the pooled within-schools standard deviation. For studies of net effects one should use the ratio of the residual variances (because that is the analog for the effect size in analysis of covariance), but since the original question is about the size of school effects as compared to other variables related to student achievement, the ratio of the residual between schools variance to the original total variance is used instead.

The multilevel model then, starting with the within-replication model, is (cf. Bryk & Raudenbush, 1992, pp. 158–161):

$$d_{rs} = \delta_{rs} + e_{rs}. \tag{3.2}$$

The effect size in replication r in study s (d_{rs}) is an estimate of the population parameter (δ_{rs}) and the associated sampling error is e_{rs} (since in each replication only a sample of schools is studied).

The between-replications model is:

$$\delta_{rs} = \delta_s + u_{rs}. \tag{3.3}$$

In this model the true replication effect size is a function of the effect size in study s and sampling error u_{rs}. Finally, the between-studies model is formulated as follows:

$$\delta_s = \delta_0 + v_s. \tag{3.4}$$

The true unknown-effect size as estimated in study s (δ_s) is a function of the effect size across studies (δ_0) with random sampling error v_s (since the studies are sampled from a population of studies).

In assessing effects of a subject domain model, equation (3.3) is extended to:

$$\delta_{rs} = \delta_s + \gamma_1 \text{subject}_{rs} + u_{rs}. \tag{3.5}$$

Effects of country and sector can modeled by extending model (3.4):

$$\delta_s = \delta_0 + \gamma_2 \text{sector}_s + \gamma_3 \text{country}_s + v_s. \tag{3.6}$$

Only a few of the studies reviewed mentioned standard errors for the estimated variance components (the size of these depends, amongst other things, on the sample size used in the study), and when they did, it was not always clear whether these standard errors related to the variance or the standard deviation (the square root of the former). For this reason the standard errors were recalculated using the following equation from Longford (1994, p. 58):

$$\text{var}(\hat{\tau}^2) = 2\sigma^4/N \times [1/(n-1) + 2\omega + n\omega^2] \tag{3.7}$$

where τ^2 is the between school variance, σ^2 is the within school variance, N is the total sample size, n is the (average) number of students per school in the sample, and ω is the variance ratio τ^2/σ^2. This approach to calculating the standard errors of the variance components is rather crude, since we have to assume balanced designs and no predictors (which is not so in the meta-analyses of the net school effects).

The sampling error variance of the effect sizes, d_{rs}, now can be calculated as:

$$\text{var}(\hat{d}_{rs}) = (1/4) \times ((1-\tau^2)/(\tau^2)) \times \text{var}(\hat{\tau}^2) \tag{3.8}$$

The analyses were done using MLn (Rasbash & Woodhouse, 1995), following and adapting a procedure suggested by Lambert & Abrams (1995).

Results

The results of the meta-analyses will be described by looking into the magnitude of the school effects, for both gross and net school effects, then trying to account for variation in results across replications and studies by modeling the potential effects of subject domain, sector and country.

Gross school effects

The results for the gross school effects are presented in Table 3.2. The results for the unconditional model are presented in the first part of the table. The estimated average effect size *d* for the gross school effect is 0.4780. There is, however, considerable variation across replications and studies: in total 0.0402. Constructing a 95% prediction interval for the average effect size for gross, uncorrected school effects results in:

$$(0.4780 - 1.96 \times \sqrt{0.0402}) = 0.0870 \text{ to } (0.4780 + 1.96 \times \sqrt{0.0402}) = 0.8730.$$

This prediction interval might be interpreted as an approximation to the population of effect sizes.

In the second part of the table the effect sizes are regressed on characteristics of the replications and studies. The intercept is the estimated effect size for language achievement of students in Dutch primary schools. The regression coefficients indicate deviations from this effect. It can readily be deduced from the results that the magnitude of gross school effects for pri-

Table 3.2 Results from the meta-analysis on gross school effects (153 replications) (Source: Bosker & Witziers, forthcoming)

	Effect	S.E.
Unconditional model		
Mean gross school effect	0.4780	0.0191
Variance across studies	0.0332	0.0056
Variance across replications	0.0070	0.0015
Conditional model		
Intercept gross school effect	0.3106	0.0638
Primary	0.0000	
Secondary	0.0732	0.0384
Language	0.0000	
Mathematics	0.0175	0.0196
Composite	0.1315	0.0481
Science	0.0001	0.0629
The Netherlands	0.0000	
United Kingdom	−0.0389	0.0614
Europe — others	0.0855	0.0503
North America	0.0829	0.0571
Other industrialized	0.0023	0.0611
Third World	0.2638	0.0859
Residual variance across studies	0.0290	0.0048
Residual variance across replications	0.0065	0.0013
% Variance accounted for	11.69%	

mary and secondary schools does not differ significantly ($z = 0.0732/0.0384 = 1.9063$, giving a p-value of 0.0566), showing larger effects for secondary schools. Since the chosen contrast is rather arbitrary with respect to the subject domains and the countries, complete contrasts were calculated for these variables. These indicate that the effect sizes for studies using composite achievement scores are significantly higher than for the other three subject domains. With respect to the countries, the contrast tests showed that the effect sizes for the Third World countries were significantly higher than those found for the other countries. Moreover, the North American studies produced effect sizes that were significantly higher than those for The Netherlands, the U.K., and the industrialized countries in the Far East and Pacific area.

In total, 11.69% of the variation in effect size estimates can be accounted for by the variables in the prediction model (sector, subject and country).

Net school effects

Net school effects are expressed in terms of the variation in school means (after adjusting for initial differences between students and schools) relative to the total initial pooled within-schools variation (i.e. without any adjustments). Table 3.3 contains the results of the meta-analysis on 89 studies or replications within studies on net school effects.

The estimated average effect size for the magnitude of net school effects is 0.3034, with the 95% prediction interval running from 0.0449 to 0.5619. The studies using composite achievement scores produce significantly higher effect size estimates than the studies on language or mathematics. These latter mathematics studies produce higher effect size estimates (0.0624 higher) than those focussing on language achievement. Third World countries have significantly larger net between-school differences than all other countries.

Discussion of the results of the meta-analyses on school effects

In this section the core question of school effectiveness research was readdressed: do schools matter? A meta-analytical approach was chosen to estimate the size of school effects quantitatively using a large number of studies from all over the world. Gross school effects had an estimated magnitude of 0.48 and net school effects had an estimated magnitude of 0.30. In other words, on average, schools account for 19% of the achievement differences between students, and for 8% when adjusting for initial differences between students. Country and subject domain affected the estimated size of school effects, whereas sector only affected gross school effects showing larger effects for secondary schools. Net school effects were higher for mathematics than for language. Using composite scores, whether looking at gross or at net effects, generally produced the highest between-school differences. Schools

Table 3.3 Results from the meta-analysis on net school effects (89 replications)

	Effect	S.E.
Unconditional model		
Mean net school effect	0.3034	0.0169
Variance across studies	0.0111	0.0031
Variance across replications	0.0063	0.0016
Conditional model		
Intercept net school effect	0.2885	0.0486
Primary	0.0000	
Secondary	-0.0116	0.0324
Language	0.0000	
Mathematics	0.0624	0.0177
Composite	0.1740	0.0597
Science	0.0820	0.0677
The Netherlands	0.0000	
United Kingdom	-0.0648	0.0391
Europe — others	-0.0788	0.0665
North America	0.0098	0.0494
Other industrialized	-0.0090	0.0537
Third World	0.1812	0.0790
Residual variance across studies	0.0078	0.0022
Residual variance across replications	0.0045	0.0011
% Variance accounted for	29.31%	

in Third World countries varied more in their output than schools in other countries, and in some instances North American secondary schools produced more differences than similar schools in other developed countries.

What , therefore, is currently known about the importance of schools? First of all, one should be aware of at least three problems concerning causality and validity.

1. The causality problem is that the term school effect suggests causation, i.e. a change in achievement level caused by the school. According to Raudenbush and Willms (1996) causation is more likely if assignment of students and schools to treatment is "strongly ignorable". This implies that adjustments for relevant covariates should be made, but either way one cannot tell whether the effects were caused by schools or by other mechanisms.

With regard to validity, there are two further problems worth mentioning, both of which concern the criterion variable used.

2. The effect size is underestimated, since measurement error in the achievement tests shows up as within-school variance; if one takes into account say, 20% "noise", the ratio of the between-school variance to the total "true" variance would improve from 8% to 11% for the net school effects and from 19% to 25% for the gross school effects (the effect sizes would increase to approximately 0.33 and 0.56, respectively).
3. The problem of curricular validity of the achievement tests used in the various studies has not yet been addressed. However, Bosker and Scheerens (1989) stress that improved curricular validity of the achievement tests will lead to higher school effects. Rowe and Hill (1994), for instance, using authentic testing procedures, present between-school differences as high as 30%of the total variance (i.e. effect sizes near 0.60).

All in all, a conservative estimate of the gross effect may be approximately 0.30, and the estimate for the net effect may be around 0.20. In terms of Cohen's effect sizes these are near medium effect sizes. The same figures can be made to look much smaller, however, when expressed as the proportion of variance in student achievement accounted for by the school attended: only 9% and 4%. Is this "much ado about nothing?". One way in which to answer this question might be to express the effects in societal merits. For the net effects, the difference between the top 10% and the bottom 10% schools is then as large as 0.65 of a standard deviation on the original test-scale. In The Netherlands this difference at the end of the primary cycle corresponds to students being placed in one of the two lower curricular tracks in secondary education versus being placed in one of the two higher tracks. Given the structure of secondary education, in general it would cost a student from the ineffective primary schools two years to attain the same certificate of achievement at the end of the secondary cycle as an equally talented student from a highly effective primary school. A technical answer to the question can be given following the conceptual idea that one school affects all of its students. The importance of the school effect then can be assessed by looking at the school total of deviations, to which the within-school variance and the between-school variance relatively contribute $1 : n\tau^2/\tau^2$ (Longford, 1994, pp. 27–28). The trick lies in the premultiplication with n. This may be the number of students per school in the sample, or a value deemed important a priori (e.g. total number of students in a cohort or, even better, the total number of students leaving the school over a number of years). The net between-school variance is then as important as the within-school variance if we consider an average class of 25 students per school, because that follows from $n \geq ((1 - 0.04)/0.04)$ It thus seems to be a matter of taste whether to judge something as important.

The Unidimensionality of the School Effect Concept

Although in the previous sections an estimate of the true magnitude of school effects was presented, one might argue that there are at least two other problems:

1. The effect size is overestimated, since the important intermediate level of the classroom is ignored. Including the intermediate level would lead to a decline in the "true" size of the school effect, since misspecification in this respect leads to a statistical artificial increase in the size of school effects (e.g. Rowe & Hill, 1994). In general, ignoring the intermediate classroom level leads to an overestimation of school effects. This overestimation amounts to variance between classes within schools divided by the average number of classes within schools.
2. We have assumed that we have investigated school effects, but there may be considerable variation in these effects within schools across subject domains, cohorts, grades and teachers. Those schools that are effective in one subject domain are not always effective in other subject domains, and stability across cohorts is sometimes problematic.

These two objections may be seen as questions regarding the unidimensionality of the school effect concept.

With the proverb "what is good for the goose is good for the gander", Walberg (1984) indicates that effective schools are effective for all groups of students. Rutter has stated that no matter which criterion is used effective schools are consistently effective (Rutter *et al.*, 1979). In other words, the schools that students enjoy attending are also the schools that make them achieve well. Is that really so?

Reviews on school effectiveness research show that the results in this field are often contradictory (Good & Brophy, 1986; Purkey & Smith, 1983; Lugthart *et al.*, 1989; Scheerens, 1989a; see Chapters Four and Five of this book) and sometimes hardly replicable (Knuver, 1989). Causes are thought to include the poor quality of the research designs and the statistical models used (e.g. Ralph & Fennessey, 1983; Kreft, 1985; Aitkin & Longford, 1986; see Chapter Nine), the different operationalizations of school factors (see Chapter Three), cultural differences between educational systems in different countries (e.g. Scheerens *et al.*, 1989; Lockheed, 1990; see Chapter Seven), and sector differences between, for instance, primary and secondary education. An even more plausible explanation for these contradictory findings may be the particular school effect measure being employed (as the ones described in Chapter Two).

In this chapter the dimensionality of the school effect concept will be investigated. In doing so we will ask ourselves the following critical questions.

1. Are school effects consistent across different modes of operationalization (i.e. value-added-based, gross, etc.)?
2. Are school effects stable over time (i.e. across grades and cohorts)?
3. Are school effects stronger than teacher effects (i.e. are school effects consistent across different subject areas)?
4. Are school effects the same for different groups of students (i.e. are there differential school effects)?

The consistency and stability of school effects comprise one of the most fundamental issues in school effectiveness research (cf. Bosker & Scheerens, 1989). The definition of consistency or stability of school effects can be operationally formulated as the correlation between two different rank orderings of schools. Consistency then refers to different criterion variables, whereas stability has to do with different time-points. So, for example, one might rank order schools on the basis of their output this year and then compare this with the rank order for the preceding year (a stability measure) or one might rank order them on the basis of their output measured as arithmetic achievement and correlate this with the rank order based on language achievement (a consistency measure).

One could argue that consistency is a key issue because the choice of the right criterion would be irrelevant if different indices for school effectiveness are empirically highly correlated, implying that school effectiveness has integrity as a comprehensive rather than a fragmented phenomenon. Furthermore, the issue of effect size might become more meaningful when schools produce the same effects year after year. In that case the argument that even a small effect size is meaningful — because it implicates an effective *school* — is underlined: these schools produce this outstanding output each year, so thousands of pupils will benefit from attending these outstanding schools. In the outlier tradition of school effectiveness research the stability question is crucial, since if the selected excellent schools are not excellent the next year, it would be hazardous to translate school processes that are seen as the causes of achievement difference into actual policy measures, such as school-improvement programs. Moreover, when parents are informed about the performance of schools to help them in choosing the right school, information as to whether the output is constantly outstanding over time, and consistent across different success domains, is decisive. The stability issue does not play a role when rewarding schools on the basis of their performance, as long as these rewards are given annually (cf. Mandeville, 1988).

The position that stability and consistency are vital issues in school effectiveness research is also based on the contention that it is a necessary condition for further theory development. Therefore one needs to know whether schools affect pupil achievement and/or attainment consistently. The consistency and stability assumption is crucial in more than one way. First of all, since organizational characteristics of schools are more or less stable over time, it is important to know whether the rank order of schools on output remains the same regardless of what criterion is used and at what point the effect is measured. The independence of effects with respect to the last specification is the most important one, because if schools affect achievement each year in a different way, and organizational features remain roughly the same, the resulting correlation between the school characteristics and the output index must be near zero.

Another aspect of stability over time is the possible existence of grade-specific effects. If school effects vary across grades this would mean — especially in primary education — that these effects are actually teacher effects.

The independence of the rank order of schools from the criterion is not a necessary assumption, but seems to be made in most school effectiveness models. These models are usually general, stating relations between school ethos, school resources, organizational features on the one hand and school success on the other, no matter what operationalization of school success is used. So, for instance, the predicted school effects for arithmetic achievement are the same as those for language performance. However, it remains to be seen whether this is really the case.

Our last contention is that if school effect theories are general and effects are to be generalized (e.g. to help parents in choosing the right school), the least one should know is whether or not schools interact with pupil background characteristics in their effects on school success. When this is true, one should consider the best school to be the school that has the highest effects for the child, considering their particular ability. Furthermore, one should know whether or not this finding is stable across years and grades and consistent across different school success criteria.

In dealing with these questions a distinction will be made between primary education and other sectors, for the following reasons.

- Educational systems in some countries are tracked in the secondary stage: when looking into stability and consistency issues here, the different hierarchically organized curricular tracks may be the cause of increased stability and consistency; this tracking may also influence the potential to discover differential school effects if students with different backgrounds are selected into different curricular tracks.
- The way in which secondary schooling is organized in virtually all countries, where specialist teachers teach each subject, raises the possibility of another layer — the subject departments in secondary schools — being effective by itself. If different results showed up for different subject areas

in primary education, at least the teacher as an individual was held constant (though not his or her mastery of that subject!).

In summary, the research will be reviewed separately for primary, secondary and tertiary educational institutions, with respect to the following stability and consistency issues:

I. Stability over time:
   (a) across years;
   (b) across grades.
II. Consistency over criteria:
   (c) across domains of success;
   (d) (in particular) across subject areas.
III. Stability across pupils with different background characteristics (also referred to in the literature as differential effectiveness).

Stability across Cohorts and Grades

Primary education

First, the issue of stability of school effects in primary education will be investigated. Mandeville and Anderson (1987) and Mandeville (1988) have presented the first results in this area. These authors use a value-added-based school effect (correcting for aptitude and socioeconomic background). Their research in elementary schools, grades 1–4 in the U.S.A. gives results for stability across different cohorts ranging from 0.34 to 0.66 (Mandeville, 1988), but these figures might be somewhat deflated because of the inadequacy of the statistical models used: Mandeville does not separate sampling variance from true parameter variance; therefore the stability is underestimated because measurement noise, sometimes amounting to nearly 80% of the observed between-year variance, confounds the effects (see Willms & Raudenbush, 1989).

Different cohorts of students leaving primary education have also been the object of study in some Dutch studies. Van Batenburg (1990) analyzed various cohorts of students who took the final Cito test at the end of primary education. Using a gross school effect measure, the stability estimate, computed as the correlation between rank orderings of schools per year or cohort, varied between 0.78 and 0.83.

Blok and Hoeksma (1993) used a database similar to that of van Batenburg (four cohorts), and their results pointed in the same direction, although the stability estimate for arithmetic (0.87) was somewhat higher than that for language (0.84) and reading (0.80).

The most recent research into this question is presented in Doolaard (1996). Using 100 primary schools where pupils were tested in 1987 and 1995, and constructing both gross and value-added-based school effect measures for both language and arithmetic, the four stability estimates were 0.53 (arithmetic, value added), 0.61 (arithmetic, gross), 0.61 (language, value added) and 0.76 (language, gross).

From a theoretical point of view stability across grades is a more interesting question. Grade-specific variation may point to teacher effects being more important than school effects. Mandeville and Anderson (1987) report correlations across grades near 0.10 (grades 1–4 in elementary schools) when using mathematics and language effects, correcting for aptitude and the socio-economic status of students. Their explanation for this inconsistency of school effects across grades is plausible. In using curriculum-specific tests, some variation may occur as a function of the grade in which the specific subject is taught to the pupils. Grade variation within primary schools has also been investigated by Bosker (1991). Comparing grade 6 and grade 8 students from the same 150 primary schools he obtained estimates running from 0.47 (value added) to 0.58 (gross) for arithmetic, and from 0.26 (value added based) to 0.69 (gross) for language.

Hill and Rowe (1996), using data from 59 schools in Victoria (Australia) and investigating grade-specific variation, report variance components that can be readily used to estimate the stability across grades 1, 3 and 5. Their results indicate a stability of 0.34 (English, gross), 0.20 (math, gross), 0.30 (English, value added) and 0.19 (math, value added).

These results suggest that teacher effectiveness is a more probable cause of differences between schools than the school's effectiveness itself. This conclusion may be corroborated when one looks at the differences between classes within grades. In this case, since the grade is constant, the only plausible explanation of potential between-class differences within schools is differences between teachers. Using the data reported by Ecob for the U.K. (in Mortimore *et al.*, 1988, p. 130, columns 1 and 2), correlations for stability across classes within grades are found, which only partly contradict the teacher effectiveness hypothesis. For reading, the correlations between the classes within grade 2 are 0.63, and 0.93 for grade 3. The figures for mathematics are 0.46 and 0.65, respectively. This topic was also investigated by Bosker (1991) in The Netherlands, who obtained results of (almost) perfect consistency for language and somewhat less consistency (0.69 for value added and 0.81 for gross) for arithmetic.

If one took the instability across grades in the Hill and Rowe (1996) study for granted, i.e. ignored this variation, and looked at the estimates of consistency across classes within the same grade, then the Australian study grades clearly points to less consistency than the U.K. and Dutch research show (0.45

for language, gross; 0.35 for math, gross; 0.39 for language, value added; 0.34 for math, value added). A strict test of the teacher effects versus school effects hypothesis is reported by Luyten and Snijders (1996), who investigated two samples of students: one with the same teacher in grades 7 and 8, and the other with two teachers for the different grades. In this way the authors could distinguish grade from teacher effects more clearly, and concluded that grade-specific variation is reduced to 20% if the teacher stayed with his or her class. In other words, 80% of the variation across grades within schools might be attributable to quality differences between teachers.

Secondary education

Research into the stability of school effects in secondary education has so far only been conducted in Europe. Evidence on this issue stems from England, France, The Netherlands and Scotland. Willms and Raudenbush (1989) used examination results data for 20 secondary schools in Scotland. The authors describe two kind of school effects, which they refer to as type A and type B effects. The type A effect deals with the question: "How well would we expect a student with average background characteristics to perform in school *j*, relative to the grand mean?" (Willms & Raudenbush, 1989, p. 213). This school effect is adjusted for student background characteristics. The type

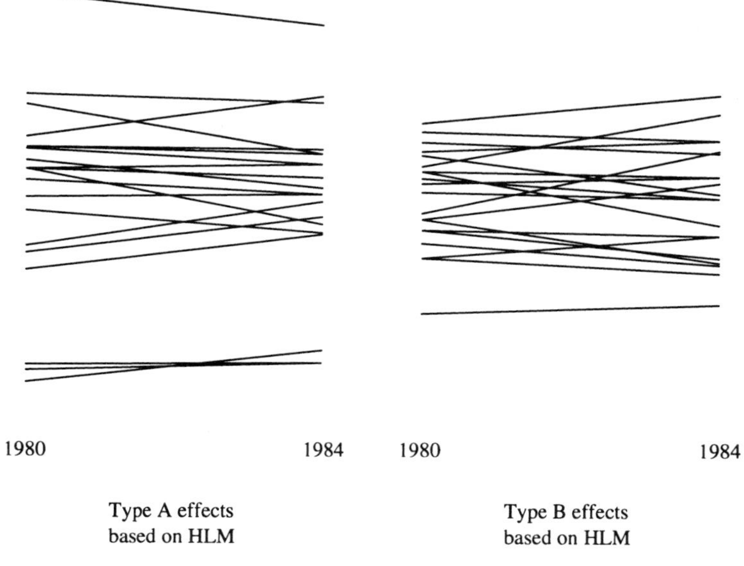

Figure 3.1 Change in school effects over time. (Source: Willms & Raudenbush, 1989.)

B effect is more restricted, since in this case corrections have also been made for the composition of the student population within a school. This type B effect approaches the "real" value added by a school, because it excludes compositional effects.

The results of their analyses are shown in Fig. 3.1. On the left of the figure one can see the change in type A school effects between 1980 and 1984. These changes are rather modest, with effective schools remaining effective, and ineffective schools remaining ineffective. There are only small changes in the group of the moderately effective schools. The correlation between the 1980 and 1984 type A school effects is 0.87. On the right of the figure one can see the change in type B school effects. Schools differ less in 1980, and changes in position are thus more likely. The correlation between 1980 and 1984 type B effects is 0.70. However, when looking into more subject-specific changes, the pattern for mathematics is far less clear (correlations between 1980 and 1984: 0.13 for type A and –0.13 for type B school effects). The results are more stable for English (0.74 and 0.79, respectively). The authors attribute the instability in subject-specific changes between 1980 and 1984 to possible teacher effects, because the 1980 students may have been taught by other teachers.

Research in The Netherlands, using examination data from the 1970s in the city of Amsterdam, has been presented by Bosker and Guldemond (1990). Their data include three cohorts of students, controlled for aptitude and gender as well as for the composition of the school populations (thus estimating the stability of type B school effects). The authors first present a stability estimate of 0.70 for gross school effects. For the value-added type B effects the stability estimate is 0.91.

Gray *et al.* (1995) presented results based on three consecutive cohorts of secondary schools' examinees in England. Correcting for prior attainment and

Table 3.4 Changes in schools' effectiveness over time (Source: Gray *et al.*, 1995)

	Position in 1992		
Position	Top quarter	Middle half	Bottom quarter
Top quarter	(1) 15%	(2) 6%	(3) 0%
Middle half	(4) 9%	(5) 35%	(6) 9%
Bottom quarter	(7) 0%	(8) 9%	(9) 18%

Fifteen per cent of the schools in the study had school-level residuals in 1990 and 1992 which placed them in the top quarter in both years (cell 1).

gender (thus using type A value-added school effects), the correlations between the three consecutive years are: 1990–1991, 0.094; 1991–1992, 0.96; and 1990–1992, 0.81. The authors also present a table to illustrate their findings (see Table 3.4). Having divided the schools in both 1990 and 1992 into three groups (the top 25%, the bottom 25%, and the middle 50%) the table shows the lack of dramatic change. Changes occur only between two adjacent groups and not otherwise.

Variation across grades within secondary schools has been the subject of only two studies. Bosker *et al.* (1989a, b) studied one cohort of students on their way through secondary schools. Since curricular tracking is predominant in the Dutch secondary school system, the authors were able to construct an interval attainment variable. This basically describes the distance between grades and tracks in terms of the years of schooling that the student in the lower track and grade would need in order to reach the higher track and grade.

The correlation matrix in Table 3.5 contains the results. These correlations represent the association between value-added-based school effects in consecutive years. Since the table is based on one cohort of students, the correlations reveal whether selection (into a track) and retention of students (in a grade) within a school show a consistent pattern over the years. On the diagonal are the correlations between consecutive years, and it can be readily seen that these are substantially higher than the off-diagonal correlations between years that are further apart.

Estimates of grade-specific variation can also be deduced from a French study (Grisay, 1996), involving 94 secondary schools in France. Gross and value-added school effects were assessed in grades 3 and 5 of secondary schooling. For mathematics the gross school effects have a correlation of 0.79, whereas for French language this correlation is 0.84. Looking at the value-added-based school effects, however, the picture is less encouraging: only 0.42 for French language and 0.27 for mathematics.

Table 3.5 Correlation between years following one cohort within secondary schools (source: Bosker *et al.*, 1989b)

1978	0.74			
1979	0.62	0.72		
1980	0.49	0.49	0.74	
1981	0.49	0.35	0.50	0.80
	1977	1978	1979	1980

The correlations in Table 3.5 are the association between value-added-based school effects in consecutive years.

Table 3.6 Consistency across subjects in primary education (partly after Luyten, 1996)

Study	Subjects	Country and region	Age groups	Number of schools and pupils	Covariates	Outcomes
Mandeville (1988)	Reading and mathematics	U.S.A., South Carolina	7-year-olds 8-year-olds 9-year-olds 10-year-olds (two cohorts each)	431 schools	School aggregates of cognitive aptitude and family background	Eight correlations, range: 0.59–0.74 median: 0.62
Bosker (1990a)	Language and mathematics	The Netherlands, national sample	12-year-olds	181 schools 2954 pupils	Cognitive aptitudes, gender and family background	One correlation: 0.73
Van Batenburg (1990)	Language, mathematics and information processing	The Netherlands, national sample	12-year-olds (three cohorts)	2382 schools	None	Nine correlations range: 0.80–0.90 median: 0.83
Sammons *et al.* (1993)	Reading and mathematics	U.K., Inner London	10-year-olds	49 schools 1240 pupils	Cognitive aptitudes, age, gender and family background	One correlation: 0.62
Hill & Rowe (1996)	Reading and mathematics	Australia, Victoria	6-year-olds 8-year-olds 10-year-olds	59 schools 6678 students	Gender, minority status, socioeconomic status	0.70 (gross) 0.51 (value added)
Luyten (1996)	Reading and mathematics	The Netherlands, national sample	12-year-olds	199 schools 2532 students	Achievement in grade 7	0.68 (gross) 0.59 (value added)

Consistency Across Subjects

Having presented stability estimates, one could conclude that school effects are stable enough to justify the concept of school effectiveness. However, they are also malleable (see the results of Doolaard, 1996) since school effects are subject to change over a longer period. Grade-specific variation within schools, however, turned out to be a predominant phenomenon, thus indicating that teacher effects may be more important than school effects.

Further evidence on the teacher versus school effectiveness hypothesis can be found by looking at subject-specific school effects, e.g. is the rank order of schools according to their effects in the mathematics domain congruent with their rank order in the language domain?

Primary education

A summary of findings from research into the consistency of school effects across subjects in primary education is presented in Table 3.6, which presents research from the U.S., the U.K., Australia and The Netherlands. In general, the research focusses on mathematics and reading (mother tongue). Mandeville (1988) looked into this consistency issue for 7–10-year-old students. The median consistency estimate was 0.62. This is exactly the same estimate that Sammons *et al.* (1993) found for 10-year-old students in the U.K.

Bosker (1990a), studying the consistency issue for 12-year-old students in The Netherlands, reports a correlation between value-added-based school effects for language and mathematics of 0.73, whereas van Batenburg (1990) reports correlations in the range of 0.80–0.90 for gross school effects.

The studies by Hill and Rowe (1996) and Luyten (1996) use a different approach in the estimation of the consistency coefficient. These authors use a multivariate multilevel model, thus explicitly modeling the covariation of mathematics and reading scores at the student as well as at the school level. Hill and Rowe report a consistency of 0.51 (value-added-based school effects). Luytens' results are of the same order of magnitude, 0.68 (gross) and 0.59 (value added).

Secondary education

The situation in secondary schools is less clear cut than in primary schools. Where inconsistency of school effects across subjects can indicate different mastery levels of the same teacher of the different subject domains, in secondary schools inconsistency indirectly points to differences between teachers or subject departments. Whereas one could argue that consistency of school effects across subject domains in primary schools may be attributable to the quality of individual teachers, this is certainly not the case in secondary schools. The research results in this area are presented in Table 3.7.

Table 3.7 Consistency across subjects in secondary education (partly after Luyten, 1996)

Study	Subjects	Country and region	Age groups	Number of schools and pupils	Covariates	Outcomes
Cuttance (1987)	English, arithmetic and overall attainment	U.K., Scotland	16-year-olds 17-year-olds 18-year-olds	456 schools 18,851 pupils	Gender and family background	Two correlations: English–overall: 0.47 Arithmetic–overall: 0.74
Willms & Raudenbush (1989)	English, arithmetic and overall attainment	U.K., Scotland	16-year-olds 17-year-olds 18-year-olds (two cohorts)	Over 6500 pupils	Cognitive aptitudes, family background (individual and school aggregate)	12 correlations: range: 0.19–0.73 median: 0.57
Thomas *et al.* (1995)	Overall attainment, mathematics, English, English literature, French, history and science	U.K., Inner London	15 years and older (three cohorts)	94 schools 17,850 pupils	Cognitive aptitudes, family background (individual and school aggregate)	21 correlations, range: 0.20–0.72 median: 0.35
Thomas & Mortimore (1996)	Overall attainment, mathematics and English	U.K., Lancashire	15 years and older	79 schools 8556 pupils	Cognitive aptitudes, age, gender and family background	Three correlations: English–maths: 0.46 English–overall: 0.65 Maths–overall: 0.68
Luyten (1996)	Mathematics, Dutch language	The Netherlands, national sample	15 years	299 schools 10,511 students	Track, achievement at age 12, family background	0.87 (gross) 0.40 (value added)

Most research in this area has been conducted in the U.K. Cuttance (1987) presented results for Scottish schools. His consistency estimates are slightly biased since the overall attainment score incorporates both mathematics and English achievement. Nevertheless, the school effects for English and the overall attainment score have a correlation of only 0.47, whereas mathematics has a correlation of 0.74 with this index.

Willms and Raudenbush (1989), also studying Scottish schools, present correlations that range from 0.19 to 0.73 (median 0.57). There appears to be even less consistency in English schools. Thomas and Mortimore (1996) found correlations of 0.65 and 0.68 for associations between English and mathematics school effects on the one hand and an overall attainment score on the other, but between the two subject domains the correlation did not reach much higher than 0.46. Further research by Thomas *et al.* (1995) using many more subject domains (French, history and science are also included) found a median correlation of only 0.35 between the subjects.

This lack of consistency is also reported in the research by Luyten (1996) on Dutch secondary schools (0.40 between mathematics and language).

Stability and Consistency: The Total Picture

It has been shown by now that school effects are stable to a certain degree, but that there appears to be a lack of consistency across subject domains. How does the picture look when the school effect is decomposed into its constituent components?

The first attempt at achieving this is shown in Bosker (1990a), who used a correlation matrix of subject and cohort-specific school effects presented by van Batenburg (1990). Analyzing this matrix, he extracted a school factor accounting for 70% of the subject and cohort-specific (gross) school effects.

The idea of unidimensionality of the school effect concept thus seems tenable. The picture, however, that arises from other research is less clear cut.

Van der Werf and Guldemond (1995) studied a series of cohorts of students in primary schools in The Netherlands. The subjects studied were arithmetic and language, and data were used on students from grades 4–8. The results clearly indicate that the unidimensionality of the value-added-based school effect concept in primary education is questionable: 31% of the school effects appeared to be year specific, 9% subject specific, and a further 21% subject and year specific, leaving only 39% to be the consistent, stable effect.

For secondary education a similar decomposition was achieved in the study by Luyten (1994). Using gross school effects, studying five cohorts of students with examination results in 17 subject domains, Luytens' results can be depicted as in Fig 3.2.

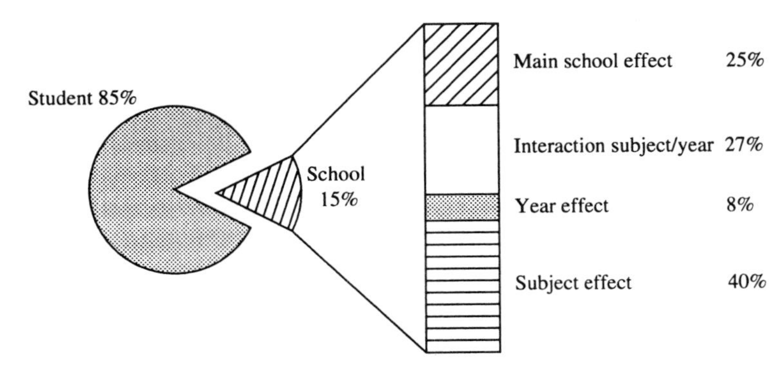

Figure 3.2 Stability of school effects. (Source: Luyten, 1994.)

Based on these findings the unidimensionality of the school effect concept in secondary schools is very questionable. Subject departments appear to be the most predominant factor (accounting for 40% of the between-school differences), whereas the consistent (across subjects) and stable (across years) school effect amounts only to 25%.

Differential Effectiveness

Until now the unidimensionality of the school effect concept has been treated along two lines: stability across different grades and/or time-points and consistency across subject domains. Another important aspect of the unidimensionality of the school effect concept is whether general effectiveness should or should not be separated from differential effectiveness. Implicit in our treatment so far has been the fact that the effectiveness of a school is its effectiveness for the average student, i.e. average with respect to aptitude, socioeconomic status, gender, etc. There are clear indications, however, that for below-average students the situation is quite different than for average or above-average students.

The first empirical research evidence in this area stems from the Coleman report (Coleman *et al.*, 1966). Coleman discovered that school effects for black students are almost twice as large as those for white students in the U.S.A. (see Tables 3.8 and 3.9). One can readily conclude from these data that if schools matter, then they matter especially for the underprivileged students.

Reanalyzing the data from *The high school and beyond* (another major piece of research by Coleman), Bryk and Raudenbush (1992) discovered that public and private secondary schools in the U.S.A. differ with respect to their relation between the socioeconomic status of a student and his/her achieve-

Table 3.8 Effects on achievement in percentages (blacks) (after Coleman, 1990, p. 82)

	School	Unknown individual causes
Grade 6	20	80
Grade 9	17	83
Grade 12	21	79

Table 3.9 Effects on achievement in percentages (whites) (after Coleman, 1990, p. 82)

	School	Unknown individual causes
Grade 6	14	86
Grade 9	10	90
Grade 12	8	92

ment level. It could be demonstrated that the differences between the public and private schools are twice as large for low socioeconomic status students as for students from middle-class families. Moreover, there are hardly any differences between the schools with respect to upper-class families.

In England, Nuttall *et al.* (1989) studied differential effectiveness for secondary schools. The criterion used was examination performance of students in 1984, 1985 and 1986. These authors conclude that school effects differ according to both aptitude and whether a student is Caribbean or British. All in all, the lower the position (either in aptitude or on the background variable) the more a school matters.

A study by Sammons *et al.* (1993) (a reanalysis of the Junior Schools data) showed that for primary schools differential effects could only be demonstrated for the prior attainment position of students, and not with respect to their gender, socioeconomic status or ethnicity status (see Fig. 3.3).

The pattern from the English study is consistent with the U.S. studies: if schools matter then this is especially so for initially low-achieving students.

Bosker (1995) studied this phenomenon for Dutch primary schools. He argues that one should study not only the within-school relation between socioeconomic status and achievement but also the between-schools relation. He then combines the stability and consistency approach with the differential effects approach, in studying Matthew effects in 150 Dutch

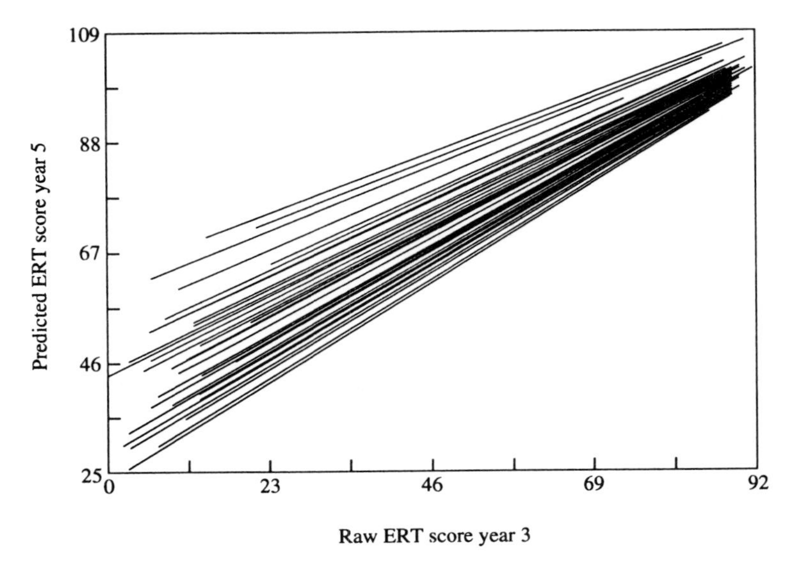

Figure 3.3 Plot of school slopes showing predicted reading score year 5. ERT: Edinburgh Reading Test (source: Sammons *et al.*, 1993).

primary schools. The Matthew effect relates to the difference in achievement levels between students of low and high socioeconomic status. His results can be summarized as shown in Table 3.10. From this correlation matrix the following conclusions can be derived.

1. Schools are highly stable in effectiveness across grades for low socioeconomic status students.
2. Schools are less stable in effectiveness across grades for high socioeconomic status students.
3. Schools are equally moderately consistent in effectiveness for both high and low status economic students across the mathematics and language domain.

Another line of research into differential school effects focusses on gender differences. Bosker and Dekkers (1994a) failed to demonstrate differential school effects for boys and girls in upper general secondary schools in The Netherlands, when analyzing differences with respect to mathematics achievement. In pre-university schools, however, they showed that schools varied widely in the difference between the numbers of girls and boys choosing mathematics. These differences were between 40 and 0%.

Table 3.10 Correlation matrix value-added school effects for low and high socioeconomic status students (Bosker, 1996)

1	1.00							
2	0.64	1.00						
3	0.99	0.65	1.00					
4	0.66	0.98	0.67	1.00				
5	0.99	0.64	0.99	0.66	1.00			
6	0.65	0.94	0.66	0.97	0.66	1.00		
7	0.64	0.49	0.68	0.56	0.68	0.55	1.00	
8	0.56	0.64	0.59	0.72	0.59	0.72	0.78	1.00
	1	2	3	4	5	6	7	8

1: language school effect for low socioeconomic status students grade 6.
2: math school effect for low socioeconomic status students grade 6.
3: language school effect for low socioeconomic status students grade 8.
4: math school effect for low socioeconomic status students grade 8.
5: language school effect for high socioeconomic status students grade 6.
6: math school effect for high socioeconomic status students grade 6.
7: language school effect for high socioeconomic status students grade 8.
8: math school effect for high socioeconomic status students grade 8.

Conclusions

In this section the dimensionality of the school effect concept was examined. It was shown that at least three foundational questions have to be investigated before deciding on this one-dimensionality issue:

1. stability of school effects across grades and cohorts of students;
2. consistency across subject domains;
3. undifferentiatedness of school effects for different groups of students.

In some cases we tried to combine these questions. The general picture that emerges from the review can be summarized as follows.

Schools are stable in effectiveness when the effects at the formal end of a schooling period are considered (e.g. the end of primary schooling) as long as the time interval is tight. Schools are far less stable in effectiveness either when the time interval is longer or in case one looks at a stage in the beginning or in the middle of a certain level of schooling. Widening the time interval as a cause for instability is not much of a concern. It shows that schools are amenable to change over a decade or so. The instability across grades within schools points to two potential causes: the curriculum and especially its timing aspect, and differences between the quality of teachers. The former aspect can be tackled in future research by looking more closely at curriculum coverage aspects when assessing school effects. The latter aspect has been studied in more detail by looking into the consistency of school effects across subjects.

The picture that emerges from this latter review is rather pessimistic from the perspective of *school* effectiveness: the almost inevitable conclusion is that teacher effects dominate over school effects.

With respect to differential school effects, i.e. differential for different groups of students, the conclusion should be that schools matter most for underprivileged and/or initially low-achieving students. Effective or ineffective schools are especially effective or ineffective for these students. A disturbing consequence of this outcome is that school choice is a most critical issue for pupils from backgrounds where it is less likely to be as much a major concern to their parents as it is for privileged students.

PART TWO: THE KNOWLEDGE BASE

CHAPTER FOUR

THE MEANING OF THE FACTORS THAT ARE CONSIDERED TO WORK IN EDUCATION

Introduction

The core of the empirically supported knowledge base on educational effectiveness is a set of factors that have been shown to be positively associated with pupils' achievement in basic school subjects. Before addressing the question of the firmness of the empirical support for these factors and the strength and direction of their association with achievement, a closer look will be taken at the conceptual meaning of the most commonly mentioned factors. Referring to the conceptual map of school effectiveness developed in Chapter One, it should be noted that the total set of factors comprises both conditions at school and conditions at classroom level and that some factors have a structural, whereas others have a more cultural nature. The central concepts in educational effectiveness are therefore only partially objective and descriptive (those related to structure). An important component relates to attitudes, perceptions and normative positions (those related to culture).

In this chapter an attempt will be made to capture the operational core of the factors that are usually mentioned in the reviews on school effectiveness research. This will be done by taking a close look at the contents of the actual instruments that have been developed within the context of empirical school effectiveness studies and as part of instruments for school self-evaluation.

The following school effectiveness studies were used as the basis for this inventory: the Junior School Project (Mortimore *et al.*, 1988), the Differential School Effectiveness Project (Sammons *et al.*, 1995a), the OECD-INES International Survey of Schools (Scheerens & ten Brummelhuis, 1996), the School Improvement and Information Service (Hill *et al.*, 1995b), the Third International Mathematics and Science Study (TIMSS) (Knuver & Doolaard,

1996; Universiteit Twente, 1995a,b), the Stability of School Effects Study (Doolaard, 1996), and the Study into School and Classroom Characteristics of Secondary Education (van der Werf & Driessen, 1993). In addition to these school effectiveness studies five Dutch school self-evaluation instruments were analyzed (Hendriks & Scheerens, 1996), as well as the Case/IMS self-evaluation system (Keefe, 1994).

Table 4.1 General effectiveness-enhancing factors

1.	Achievement orientation/high expectations/teacher expectations
2.	Educational leadership
3.	Consensus and cohesion among staff
4.	Curriculum quality/opportunity to learn
5.	School climate
6.	Evaluative potential
7.	Parental involvement
8.	Classroom climate
9.	Effective learning time (classroom management)
10.	Structured instruction
11.	Independent learning
12.	Differentiation, adaptive instruction
13.	Feedback and reinforcement

The general factors summarized in Table 4.1 were analyzed, and the elements found in the operational definitions and instruments concerning these factors will be summarized for each factor. In addition, an impressionistic view on the core of each factor will be given.

It should be noted that the selection of effectiveness-enhancing factors closely resembles the factors included in Scheerens' (1992) model, presented in Chapter Two: factors that were represented in the set of instruments that was analyzed and which are not included in Scheerens' model are: adaptive instruction, classroom climate and independent learning. When comparing the factors in Table 4.1 to the modes of schooling in Table 1.5, it is clear that these factors are only a subset of the modes, the major distinction being the omission of environmental conditions in Table 4.1.

Achievement Orientation/High Expectations

Within the set of operational definitions that was considered the following main components could be distinguished:

1. a clear focus on the mastery of basic subjects;
2. fostering high expectations on pupils' achievement, at school and teacher level;
3. the use of records of pupils' progress.

Table 4.2 contains an overview of elements that were distinguished as a further specification of these three major components.

Elements of achievement orientation or pressure for achievement that are not contained in this overview, but have been mentioned in the literature, are:

1. "placing 'attainment' on the agenda of staff meetings and in talks between the school head and individual staff";
2. "employing achievement pressure as a criterion when recruiting new teaching staff";
3. "implementing resources, including testing systems, that make it easier to introduce an achievement-oriented policy" (Scheerens, 1992, p. 87).

It is clear that the general concept of achievement orientation and fostering high expectations comprises overt policy choices, attitudes, behaviors and structural facilities. The core idea is the determination to get from pupils what they are worth, in terms of aptitude and home environment. Standard setting in such a way that pupils are challenged, but not demotivated because the standards are either too high or too low, appears to be the main structural measure in a balanced interpretation of achievement orientation — balanced in the sense that no mono-maniacal preoccupation with achievement, regardless of ability levels, is implied, but care is taken of individual differences between pupils.

Educational Leadership

In the operational definitions and instruments that were analyzed a first general division in conceptions of educational leadership can be made between:

(a) general leadership skills applied to educational organizations:
  - articulated leadership
  - information provision
  - orchestration of participative decision making
  - coordination; and

(b) instructional/educational leadership in a narrower sense, i.e. leadership directed at the school's primary process and its immediate facilitative conditions:
  - time devoted to educational versus administrative tasks
  - the head teacher as a meta-controller of classroom processes
  - the head teacher as a quality controller of classroom teachers
  - the head teacher as a facilitator of work-oriented teams
  - the head teacher as an initiator and facilitator of staff professionalization.

Table 4.2 Components and elements of achievement orientation/high expectations

Achievement-oriented school policy/high expectations
A clear focus on the mastery of basic subjects • A stronger curricular emphasis on basic subjects than on other subjects • A stronger curricular emphasis on basic subjects than on general pedagogical aims such as personal, cultural and social development • More emphasis on basic subjects now than five years earlier • Emphasis on value added or progress • In which areas has progress been made during the last five years? • Knowledge transfer and academic development have precedence over general development • Explicit statement of minimum competency levels in basic subjects • Explicit measures to improve quality of education in basic subjects
High expectations (school level) • School policy is aimed at reaching minimum competency objectives for all pupils • All teachers stimulate pupils to reach the highest possible score on an assessment test in the highest grade • Pupils do as well today as previously • Stating relatively ambitious achievement levels motivates teachers and pupils • Explicit statement of high expectations on pupils' achievement in policy plans, in communications between head teachers and teachers and by means of rewarding pupils for outstanding performance or good progress at each level of achievement • Becoming an effective school is the central mission of the school
High expectations (teacher level) • Teachers believe that high expectations on pupils' achievement stimulate school effectiveness • The degree to which teachers strive for pupils' high achievement • The degree to which teachers believe that their own perceptions influence achievement • Teachers' attitude towards the degree to which pupils' performance can be improved • The degree to which teachers strive for minimum competency levels • The degree to which teachers require high achievement of each pupil • The degree to which teachers believe that objectives and standards can be reached • Teachers emphasize that performance can always be improved • Teachers stimulate pupils to work harder • Teachers pay attention to good performance and reward good achievement • The degree to which pupils experience that teachers have high expectations of their performance
Keeping and using records on pupils' achievement • The school keeps achievement records on all pupils • The school uses achievement records to compare itself with other schools and with earlier performance

Table 4.3 provides an overview of elements belonging to these nine subcategories of educational leadership.

Of the two dimensions that were distinguished as part of the general concept of educational leadership, the second, namely leadership, focussing on the school's primary process, should be considered as central. The other dimension addresses the specific demands required for leading and controlling organizations in which professionals at the operating core need to have a considerable degree of autonomy.

Table 4.3 Components and elements of educational leadership

(a) General leadership skills

Articulated leadership

- The school leader has a clear and explicit view on how the school has to be managed
- The school leader provides clear and unambiguous leadership
- The degree to which head teachers take the lead
- The school leader has considerable discretion
- The school leader plays a major role in hiring new teachers, initiating new policy, initiating new curricular options and teaching methods

The school leader as an information provider

- Degree, timeliness and quality of information provision
- Adequate dissemination of information
- The head teacher regularly informs parents, parents' association and board
- The head teacher channels information so that it reaches the relevant people involved
- The head teacher ensures that there is enough information on the work of colleagues in order to reach sufficient coordination of tasks
- The school leader informs the teaching staff about the board's decisions

The school leader as an orchestrator of participative decision making

- The school leader uses a clear decision-making procedure
- Decisions are taken on the basis of sound and well-grounded information
- Decisions are supported by a sufficient number of staff
- The time needed to take decisions is fair
- It is clear in our school who decides on what subject
- Decisions are taken by the whole team
- Head teachers feel they can control matters at school
- The school leader engages teachers in the choice of new subject matter and teaching methods
- The classroom teacher has a say in decisions about his/her classroom
- The school leader engages personnel in the school's policy making

(cont'd)

Table 4.3 (continued)

- The school leader engages parents in decision making
- The school leader ensures that decisions taken are carried through
- Innovation is not hindered by decision making
- The head teacher ensures that clear decisions are made in staff meetings
- The school leader is firm in adhering to rules and agreements
- The school leader feels that engaging teachers in decision making stimulates school effectiveness
- The school leader engages the staff in drawing up the guidelines for running the school
- The school leader engages department heads in matching teachers and classes, staff appraisal, and policy decisions
- The school leader engages teachers in decisions on matching teachers and classes, provision of teaching aids and materials, the development of school guidelines, and the recruitment of new personnel
- There are forums in the school to express views and opinions
- Procedures for teacher appraisal are developed in conjunction with the staff
- Ease of communication with the school leader as seen from the perspective of the staff

The school leader as a coordinator

- The school leader as an initiator of staff meetings

(b) instructional leadership

Time devoted to educational versus administrative tasks

- The number of hours a head teacher teaches
- Total number of hours for managerial, non-teaching activities
- Division of school leader activities over administrative/organizational, instructional leadership, contacts with parents, own professional development
- The number of times per year/month a head teacher attends lessons, and discusses pupils' functioning with teachers
- Teachers are content with the relative emphasis the head teacher places on instructional versus other leadership tasks
- The degree to which teachers are satisfied with stimulating effectiveness-enhancing leadership

The school leader as a meta-controller of classroom processes

- The school leader is aware of pupils' progress
- The school leader initiates consultations about the progress of individual pupils
- The school leader uses records on pupils' progress as a basis to set teaching priorities, modification of curricula and methods, adaptation of teaching methods and placing pupils in ability groups
- The school leader stimulates the systematic counseling of pupils with learning and behavioral problems throughout the school

Table 4.3 (continued)

- The degree to which the school leader takes corrective action on the basis of test results
- The degree to which the school leader emphasizes specific attention to be given to weak pupils
- The school leader requires that teachers keep records on pupils' progress

The head teacher as a counsellor and quality controller of classroom teachers

- Teachers are happy with their relationship with the school leader
- Teachers experience support, appreciation, counseling and feedback from the school leader
- The school leader knows about educational practice in each classroom
- The school leader regularly asks teachers about their work
- The school leader attends lessons and talks about them with teachers
- The school leader appraises teachers
- The school leader shows his/her appreciation if teachers do a particularly good job
- The school leader encourages teachers to exploit their talents
- The school leader supports teachers who need help in carrying out improvement measures
- The school leader guides and counsels teachers during staff meetings by enquiring about how things go in classrooms in a detailed way, by discussing strong and weak points with teachers, by advising them on how to optimize instruction, by setting successful teachers as examples, and by stimulating the further development of teachers
- The school leader stimulates teachers to improve their professional craftsmanship
- The school leader may try to modify teaching strategies
- The degree to which the school leader encourages teachers and gives them feedback and recognition
- The number of times the head teacher informally communicates with one or more staff members
- Frequency of counseling contacts with beginning teachers
- The school leader uses records on pupils' achievement in appraisal interviews with teachers
- Frequency of the school leader attending lessons
- Any type of information gathering with respect to the quality of teachers

The school leader as a facilitator of work-oriented teams

- The school leader encourages the staff to work as a team
- The school leader encourages a clearly established division of tasks among staff
- Special skills of teachers are taken into account when tasks are divided among staff
- The school leader monitors the general orientation of the various subject matter areas

(cont'd)

Table 4.3 (continued)

- The school leader ensures that different learning routes are aligned
- The school leader monitors the attainment of educational objectives
- The school leader has an open mind with respect to initiatives to improve the quality of education
- The school leader takes appropriate action when desired educational and organizational aspects are not fulfilled
- The school leader and team talk about desired changes at school
- The school team is invited to put forward improvement proposals
- A supportive attitude of the head teacher with respect to the implementation of new methods of work

The school leader as an initiator and facilitator of staff professionalization

- The school leader emphasizes the importance of team development and further education
- The school leader tries to educate him/herself further by means of courses and study of literature
- The head teacher encourages further education of teachers in a selective, targeted way
- There is an explicit policy for furthering training of teachers
- Who decides about further training of teachers?
- Percentage of staff that has followed courses for further training as a teacher
- Percentage of staff that has followed courses during out of school hours/during school hours
- Has the school leader taken part in courses aimed at his/her own professionalization?

As a whole, educational leadership can be seen as a phenomenon that needs to strike a balance between several extremes: direction versus giving leeway to autonomous professionals, monitoring versus counseling and using structures and procedures versus creating a shared (achievement-oriented) culture. In this context, Sammons *et al.* (1995a) refer to the "*leading professional*".

The system-theoretical concept of meta-control is perhaps the most suitable for expressing the indirect control and influence exercised by an educationally or instructionally oriented school leader on the school's primary process. This does not imply that the head teacher is looking over the teachers' shoulder all the time, but rather that he or she is involved in important decisions on objectives and methods, and visibly cares about overall achievement levels and individual pupils' progress. It is evident from the set of components listed in Table 4.3 that the meta-control of the school leader is exercised in a non-authoritarian way, expressing concern about pupils, individual staff members and team work.

Some authors who define educational leadership say more about structural conditions surrounding the instructional process, whereas others focus more on cultural aspects. Irwin (1986, p. 126) belongs to the former category in mentioning the following aspects of educational leadership:

the school leader:

- functions as an initiator and coordinator of the improvement of the instructional program;
- states a clear mission of the school;
- has a task-oriented attitude;
- establishes clear objectives;
- supports innovation strategies;
- stimulates effective instruction;
- is quite visible in the organization;
- ensures that pupils' progress is monitored regularly;
- delegates routine tasks to others;
- regularly observes the work of both teachers and pupils.

Leithwood and Montgomery (1982, p. 334) mention the following, more cultural aspects of educational leadership:

- stimulation of an achievement-oriented school policy;
- commitment to all types of educational decisions in the school;
- stimulating cooperative relationship between teachers, in order to realize a joint commitment to the achievement-oriented school mission;
- advertising the central mission of the school and obtaining the support of external stakeholders.

In more recent views on educational leadership, inspired by the concept of the learning organization, emphasis is placed on motivating staff by providing incentives and creating consensus on goals. Mitchell and Tucker's (1992) concepts of transactional leadership and transformational leadership form a case in point. Staff development and the human resource factor are further underlined in these approaches. These newer perspectives do not create a sharp break from the longer existing conceptualizations of educational leadership, but emphasize the cultural and the staffing mode of schooling.

Scheerens (1992, p. 89) draws attention to the point that the rather heavy requirements of an educational leader do not necessarily rest on the shoulders of just one individual:

> At first glance the description of "educational leadership" conjures up an image of a show of management strength: not only the routine work necessary for the smooth running of a school, but also active involvement with what is traditionally regarded as the work sphere of the routine assignments leave sufficient time for the more pedagogic tasks. Nevertheless, this leadership does not always have to come down to

the efforts of one main leader. From the school effectiveness research of Mortimore *et al.* (1988) it emerges that deputy heads in particular fulfil educational leadership duties. Delegation can go further than this level: it is desirable that, given the consensus of a basic mission for the school, there is as broad as possible a participation in the decision making. In the end certain effects of pedagogic leadership such as a homogeneous team, will fulfil a self-generating function and act as a substitute for school leadership (according to Kerr's (1977) idea of "substitutes for leadership".

Consensus and Cohesion Among Staff

Given the traditional autonomy of teachers it is clear that consensus, cohesion and sufficient continuity for pupils, when they pass from one teacher to the next, should not be taken for granted in schools. Therefore, in many school effectiveness studies, the degree to which schools succeed in building coherence and consistency is seen as a hypothetical explanation for the fact that some schools do better than others.

In the operational definitions and instruments that were analyzed the following components of consensus and cooperation were distinguished:

- types and frequency of meetings and consultations;
- the contents of cooperation;
- satisfaction about cooperation;
- the importance attributed to cooperation;
- other indicators of successful cooperation.

Table 4.4 contains an overview of the elements that were distinguished as part of each of these five components of consensus and cooperation.

In the way in which consensus and cooperation are measured, facts, actual cooperation and frequency of sessions where staff meet and cooperate, as well as perceptions and attitudes on cooperation are included. With respect to the substance of cooperation, both agreement on overall mission and educational philosophy, and consultation on technical aspects of teaching and instruction are measured.

There appears to be no agreement on areas of cooperation that are thought to be particularly relevant. Across studies, a broad range of cooperation activities and topics on which to cooperate is chosen.

Curriculum Quality and Opportunity to Learn

The curriculum has been described as the blueprint for the functioning of the primary process in education. In articulating the curriculum and by indicating clear targets, the curriculum could function as a powerful coordination mechanism (i.e. a form of standardization). Such standardization, however, is usually balanced by the opportunity for teachers to exercise their own professional autonomy.

Table 4.4 Components and elements of consensus and cooperation in schools

Types and frequency of meetings and consultations

- Number of formal staff meetings with the head teacher
- Frequency of informal meetings among groups of teachers
- Informal contacts between staff

The contents of cooperation

Items considered important in cooperation at school:

- Pedagogical mission
- Educational concept
- School aims, objectives
- Pedagogic actions
- Planning and implementation of lessons
- Acquiring teaching methods and materials
- Discussing pupils' achievement
- Establishing entrance behavior at the beginning of the school year
- Treatment of pupils with learning difficulties
- Educational change and innovation
- Subject matter choice, assignments, achievement test, homework, preparation of lessons, observation of lessons
- Counseling of beginning teachers

Satisfaction about cooperation

- Satisfaction in relation to colleagues with respect to allocation of duties and coordination, concerning:
  - variety of interests
  - professional competence
  - supporting school improvement
  - involvement in pupils' learning and satisfaction
  - the amount of curriculum/techniques discussion in team meetings
  - acceptance, support and opportunity to cooperate
  - cooperation at school and within the team

The importance attributed to cooperation

- To what degree do head teachers agree on the importance of the following activities as effectiveness-enhancing conditions?
  - the necessity of aligning the curriculum of subsequent grade levels
  - similarity in teaching approach among grades and classrooms
  - a common policy with respect to pupils with special learning and behavioral problems
  - the use of pupil records to be passed from one grade level teacher to the next
  - the importance of cooperation within departments

Other indicators of successful cooperation

- Explicit policy aimed at furthering cooperation among staff
- Encouragement of consultations on lesson goals, teaching strategies and use of equipment
- Explicit division of tasks and coordination activities
- An established practice of team teaching
- Consensus among staff, within departments
- Frequent discussions about curriculum and teaching approach

The degree to which content that is actually taught (sometimes described as the implemented curriculum) corresponds to the test or examination of items used to assess achievement (the achieved curriculum) is usually taken into account in international comparative studies under the label "opportunity to learn".

Examination of the instruments in this area led to the following categories:

- the way curricular priorities are set;
- choice of methods and textbooks;
- application of methods and textbooks;
- opportunity to learn;
- satisfaction with the curriculum.

When overviewing the elements in the instruments, summarized in Table 4.5, the core elements appear to be:

- a clear focus of the curriculum;
- coordination and alignment of the curriculum (relationship goals and curricular choices, correspondence among grade levels, classes and teachers);
- test–curriculum overlap, or opportunity to learn.

Table 4.5 Components and elements of curriculum quality and opportunity to learn

How curricular priorities are set

- The extent to which subject matter provision is determined (i.e. guidelines are developed) by the ministry, the school board and the school team
- Knowledge about core objectives, arithmetic/math and science, the school work plan
- The importance of a good range of extracurricular activities for the school's effectiveness
- The importance of:
  - provision improvement for extending special needs in ordinary schools
  - improving preparation for the postgraduate course/profession-oriented education
- Attention to:
  - acquiring unconventional behavior
  - subject integration, factual subjects
  - realistic math education
  - introducing computers
  - the attainment targets
- Attention to learning study skills

Choice of methods and textbooks

- Availability of books for language and math
- Well-functioning methods for spelling, decoding, reading comprehension, composition writing and math, meaning:

Table 4.5 (continued)

- a clear line with regard to subject matter content
- clear directives for instruction and testing
- a step-by-step approach for the low achievers
- a clear distribution of minimum competency goals over school years

• Which language methods (in which group)
• Which arithmetic/math methods (in which group)
• Method for science

Application of methods and textbooks

• Knowledge of the manual for arithmetic/math/science methods
• The time the method is used
• Considering transfer to other methods
• Which part and which chapter in the beginning of the school year
• Which part and which chapter now
• Keeping sequence in the method
• Percentage of subject matter dealt with at the end of the school year
• Progress in method at the end of the school year
• Material for arithmetic/math/language/science other than prescribed in method
• Use of a calculator
• Percentage of pupils in a position to use a calculator

Opportunity to learn

• Percentage of time for arithmetic/math/science spent on method
• Division of lessons to subject matter components
• Other subject matter areas (within the subject)
• Number of lessons per subject matter area
• Which test items link up with education taught so far (for arithmetic/math and science)

Satisfaction with the curriculum

• Education gets shape and content in accordance with the school's vision and goals
• The extent of satisfaction with the curriculum now and five years ago
• Satisfaction with the curriculum and the teaching materials
• Satisfaction with the choice of subjects offered
• Effectiveness of the curriculum's coordination within in the school
• Successes with respect to extracurricular activities and curriculum development over the past five years
• The degree to which the work at school is considered interesting
• The extent to which a curriculum is modern
• Lessons:
  - number of lessons that stir the imagination
  - diversity of subjects

School Climate

The concept of school climate can be seen as a synonym of school culture. In the history of school effectiveness research two aspects of culture and climate have received emphasis: orderliness and achievement orientation. In the former presentation achievement orientation was treated as a characteristic of explicit, or even official policy. Achievement-oriented climate refers more to internalized norms and views of individual staff members shared with their colleagues, also in less formal relationships. A third aspect of school climate is the experience of the general "goodness" of all kinds of internal relationships and the satisfaction this gives to staff and pupils.

Indicators on the school climate range from perceptions and normative views to behavioral characteristics and factual circumstances such as a set of explicit behavioral rules, absenteeism statistics and characteristics of the school building (Table 4.6).

Rules about proper behavior and discipline express the conviction and effort of schools to suppress disruptive and negative, non-task-related activities as much as possible. In school effectiveness thinking, good relationships and satisfaction are considered instrumental to enhanced school effectiveness, and not just as aims in themselves.

The main subcategories express the breadth of scope of the school climate concept.

Table 4.6 Components and elements of school climate

(a) Orderly atmosphere

The importance given to an orderly climate

- Good discipline, pupil behavior and an orderly and safe learning environment are effectiveness-enhancing conditions
- Inconsistent approach of pupil behavior and discipline and bad pupil behavior impede the school's effectiveness
- The school has a corresponding philosophy with respect to an orderly climate
- The school head finds it important to create a quiet, orderly environment
- The extent to which a school head attaches importance to a task-oriented atmosphere
- The extent to which a teacher pursues an orderly climate

Rules and regulations

- Clear rules for pupils; pupils know where they stand
- Clear (written) rules for:
  - clothing and physical care of pupils
  - pupils doing paid jobs

Table 4.6 (continued)

- Formally recording and applying rules with respect to a.o. lateness, disturbing the lesson, absenteeism
- The extent to which school rules are recorded per subject
- Rules and sanctions with respect to discipline are well understood by staff and pupils and are not consistently broken
- The extent to which behavioral rules are honest and are being maintained
- Proportion of teachers using the following behavioral rules (a.o. looking after pupils, leaving the classroom orderly, seeing to it that the classroom is left behind clean)
- The way rules are applied in case of lateness, disturbing the lesson, cheating and truancy
- Improving and maintaining behavioral rules is an important objective for the school

Punishment and rewards

- Percentage of pupils receiving disciplinary punishment last year
- Number of rewards mentioned by the school head
- Number of punishments mentioned by the school head
- Rewards/punishments ratio
- Teacher rewards work more than punishment
- Teacher rewards behavior more than punishment
- Forms of rewards by school head (e.g. praise)
- Forms of punishments by school head (e.g. verbal warnings, confinement)
- A clearly applied system of punishment and rewards at the school

Absenteeism and drop-out

- Registration of pupils' presence/absenteeism
- Control of absentee registration by teachers
- The frequency school heads or teams are being confronted with the following behavior (of grade 6):
  - being late at school
  - being illegally absent
  - staying away from a lesson
- Measures to avoid structural canceling of lessons as much as possible
- Policy in case a teacher is absent
- Measures with respect to truancy
- Policy aimed at preventing early school leaving
- Measures to prevent early school leaving
- Measures taken when a pupil seems to become an early school-leaver

Good conduct and behavior of pupils

- Other pupils do not encourage a child teasing another child
- Teachers and pupils ensure that teaching–learning processes are undisturbed

(cont'd)

Table 4.6 (continued)

- Teachers create a learning environment in which pupils can work in a task-oriented way
- Ensure that nobody disturbs a teacher during the lesson
- The pupils behave well when the teacher leaves the classroom
- The lessons are not often disturbed by noise down the hall
- Level of pupil noise in the classroom
- Level of pupil movement in the classroom
- Teachers' audibility in the classroom
- Pupils' behavior around the school
- Strengthening pupils' behavior
- The level of pupils' unaccepted behavior now and 5 years ago
- Important successes and problems with respect to pupils' behavior and discipline now and 5 years ago
- The school's high standards of pupil behavior
- The frequency school heads or team are being confronted with the following behavior (of grade 6):
  - vandalism
  - theft

Satisfaction with orderly school climate

- A quiet, orderly learning environment at school
- The school yard, the group classrooms and the common rooms form an orderly and attractive playing/learning environment for the pupils
- The school supplies a supporting and secure environment
- Pupils and teachers feel secure at school
- There is a safe and orderly climate in my group
- Satisfaction with respect to safety at school, behavior in the classroom, the school and teachers being attentive
- Satisfaction with respect to pupils' behavior
- Degree of satisfaction with pupils' behavior now and 5 years ago
- The extent to which teachers set an example in their behavior to pupils
- Satisfaction with respect to precautions/the way the school handles vandalism, drugs, alcohol and tobacco

(b) Climate in terms of effectiveness orientation and good internal relationships

Priorities in an effectiveness-enhancing school climate

- Effectiveness-enhancing conditions of a school:
  - a caring pastoral environment
  - positive interpersonal relationships for staff and students
  - the encouragement of a positive attitude to school (pride in school)
  - shared goals and values by staff and students
  - high level of pupil motivation
  - student satisfaction

Table 4.6 (continued)

- Effectiveness-enhancing conditions of a school:
  students feel valued as people
  encouragement of student responsibility

Perceptions on general effectiveness-enhancing conditions

- Effectiveness-enhancing conditions of a school:
  - teacher motivation
  - teacher commitment/effort
  - personal effectiveness of teaching staff
  - commitment/enthusiasm of teaching staff
- Effectiveness restricting conditions of a school:
  - heavy workload
  - low staff morale
  - lack of commitment and enthusiasm by some staff
  - high teaching staff absence rates

Relationships between pupils

- How do you feel about relationships between pupils?
- Communication between pupils
- Pupils want to belong to the school and to each other

Relationships between teachers and pupils

- How do you feel about relationships between pupils and teachers?
- Contacts with pupils are open and pleasant
- The teacher/pupil social relations are good
- The team tries to understand pupils' needs
- Communication with teachers
- Teachers like pupils, support them, want them to associate nicely, know what every pupil wants, treat them fairly, etc.
- Did the school have success with respect to better relationships between teachers and pupils the past 5 years?
- Team functioning with respect to controlling pupils (firm but friendly relations)

Relationship between head teacher and pupils

- Communication between head teacher and pupils
- Head teacher listens to ideas/opinions/complaints from pupils about the climate and atmosphere

Relationships between members of staff

- Relationships between teachers
- Feeling like a member of a group
- Joint informal meetings occur a few times per year
- Colleagues lending a ready ear for personal problems

(cont'd)

Table 4.6 (continued)

- Feeling a joint/shared responsibility
- Break new colleagues in to make them feel at home
- Mutual relations aimed at learning from each other
- Staff behave according to jointly agreed rules
- Deviation of habits/rules is allowed
- Not blaming colleagues for mistakes
- Meetings are characterized by openness and commitment
- The extent of mutual confidence allows public expression of feelings of (dis)pleasure
- Considering people's wishes when taking decisions
- Colleagues pressing to put up with a decision in case of a different point of view
- A minority opposing a majority in team meetings
- Problems are usually dealt with by the team within a reasonable time
- There are conflicts
- Team morale (degree of energy, enthusiasm and spirit)
- Team functioning now and 5 years ago with respect to:
  - solidity of work relationships
  - commitment
  - shared support
  - morale
  - working hard
  - stress level

Relationships: the role of the school head

- Relationships between school head and teachers
- The school head:
  - trusts his/her team members
  - can easily be approached
  - progresses job satisfaction
  - takes suggestions and ideas of teachers with respect to work climate and atmosphere seriously
  - pays attention to solving/improving mutual relations in cases of conflict
- The behavior of school head evokes conflict

Engagement of pupils

- Pupils have a say in what happens at school
- Pupils co-decide about what happens at school
- Pupils are proud of the school and show responsibility
- Did the school have success with respect to pupils' responsibility in the past 5 years?

Table 4.6 (continued)

Appraisal of roles and tasks

- Teaching/other tasks
- Role clarity (clearly described tasks)
- Job variety
- Degree of job satisfaction

Job appraisal in terms of facilities, conditions of labor, task load and general satisfaction

- Sufficient facilities (methods/materials) to carry out work efficiently
- Salary and (secondary) conditions of labor
- Competent authority passing on to a rewarding system based at personal commitment and motivation of teachers
- Importance of part-time appointments
- Opportunities for career enhancement
- Task load (general anticipatory and perceived psychosocial mental strain):
  - in general
  - own task load
- Satisfaction with respect to working hours
- Teachers believe they are overworked and under pressure
- Average absenteeism of team members now and 5 years ago
- Quality of working life
- Satisfaction with respect to working with pupils
- Enthusiasm for the work/the school (now and 5 years ago)
- Attention for extracurricular activities
- Feeling valued in functioning as a teacher
- Opinion with respect to teachers' motivation
- Successes/problems with respect to teachers' motivation during the past 5 years

Facilities and building

- Classrooms/school building/playground clean, neat and well equipped
- Sufficient space in and around the school
- Sufficiently good facilities in and around the school
- No problems with respect to either the school's entrance or the stairs and halls in the school
- Service quality in the area of safety, advice, care, health and canteen/stay-over facilities

Evaluative Potential

The concept of "evaluation potential" (Scheerens, 1987) expresses the aspirations and possibilities of schools to use evaluation as a basis for learning and feedback at the various levels within the organization, also taking into account limitations and constraints. Aspects of this concept are:

- priority given to assessment and monitoring;
- evaluation technology (e.g. standardized pupil monitoring systems or computerized test service systems);
- use of evaluation results and records at the school level.

The main components and elements of evaluative potential as part of available instruments are listed in Table 4.7.

One of the problems in measuring schools' involvement in evaluation is the diversity in evaluation methods, which range from very informal procedures such as marking assignments to the regular use of standardized achievement tests. Also, there are several objectives of school-based evaluation:

- monitoring "normal" progress in pupils' achievement;
- diagnosing learning difficulties;
- assessment of whole school, department or classroom/teacher performance;
- school diagnosis as a basis for prospective innovations and school improvement activities;
- assessment to meet external accountability requirements;
- assessment to be used as a basis for marketing the school and informing parents and other stakeholders.

The main aspects of evaluative potential, distinguished in the introductory section on this factor, orientation, technique and use, are clearly reproduced in the instruments that were analyzed.

Table 4.7 Components and elements of evaluative potential

Evaluation emphasis
• School-wide policy with respect to marketing/assessment and regularly monitoring pupils' progress are effectiveness-enhancing conditions
• An inconsistent approach to student assessment restricts effectiveness
• The quality of education is regularly put on the agenda
• The quality of education is a central factor when discussing possible changes
• The majority of the staff is very committed and prepared to deal with quality issues

Table 4.7 (continued)

Monitoring pupils' progress

- A strong emphasis on the evaluation of test results
- Agreements and/or rules at school level with respect to testing/registration
- At our school pupils' progress is regularly tested/we handle a good testing system for progress registration to register problems with pupils in time and to take appropriate measures
- The extent to which a department head evaluates the learning progress in the department
- In groups 1 and 2 attention is paid to early signaling of pupils at risk with regard to speech-language, social-emotional, auditive, visual-spatial and motor development, concern for more cognitive activities and the task and work attitude
- The extent to which reading and arithmetic are tested
- Evaluation of pupils' progress takes place by means of standardized progress tests
- What is pupils' assessment based on (national standards, comparison with other schools, progress of the child)?
- Does the school handle achievement standards for individual pupils/standards at school level?
- (Written) rules for promotion to the next year/retention, yes/no
- Decision on promotion/retention based on opinion of teacher
- Is the school posted on pupils' functioning in further education?

The use of pupil monitoring systems

- Pupils' progress being administered in a pupil monitoring system at school level
- Evaluating pupils' progress in basic skills at least twice a year by means of a pupil monitoring system
- Registration of pupils' progress in individual pupil files, in group surveys, in central pupil monitoring system
- Which pupil monitoring system is being used and do all teachers use the same pupil monitoring system?

School process evaluation

- Has the school been assessed during the past 5 years by means of an instrument for school self-evaluation?
- Which aspects are structurally tested/evaluated, analyzed and, if necessary, improved:
  - pupil satisfaction
  - teacher achievement on the basis of pupil data
  - teacher satisfaction on the basis of:
    - functioning of the school management
    - resource expenditure

(cont'd)

Table 4.7 (continued)

- courses and teaching
- provision of education
- new teaching methods
- dissemination of innovations
- the process of educational improvement
- implemented changes
- policy formation

• Comment on each other's functioning in a positive way

Use of evaluation results

• The school being aware of possible level of changes in pupil performance during the past 5 years
• The school being aware of its position with respect to pupil performance with regard to other schools having a comparable pupil population
• For how many subjects is it possible to compare the present average achievement level with that 5 years ago?
• For how many subjects does the school compare pupil progress with other schools?
• Discussing pupils' progress and development regularly and systematically
• Evaluation of pupil performance:
  - leads to adjustment of instruction and learning strategies
  - supports assignment to ability groups
  - changes in teaching strategies
• Comparisons in achievement are used for educational improvement
• Using former pupil data for educational improvement

Keeping records on pupils' performance

• Is keeping records on pupils' performance dealt with in the school work plan?
• If yes, indications for keeping records on pupils' performance concern its recording
• Teachers keep records on pupils' development and progress
• Does the teacher keep records on language progress?
• Total number of registrations by teacher
• How often does keeping records on individual pupil's progress in documents open to the school head occur?
• Method of registration of learning progress:
  a. Standardized data
  b. Judgment by individual teacher
  c. Both a and b
  d. There is no registration
• Registration school progress:
  - none

Table 4.7 (continued)

- in individual pupil file
- in group summary
- in central pupil monitoring system

- Are pupils' data maintained through the entire school career?
- If yes, by means of automatized computer system?
- Frequency with which summaries of registration data are presented:
  - per pupil
  - per teacher
- Group summaries of pupils' achievement are made
- Use summaries per pupil/teacher to:
  - record results of written assignment
  - record test results
  - execute an error analysis
  - process pupils' achievement in pupil monitoring system at school level
- Frequency of written reports to parents (per school year/group)
- Quality of reporting of pupils' progress (all-embracing, exploratory and valuable information on pupils' progress)
- The school pays a lot of attention to reporting towards pupils and parents
- Written pupils' report when pupils pass to next school year

Satisfaction with evaluation activities

- Degree of satisfaction with the student assessment/monitoring system now and 5 years ago
- During the past 5 years, did the school succeed in establishing:
  - improved record-keeping/student profiles
  - improved monitoring of pupils' progress?
- The team's satisfaction with respect to the amount of attention paid to improving education

Parental Involvement

Continuity in home and school learning and an active involvement of parents in school matters are considered relevant in various strands of school effectiveness research. Both actual involvement and effort of the school to facilitate involvement are usually included in instruments for measuring this alleged effectiveness-enhancing factor.

A feature that does not receive much attention in the list of elements cited in Table 4.8 is a particular emphasis on contacts with parents from cultural minority and lower socioeconomic status backgrounds (compare, e.g. Cotton, 1995). Stringfield and Teddlie (1990) indicate that in neighborhoods where the majority of parents has a generally uninterested or negative attitude to schooling, buffering against parental influence rather than seeking involvement might be a more suitable policy.

Table 4.8 Components and elements of parental involvement

Emphasis in school policy

- Strong parental support as an important condition for school effectiveness
- Little parental support impedes effectiveness
- School heads and teachers are open to suggestions from parents
- The school emphasizes the importance of parental involvement with respect to education and pedagogical affairs
- The school is open for parents attending lessons
- The school has a parents' association of which parents can become a member on a voluntary basis
- Are parents in parents' committees, parents' councils or participation councils reflecting the pupils' population, and is this aimed for?
- Agreements with respect to home visits
- Facilities for parents to be present in the school
- Parents' complaints are taken seriously
- Agreement with the following pronouncements:
  - parental involvement is considered positive
  - parents are allowed to influence education's organizational structure
  - parents are allowed to influence educational contents
  - the school's and parents' responsibilities should be clearly defined
  - disappointing achievement is often due to parents not supporting the school
- A parent activity program is drafted yearly
- The school stipulates that as many parents as possible attend the individual talks about their child's progress
- The school pays specific attention to parents who are hard to reach
- The school encourages parents to help and support children at home

Contact with parents

- A good written information exchange between school and parents (school newspaper, monthly bulletin, etc.)
- Does the school inform parents about:
  - progress, yes/no
  - educational content aims
  - pedagogical/educational starting-points of the school
  - important changes with respect to content/structure of education
  - subjects dealt with in participation council meetings
- Parental involvement when deciding on:
  - policy
  - curriculum

Table 4.8 (continued)

- school planning
- finances
- personnel
- school organization

- The school head is available for parents at fixed times
- Parents can drop in with the school head any time
- Number of parents' evenings for discussing individual pupils' progress
- Number of parents' evenings for discussing general subjects
- Parents' attendance at parents' evenings about learning progress/general subjects
- All parents are visited at home at least once a year
- The extent to which parents seek for/want information with respect to their child's progress
- Teachers give parents concrete instructions with respect to supporting learning and developing skills of the children
- Percentage of parents involved in:
  - instructional/learning process
  - other school activities (e.g. library/documentation center policy)
  - out-of-school activities
  - other supporting activities
  - homework and homework conditions

Satisfaction with parental involvement

- The school's satisfaction with contacts with parents
- The school's satisfaction with respect to parents' assistance with school activities
- Teachers feel they can rely on parents
- The team can be approached easily by parents
- Parents' satisfaction with respect to the speed they are informed about pupils' progress
- Parents' satisfaction with respect to quality of report cards
- Improving parental involvement is an important goal of the school

Classroom Climate

As in school climate, the main components of classroom climate are orderliness, good relationships and satisfaction (Table 4.9). In comparison to the components that were distinguished for school climate the achievement orientation component is missing. This aspect, however, is more or less covered by another factor, namely, teacher expectations.

Pedagogical aspects, such as the enactment of moral values, in classroom interaction are also underrepresented in the set of instruments that was analyzed.

Table 4.9 Components and elements of classroom climate

Relationships within the classroom

- Classroom scores on:
  - relationships between pupils
  - relationships between teacher and pupil
- Appreciation for teacher as a companion
- Situation with respect to relationships between teacher and pupil now and 5 years ago
- Warmth towards pupils (a more rewarding than punishing position)
- Attitude teacher towards pupils (treat pupils as responsible, let pupils experience success)
- Empathy (the extent to which a teacher comprehends the pupils and take care of them)

Order

- Fairness/firmness (control in the classroom)
- Classroom scores on:
  - order in the classroom
- Rules in the group are clear for each pupil
- Creation of an orderly, quiet work environment
- Situation with respect to control (firm but friendly relations) of pupils now and 5 years ago

Work attitude

- Work attitude in the classroom
- In the group there is a (serious) atmosphere, aimed at learning
- Ensure that pupils are working on their assignment in a task-oriented manner
- Teacher energy/enthusiasm (teacher interested and enthusiastic with respect to the curriculum offered
- Pleasure in mathematics
- The use of mathematics
- Fear and difficulty

Satisfaction

- Classroom fun factor
  The fun factor is used to give an indication of whether or not it was an enjoyable experience to be a pupil in a particular teacher's class. The fun factor is the sum of all 'yes' responses to the eight items that follow:
  - Did the teacher smile often?
  - Was there positive physical contact with pupils?
  - Did the teacher show a sympathetic interest in the children other than as learners?

Table 4.9 (continued)

- Did the teacher chat to the pupils about non-work matters on any occasion during the day (whether pupil or teacher initiated)?
- Was communication between children generally cheerful?
- Was the children's behavior generally relaxed?
- Were there any jokes and/or was there any laughter in which the teacher was involved (this does not include jokes at the expense of other pupils)?
- Was there any sign that pupils wanted to be in the classroom outside class teaching time, either before or after sessions?

Effective Learning Time

Learning time can be interpreted as a measure of the quantity of exposure to "educational treatment" at school. Time can be assessed at both school and classroom level, and a distinction can be made between "planned time" (e.g. the time per subject matter area in the timetable) and "implemented time" or "time on task".
When summarizing the elements found in the set of instruments that were analyzed the following components were distinguished:

- importance of effective learning time;
- monitoring of absenteeism;
- time at school level;
- time at classroom level;
- classroom management (avoiding and minimizing ineffective time-consumers);
- homework.

These elements are further distinguished in Table 4.10. The attitudinal component "importance of effective learning time" might as well be considered as an aspect of "achievement orientation" or an "achievement-oriented climate". The rest of the components and elements appear to be amenable to construction of a one-dimensional index in which official time per subject constitutes the upper limit, and actual observed time on task during lesson hours, to which time for homework may be added, constitutes the lower limit.

Structured Instruction

Although, as will be discussed in more detail in subsequent chapters, there are diverging instruction-theoretical and pedagogical perspectives on "good instruction", in school effectiveness research the view predominates that instruction should be well structured and closely monitored. In the set of instruments that were analyzed the following components could be distinguished:

- importance of structured instruction;
- structure of lessons;
- preparation of lessons;
- direct instruction.

The main subfactors in "structured instruction" are basic requirements of well-prepared and well-controlled teaching on the one hand and aspects of direct instruction on the other (Table 4.11).

Table 4.10 Components and elements of effective learning time

Importance of effective learning time

- Emphasis on:
  - developing better policy and better procedures to lengthen instruction time
- Impeding/progressing school effectiveness:
  - good registration of presence and absenteeism
  - good class management
  - give high priority to homework

Monitoring of absenteeism

- Percentage of pupils truanting
- The way the school handles absenteeism and lateness
- Satisfaction with respect to pupils' presence now and 5 years ago

Time at school level

- Number of school days
- Number of teaching days/hours
  - Number of teaching days per school year
  - Number of full teaching days per school week
  - Number of semiteaching days per school week
  - Total number of hours per school week
  - Length of a school day
- Percentage canceling of lessons
- Number of days with no lessons due to structural causes
- Percebtage of total number of hours indicated on the table
- Measures to restrict canceling of lessons as much as possible
- Policy with respect to unexpected absenteeism of a teacher
- (In school work plan) agreements on substituting teachers

Time at classroom level

- Number of lessons on timetable per school year
- A lesson consists of how many minutes?
- Number of teaching hours for language/arithmetic
- Number of minutes for arithmetic/physics per week

Table 4.10 (continued)

- Duration of last arithmetic lesson in minutes
- Accuracy with respect to starting and finishing lessons in time now and 5 years ago
- Number of lessons that are canceled
- Satisfaction with respect to available amount of time for working in the classroom

Classroom management

- Attention to classroom management in the school work plan:
  - with respect to lesson preparation
  - rules and procedures for the lesson's course
- Situation with respect to aiming at work in the classroom (now and 5 years ago)
- Average percentage of teachers spending time on:
  - organization of the lesson
  - conversation (small talk)
  - interaction with respect to the work
  - supervision (pupil activities/behavior)
  - feedback/acknowledgement
- Average time during lesson spent on discussing homework, explaining new subject matter, maintaining order
- Sources of loss of time during lessons:
  - pupils do not know where to find equipment
  - disturbances due to bad behavior of pupils
  - frequent interruptions
  - loss of time due to lengthy transitions from one activity to the next
  - unnecessary alterations in seating arrangements
  - frequent temporarily absence of pupils during lessons
  - waiting time for individual guidance
  - many (more than three) teacher interventions to keep order
  - lack of control on pupils' task-related work

Homework

- Attention to assigning homework at school/agreements in school work plan
- Homework after last (arithmetic) lesson, yes/no
- Number of homework assignments per week
- Type of homework (arithmetic/language) (reading/composition writing)
- Amount of homework
- Amount of time needed for homework (per day)
- Extra homework for low-achieving pupils
- Successes and problems now and 5 years ago with respect to:
  - prioritizing homework
  - a consistent homework policy

(cont'd)

Table 4.11 Components and elements of structured instruction

Importance of structured instruction

- Emphasis in school's policy on:
  - the quality of teaching
  - encouraging pupils to take responsibility for their own learning process (teacher-independent learning)
  - emphasizing exam preparation
  - sufficient challenge for both high- and low-achieving pupils
- To what extent agreed upon:
  - whole class instruction gives the best results
  - discovery learning mainly needs to happen outside the school
  - pupils acquire less knowledge when different pupils do different tasks
  - repeating a year often benefits pupils' development
  - the high-achieving pupil in particular is the victim of individualizededucation
  - individualized education benefits all pupils
  - achievement is an adequate criterion when dividing pupils into groups

Structure of lessons

- Direct instruction divided into:
  - looking back daily
  - presenting subject matter
  - guided practice
  - giving feedback and correction
  - independent practice
  - looking back weekly/monthly
- Teacher uses a lesson plan

Preparation of lessons

- Lesson preparation building upon:
  - lessons previously taught
  - written plan
  - other teachers/math specialists
  - textbooks
  - standardized tests
- Most important information source for planning arithmetic/math lessons (lesson content, presentation, homework, tests):
  - core objectives
  - school work plan
  - manual
  - textbooks
  - other source books
- The subject matter is the central factor when teaching

Table 4.11 (continued)

Direct instruction

- Attention to instruction in the school work plan
- Indications in school work plan with respect to:
  - clear objectives of the instruction
  - construction of the instruction
  - method of presenting subject matter
  - the use of instructional materials
- Explanation or help to individual/groups of pupils in or outside the lesson
- Teachers deal with subject matter that corresponds to the lesson's aim
- Teacher explains at the beginning of the lesson to what prior knowledge the subject matter corresponds
- Teacher gives pupils the chance to raise questions about the last lesson
- Teacher explains beforehand what pupils should know at the end of the lesson
- Teacher knows what to achieve with the lesson
- Lesson objectives are clear to pupils
- Teacher applies instructional methods to increase pupils' achievement
- Teacher deals with only one subject matter component at a time
- Explanation in small successive steps
- Teacher takes next step when preceding step is understood
- Teacher gives concrete examples
- It appears from pupils' reactions that the teacher explains the subject matter clearly
- Teacher poses intellectual questions that invite pupils to participate actively
- After posing a question the teacher waits to let the pupils think
- Teacher gives many pupils a turn
- A lot of interaction between teacher and pupils
- Pupils respond well to questions posed by the teacher
- Teacher have pupils practiced under guidance
- Teacher continues until all pupils have mastered the subject matter
- Explanation is clear
- Teacher involves pupils in instruction
- Teacher takes care that pupils are concentrating during instruction
- During instruction immediate feedback to pupils' answers
- The lesson displays a clear structure
- Summary of subject matter at the end of instruction (by teacher/pupils)
- Pupils are given tasks they can handle
- Group work, if appropriate
- Teacher's activities (controlling) when pupils work on assignments
- Teachers take time to help pupils with tasks
- Pupils know which tasks are to be carried out
- Teacher ensures that pupils work in a concentrated way during assignments
- Teacher ensures that pupils work in a task-oriented manner during assignments

Table 4.11 (continued)

- From pupils' reactions it appears that everyone knows what he or she has to do
- There is sufficient control on pupils doing the assignments they are supposed to do
- Pupils work at a good pace
- Percentage of time during lessons in which assignments are discussed
- Analysis of mistakes
- Checks on homework

Monitoring

- Is monitoring of pupils' achievement mentioned in the school work plan?
- Indications concerning:
  - pupils' written assignment
  - the use of tests
- Percentage of lessons containing tests
- The number of tests, hearings
- Types of tests per school year (e.g. posng questions in class, own tests, curriculum-embedded tests)
- Which procedures are used to assess pupils' achievement with respect to arithmetic?
- Progress in pupil learning outcomes is measured by means of (curriculum-embedded) tests
- Teacher uses checklist for oral hearing of pupils
- The way the teacher prepares pupils for tests
- Teacher checks whether all pupils have reached the minimum goals
- Teacher checks up on difference between expected and actual pupil achievement
- Compare pupil achievement to:
  - former pupil achievement
  - fellow pupil achievement
  - norms and standards
- In what way is arithmetic/math work of a pupil judged (absolute criterion, class average, etc.)?
- Are test results used for individual help, extra explanation?
- Take action in connection with test results
- Use learning progress for:
  - preparing a program for an individual pupil
  - reporting to parents
  - informing teacher about next group
  - evaluating the school's functioning
  - putting pupils into (parallel) classes
  - selecting pupils for teaching programs (enrichment/remediation)
  - grouping pupils within classes
  - other

Table 4.11 (continued)

- The degree of pupils' progress has an effect on class level (e.g. other grouping patterns, more or less instruction)
- Successes/problems with respect to preparation for tests over the past 5 years
- Review and correct written assignment of pupils
- Use of curriculum-embedded tests
- Use of curriculum-independent tests
- Use of self-made tests

Independent Learning

Next to the direct instruction perspective, a new instructional paradigm based on constructivism has emerged. It emphasizes independent learning, use of meta-cognitive skills and learning embedded in authentic assignments and "real-life" situations. The elements of independent learning that were found in some of the instruments analyzed are summarized in Table 4.12.

Table 4.12 Components and elements of independent learning

- Attention for independent learning in school work plan
- Teacher-independent learning is encouraged, yes/no
- If yes, indications concerning:
  - relation between instruction/processing time
  - organization of independent learning
  - other types of differentiation
- State of affairs with respect to teacher-independent learning/independent learning
- The extent to which pupils are responsible for their own work
- The extent to which pupils are responsible for their own work during a longer period
- The extent to which pupils are able to choose their own assignments
- The extent to which pupils' cooperation is encouraged by teachers
- In the case of independent learning, do pupils work:
  - on the same subject
  - on various subjects per group of same level
  - on the same subject at own level
  - on various subjects at various levels?
- Opportunity for pupils to plan the school day themselves
- Successes and problems with respect to teacher-independent learning/ independent learning

Table 4.13 Elements of differentiation

- Attention for differentiation in school work plan
- Indications for differentiation concerning:
  - instruction
  - processing
- Minimum goals per class for all pupils
- Use of differentiation model: if yes, which one
- Application of setting/streaming with respect to capacities in the school/department
- How to deal with differences between pupils in arithmetic/math attainment levels during lessons (all pupils have the same subject matter ...)
- Percentage of lessons in which pupils:
  - work on the same subject
  - work on two subjects
  - work on three or more subjects
- How often do pupils work individually or in pairs?
- Percentage of teacher time spent on communication with the class, groups and individuals
- Criteria with respect to subject matter provision/grouping:
  - achievement
  - results of standardized test
  - results of diagnostic test
  - results of oral test
  - teachers' recommendations
  - parents' wishes
  - pupils' wishes
  - demands of the methods
- Pupil grouping within the class:
  - no grouping
  - age groups
  - level group
  - interest groups
  - other
- Frequency of regrouping pupils (possibly of more classes) on behalf of level groups
- Problems and successes with respect to differentiation in the past five years
- Mastery of subject matter adapted to slow and fast learners

Special attention for pupils at risk

- Policy with regard to low-achieving pupils
- School policy is explicitly aimed at catering for a wide range of educational needs: in other words, clear directives and structural attention for pupils with problems

Table 4.13 (continued)

- Catering for special individual educational needs concerning:
  - diagnosing at-risk pupils
  - remedial teaching
  - cooperation with special education
  - drafting intervention plans
  - drafting group plans
- Amount of extra time teachers are prepared to spend on problem pupils
- Extra provisions for problem pupils
- Low-achieving pupils are given more time for reflection, extra attention, instruction, help, material and exercise material
- Provisions/approved methods for preventing (teaching) problems
- Check systematically which subject matter is not being mastered
- Group teachers having expertise with regard to diagnostic test administration
- Group teachers are able to translate test data into intervention plans

Table 4.14 Elements of reinforcement and feedback

Reinforcement

- Is feedback in connection with pupils' achievement discussed in the school work plan?
- Indications for feedback in connection with pupils' achievement are related to discussion by the teacher?
- How often, in arithmetic/math lessons, do you take the following action when pupils answer incorrectly (e.g. correct wrong answer, pose different question)?
- During the lesson feedback is given and pupils' mistakes are corrected
- When pupils have carried out an assignment it is discussed immediately
- The teacher explains what was wrong when he/she returns the tests
- Teacher gives pupil as much as possible real and positive feedback on achieved results
- Frequency of discussing learning progress with pupils
- Low-achieving pupils get extra feedback

Feedback

- Results of written assignment are discussed with pupil if necessary
- Results of curriculum-embedded test are discussed with pupil if necessary
- Results of method-independent tests are discussed with pupil if necessary
- Results of self-made tests are discussed with pupil if necessary
- A differentiated supply based on tests is offered
- Quality/suitability of feedback
- State of affairs with respect to giving constructive feedback now and five years ago
- Problems with respect to inadequate feedback

Differentiation

Differentiation is aimed at instruction that is adaptive to the specific needs of subgroups of pupils. The success of differentiation is to a large extent dependent on school and classroom organization. Crucial intervening variables include time on task, and the quality of tuition during group work. Elements of differentiation are summarized in Table 4.13. The selection of elements is somewhat colored by the strong current focus on taking care of pupils with learning and behavioral problems in regular (as opposed to special) primary schools in The Netherlands.

Reinforcement and Feedback

Reinforcement and feedback are important basic conditions for learning. Elements of these instruments are summarized in Table 4.14. It should be noted that reinforcement and feedback have both cognitive and motivational implications, as a basic requirement in learning and in rewarding exertion and good performance.

Summary and Conclusions

The main components of each of the 13 general effectiveness-enhancing factors are summarized in Table 4.15. The range of components within factors in several cases shows that effectiveness-enhancing conditions are measured in terms of:

1. priorities assigned to factors and components, i.e. attitudes, beliefs, goal statements;
2. the factual state of affairs relevant to factors and components;
3. appraisal and judgment on the degree to which factors and components are realized.

There is a danger of reactivity in the measurement of (hypothetical) effectiveness-enhancing conditions, particularly with respect to the final category (appraisal), because judgment on processes and antecedent conditions may be colored by knowledge about outcomes and dependent variables.

The divergence in choice of elements for instruments across sources (i.e. instruments used in school effectiveness studies and school diagnosis instruments) is somewhat inflated, because there are sometimes slight differences between elements. It should also be noted that divergence at item level does not preclude that elements will be correlated and be shown to be subsumable under common headings, also by means of data-analytical procedures such as factor analyses. It is quite clear, however, that there is little agreement, at the operational level, on the substance of the key factors that are supposed to determine school effectiveness.

Table 4.15 Components of 13 effectiveness-enhancing factors

Factors	Components
Achievement, orientation, high expectations	Clear focus on mastering basic subjects High expectations (school level) High expectations (teacher level) Records of pupils' achievement
Educational leadership	General leadership skills School leader as information provider Orchestrator of participative decision making School leader as coordinator Meta-controller of classroom processes Time educational/administrative leadership Counsellor and quality controller of classroom teachers Initiator and facilitator of staff professionalization
Consensus and cohesion among staff	Types and frequency of meetings and consultations Contents of cooperation Satisfaction about cooperation Importance attributed to cooperation Indicators of successful cooperation
Curriculum quality/ opportunity to learn	The way curricular priorities are set Choice of methods and textbooks Application of methods and textbooks Opportunity to learn Satisfaction with the curriculum
School climate	(a) Orderly atmospheres The importance given to an orderly climate Rules and regulations Punishment and rewards Absenteeism and drop-out Good conduct and behavior of pupils Satisfaction with orderly school climate (b) Climate in terms of effectiveness orientation and good internal relationships Priorities in an effectiveness-enhancing school climate Perceptions on effectiveness-enhancing conditions Relationships between pupils

Table 4.15 (continued)

	Relationships between teacher and pupils Relationships between staff Relationships: the role of the head teacher Engagement of pupils Appraisal of roles and tasks Job appraisal in terms of facilities, conditions of labor, task load and general satisfaction Facilities and building
Evaluative potential	Evaluation emphasis Monitoring pupils' progress Use of pupil monitoring systems School process evaluation Use of evaluation results Keeping records of pupils' performance Satisfaction with evaluation activities
Parental involvement	Emphasis on parental involvement in school policy Contacts with parents Satisfaction with parental involvement
Classroom climate	Relationships within the classroom Order Work attitude Satisfaction
Effective learning time	Importance of effective learning Time Monitoring of absenteeism Time at school Time at classroom level Classroom management Homework
Structured instruction	Importance of structured instruction Structure of lessons Preparation of lessons Direct instruction Monitoring
Independent learning	(No subcomponents)
Differentiation	General orientation Special attention for pupils at risk
Reinforcement and feedback	(No subcomponents)

A further observation is that most of the factors are broad, in the sense that there is a wide range of components and elements. This is particularly the case for educational leadership and school climate. The broadness of the factors makes it hard to decide which of the set of elements is supposed to be crucial in enhancing effectiveness. Both the divergence and the broadness of the factors make summary review and quantitative research synthesis hazardous, because operationalizations of the same general factor may differ across studies.

A third and final observation is that the factors are not mutually exclusive. Zones of overlap exist between:

- achievement orientation in policy and climate;
- evaluative potential and monitoring as an aspect of structured teaching;
- curriculum aspects and coordination and consensus;
- educational leadership and use of students' records (also an aspect of evaluative potential);
- participatory decision making and consensus.

The modes of schooling that are most strongly represented in the set of instruments that was analyzed are:

- school policy;
- management/leadership;
- climate;
- curriculum;
- instruction;
- relations with parents.

Modes such as financial inputs and professional development of teachers were underrepresented in the set of instruments that was analyzed.

The main data provider for the instruments that were analyzed is the school leader, followed by the teacher. For some instruments inspectors or head of departments were the data provider. Most instruments are written questionnaires asking for self-reports from head teachers and teachers. Direct observation and structured content analysis of documents occurred in a small minority of cases.

The main conclusion from this analysis of instruments used in school effectiveness research is that there is great divergence among studies, that each project leader appears to be reinventing the wheel in the area of instrument development for measuring effectiveness-enhancing school and classroom variables and that there are no commonly used standardized research instruments to measure factors that are supposed to be the core of effectiveness-enhancing conditions.

Despite the need to adapt the choice of measurements and instruments to local circumstances, it appears worthwhile to try and develop a set of core indicators on the most promising antecedent conditions of school effects. The development of process indicators within the framework of the current OECD education indicator project can be seen as a first attempt to achieve this task in an international comparative context (Scheerens & Ten Brummelhuis, 1996).

CHAPTER FIVE

THE KNOWLEDGE BASE ON EFFECTIVENESS-ENHANCING CONDITIONS, PART 1: QUALITATIVE REVIEWS[1]

Introduction: Research Traditions in Educational Effectiveness

There are several bodies of empirical educational research that are focussed on the overall question of "what works" in education. Scheerens (1992, p. 33) distinguishes five effectiveness-related areas of research:

1. Research on equality of opportunities in education and the significance of the school in this.
2. Economic studies on education production functions.
3. The evaluation of compensatory programs.
4. Studies on effective schools and the evaluation of school improvement programs.
5. Studies on the effectiveness of teachers, classes and instructional procedures.

---

[1]The first section of this chapter and the first paragraph of the second section are adapted versions of parts of Chapter 2 of Scheerens, J. (1992). Effective schooling. Research, theory and practice. London: Cassell, plc. This material is used with permission of the publisher.

Table 5.1 General characteristics of types of school effectiveness research

	Independent variable type	Dependent variable type	Discipline	Main study type
(a) (Un)equal opportunities	Socioeconomic status and IQ of pupil, material school characteristics	Attainment	Sociology	Survey
(b) Production function	Material school characteristics	Achievement level	Economics	Survey
(c) Evaluation compensatory programs	Specific curricula	Achievement level	Interdisciplinary pedagogy	Quasi-experiment
(d) Effective schools	"Process" characteristics of schools	Achievement level	Interdisciplinary pedagogy	Case study
(e) Effective instruction	Characteristics of teachers, instruction, class organization	Achievement level	Educational psychology	Experiment, observation

Certain general characteristics of these areas of research are reproduced in Table 5.1. In the discussion of these five areas of research a general summary is given of the way in which the research was carried out and the most important study findings are broadly indicated.

School effectiveness in equal educational opportunity research

Coleman's research into equal educational opportunity, about which a final report known as the Coleman report was published in 1966, forms the cornerstone for school effectiveness studies (Coleman *et al.*, 1966). His Equal Educational Opportunity survey took place in no fewer than 4000 primary and secondary schools, with data collected by 60,000 teachers and 600,000 pupils. While this study was intended to give a picture of the extent to which school achievement is related to pupils' ethnic and social background, the possible influence of the "school" factor on learning attainment was also examined.

In the survey three clusters of school characteristics were measured: (a) teacher characteristics; (b) material facilities and curriculum; and (c) characteristics of the groups or classes in which the pupils were placed. After the influence of ethnic origin and socioeconomic status of the pupils had been statistically eliminated, it appeared that these three clusters of school characteristics together accounted for 10% of the variance in pupil performance. Moreover, the greater part of this 10% variance was due to the third cluster that was operationalized as the average background characteristics of pupils, which means that the socioeconomic and ethnic origin — now defined at the level of the school — again played a central role. In reactions to the Coleman report there was general criticism on the limited interpretation of the school characteristics. Usually, only the material characteristics were referred to, such as the number of books in the school library, the age of the building, the training of the teachers, their salaries and expenditure per pupil. Nevertheless there were other characteristics included in Coleman's survey, such as the attitude of school heads and teachers towards pupils and the attitude of teachers towards integrated education, i.e. multiracial and classless teaching.

A study just as large was carried out after the publication of Coleman's report by Jencks (Jencks *et al.*, 1972). He reanalysed statistical data at national level (U.S.A.): the Coleman findings, findings from a longitudinal study of more than 100 high schools and data from numerous smaller studies. His conclusions read as follows:

1. Schools contribute little towards bridging the gap between rich and poor, able and less able pupils.

2. The quality of education received hardly affects the postschool career of pupils, especially their future income.
3. School achievement is largely determined by one particular input factor and that is the family circumstances of a pupil. All other factors are of secondary importance and even irrelevant.
4. There are few indications that educational reforms like compensatory programs can greatly redress cognitive inequality.
5. Taking the above into account, total economic equality irrespective of intellectual capacity, can only be realized by means of a redistribution of incomes.

After reanalysing 11 studies which included data on the relationship between schooling and career, Jencks came to a similar conclusion seven years later (Jencks *et al.*, 1979). In addition to Coleman's and Jencks' surveys into unequal opportunity, there are two other important American school career studies that should be mentioned in which attention was given, although indirectly, to school characteristics. While these were commonly billed as educational attainment studies, here too the influence of social class and ethnic origin on school career was closely looked at. The surveys were centered around two longitudinal data sets: the Explorations in Equality of Opportunity Survey and the Wisconsin longitudinal study of social and psychological factors. The first-mentioned data set included school characteristics related to differentiation and internal selection procedures (track placement). According to the authors, these internal selection policies were important for pupils' careers (Alexander & Eckland, 1980). Analysis of the second file was focussed on the influence of school characteristics in the sense of contextual factors, such as the average socioeconomic status of pupils. Hauser *et al.* (1976) concluded that the influence of these contextual effects, after pupils' background characteristics such as socioeconomic status and intelligence had been taken into account, was slight.

One other survey from Thorndike falls into this category of large-scale studies on school career and environmental background of pupils (Thorndike, 1973). This was an international comparative study (15 countries) in which reading comprehension was the main achievement variable. The survey was conducted as one of the projects of the International Association for the Evaluation of Educational Achievement (IEA). The school and class characteristics in this study included not only the more material factors of the Coleman survey which had aroused criticism, but also variables such as individualized instruction and the use of reading tests. The results of this survey were, however, to a great extent in keeping with the previously mentioned findings: a relatively high correlation between socioeconomic-related family characteristics and learning attainment, and an insignificant influence from school and instruction characteristics.

In summary, the following points may be raised about this research tradition:

1. Unequal opportunity, that is to say the limitations imposed by environment on learning attainment, was central, and this explanation for differences in achievement among pupils and schools was strongly borne out by the findings.
2. School characteristics were especially related, though not in all cases, to material input factors and appeared to account for little variance in individual pupil attainment.
3. Questions which should be asked about this study area, especially when one is particularly interested in school characteristics, are:
   - how should one interpret the range of school effects found;
   - whether including school characteristics which would have been more like throughputs would have led to different results;
   - critical questions regarding the analysis techniques employed (see, for example, Aitkin & Longford, 1986).

The above-mentioned questions will be addressed in this and other chapters. Incidentally, it should be mentioned that studies on inequality of opportunity largely similar to the Wisconsin longitudinal survey have been carried out in The Netherlands (e.g. Dronkers, 1978).

Economic research into education production functions

The focus of economic approaches towards school effectiveness is the question of what manipulative inputs can increase outputs. If stable knowledge was available on the extent to which variety of inputs is related to variety of outputs it would also be possible to specify a function which is characteristic of the production process in schools: in other words, a function which could accurately indicate how a change in the inputs would affect the outputs.
This leads to a research tradition that is identified both by the term input–output studies and by the term research into education production functions. In theory, the research model for economic-related production studies hardly differs from that for other types of effectiveness research: the relationship between manipulative school characteristics and attainment is studied while the influence of background conditions such as social class and pupils' intelligence is eliminated as far as possible. The specific nature of economic-related production research is the concentration on what can be interpreted in a more literal sense as input characteristics: the teacher–pupil relationship, teacher training, teacher experience, teachers' salaries and expenditure per pupil. In more recent observations of this research type one comes across the suggestion of taking effectiveness predictors known from

educational psychology research into account (Hanushek, 1986). It should be noted that the Coleman report is often also included in the category of input–output studies. In view of its emphasis on the more material school characteristics, the association is an obvious one.

The findings of this type of research may be seen as disappointing. Review studies like those from Mosteller and Moynihan (1972), Averch *et al.* (1974), Glasman and Biniaminov (1981) and Hanushek (1979, 1986) always produce the same conclusions: inconsistent findings throughout the entire available research and scant effect at most from the relevant input variables. More comments on this issue will be provided in the next chapter, when considering the results of meta-analyses, and reanalyses of meta-analyses (cf. Hedges *et al.*, 1994).

The evaluation of compensatory programs

Compensatory programs may be seen as the active branch in the field of equal educational opportunity. In the United States compensatory programs such as Head Start were part of President Johnson's "war on poverty". Other large-scale American programs were Follow-Through — the sequel to Head Start — and special national development programs that resulted from Title 1 of the Elementary and Secondary Education Act, enacted in 1965. Compensatory programs were intended to improve the levels of performance of the educationally disadvantaged. In the late 1960s and early 1970s there were also similar programs in The Netherlands like the Amsterdam Innovation project, the Playgroup Experiment project, Rotterdam's Education and Social Environment (OSM) project and the Differentiated Education project (GEON) of the city of Utrecht.

Compensatory programs manipulate school conditions in order to raise achievement levels of disadvantaged groups of pupils. The level at which this is achieved demonstrates the importance of the school factor, and in particular the conditions and educational provisions within it.

However, it proved to be not that simple to redress the balance with effective compensatory programs. In fact, no overwhelming successes could be established. There was heated debate on the way in which available evaluation studies should be interpreted.

The key question is: what results can be realistically expected from compensatory education given the dominant influence in the long run of family background and cognitive aptitudes on pupils' attainment level? Scheerens (1987, p. 95) concluded that the general image provided by the evaluation of compensatory programs reveals that relatively small progress in performance and cognitive development can be established immediately after a program finishes. Long-term effects of compensatory programs cannot usually be established. Moreover, it has occasionally been demonstrated that it was the moderately disadvantaged in particular that benefited from the programs, while the most educationally disadvantaged made the least progress, relatively speaking.

In view of the variety of compensatory programs, the evaluation studies gave some insight into the best type of educational provision. When comparing the various components of Follow Through, programs aimed at developing elementary skills such as language and mathematics and which used highly structured methods turned out to be winners (Stebbins *et al.*, 1977; Bereiter & Kurland, 1982; Haywood, 1982).

As will be shown later, there is a remarkable similarity between these characteristics and the findings of other types of effectiveness research. In any case, when interpreting the results of evaluations of compensatory programs one should be aware that the findings have been established among a specific pupil population: very young children (infants or first years of junior school) from predominantly working-class and ethnic-minority families.

Effective schools research and the evaluation of school improvement programs

Research known under labels such as "identifying unusually effective schools" or the "effective schools movement" can be regarded as the type of research that most touches the core of school effectiveness research. All other research areas mentioned here have essentially a different focus of interest than the effectiveness of school as a whole, i.e. the effectiveness of characteristics that can be defined at school level. In Coleman's and Jencks' surveys the inequality of educational opportunity was the central problem. In economic-related input-output studies the school was even conceived as a black box. During the evaluation of compensatory programs the concern was with the effect of these particular programs and, with the still-to-be discussed research on the effectiveness of classes, teachers and instruction methods, education characteristics on a lower aggregation level than the school are the primary research object.

Effective schools research is generally regarded as a response to the results of studies like Coleman's and Jencks' from which it was concluded that schools did not matter very much when it came down to differences in levels of achievement. From titles such as "Schools can make a difference" (Brookover *et al.*, 1979) and "School matters" (Mortimore *et al.*, 1988), it appears that refuting this message was an important source of inspiration for this type of research. The most important distinguishing feature of effective schools research was that it attempted to break open the black box of the school by studying characteristics related to organization and curriculum. As far as practical research goes a few main types of effective school studies may be distinguished.

First, there are studies in which schools are identified which, after controlling for prior achievement of pupils, display an exceptionally favorable output. These schools are then looked at to determine what distinguishes them from schools with an unfavorable output. A second category of effective schools research is made possible by the influence of the first category on education practice: the findings of studies of exceptionally effective schools were adopted with great dispatch for school improvement programs. Against this background, evaluating the programs naturally produces extremely useful data on the question of which school characteristics are important for improving performance levels.

Finally, a third, fairly recent, research category can be distinguished whereby a larger scale study is made of the school characteristics related to achievement level. This last category integrates the models of the various research areas discussed in this chapter, e.g. Mortimore *et al.* (1988) and Brandsma and Knuver (1988).

Effective schools research findings appear to converge more or less around five factors:

- strong educational leadership;
- emphasis on the acquiring of basic skills;
- an orderly and secure environment;
- high expectations of pupil attainment;
- frequent assessment of pupil progress.

In the literature this summary is sometimes identified as the "five-factor model of school effectiveness". It should be mentioned that effective schools research has been largely carried out in primary schools, while at the same time studies have been largely conducted in inner cities and predominantly working-class and ethnic-minority neighborhoods.

Studies on the effectiveness of teachers and teaching methods

The topic of this book is school effectiveness. The teaching techniques that occur within a school can best be studied at the teacher and classroom level. Studying the effectiveness of teaching techniques thus takes place on a different level then studying school effectiveness. Nevertheless, attention is given here to studies on effective instruction because the impact of effectiveness-promoting school characteristics on pupil performance largely happens via class teaching techniques. This can be represented diagrammatically as a step-by-step, causal process (see Fig. 5.1).

In Fig. 5.1, C indicates that school characteristics can also directly influence performance. As an illustration, the often found research result that the favorable effect of frequent assessment of pupils' progress is established can be referred to. In the first instance this is a matter that takes place at class level (the influencing connection is arrow B in Fig. 5.1). Evaluating at class level, however, can be highly stimulated when much importance is ascribed to assessment at school level; in other words, when a certain policy priority is adhered to. Another evaluation-promoting condition at school level could be a pupil monitoring system or a "test-service system" for the entire school (represented by arrow A in Fig. 5.1). Finally, it is conceivable that as part of the management duties of running a school, the school head, with or without the help of a testing system, checks on the progress of pupils in various classes from time to time (this is an example of relation C in Fig. 5.1).

A review of the study findings within the field of effective teaching could easily fill a separate book. Here, only a short review of various approaches of research on teaching will be given.

A review of approaches in research on teaching and instructional effectiveness.

In the 1960s and 1970s the effectiveness of certain personal characteristics of teachers was particularly studied. Medley and Mitzel (1963), Rosenshine and Furst (1973) and Gage (1965) are among those who reviewed the research findings. From these studies it emerged that there was hardly any consistency

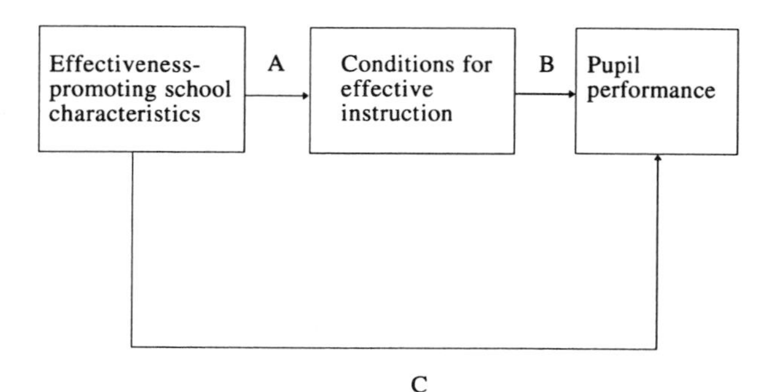

Figure 5.1 Step-by-step causal process with school and instruction conditions as malleable factors.

between personal characteristics of the teacher, such as warmheartedness or inflexibility, on the one hand, and pupil achievement on the other. When studying teaching styles (Davies, 1972), the behavioral repertoire of teachers was generally looked at more than the deeply rooted aspects of their personality. Within the framework of "research on teaching" there followed a period in which much attention was given to observing teacher behavior during lessons. The results of these observations, however, in as far as they were related to pupil achievement, seldom revealed a link with pupil performance (e.g. see Lortie, 1973). In a following phase more explicit attention was given to the relation between observed teacher behavior and pupil achievement. This research is identified in the literature as "process-product studies". Lowyck, quoted by Weeda (1986, p. 68), summarizes variables which emerged strongly in the various studies:

1. *Clarity:* clear presentation adapted to suit the cognitive level of pupils.
2. *Flexibility:* varying teaching behavior and teaching aids, organizing different activities, etc.
3. *Enthusiasm:* expressed in verbal and non-verbal behavior of the teacher.
4. *Task-related and/or businesslike behavior:* directing the pupils to complete tasks, duties, exercises, etc., in a businesslike manner.
5. *Criticism*: much negative criticism has a negative effect on pupil achievement.
6. *Indirect activity*: taking up ideas, accepting pupils feelings and stimulating self-activity.
7. *Providing the pupils with an opportunity to learn criterion material*, i.e. a clear correspondence between what is taught in class and what is tested in examinations and assessments.
8. Making use of *stimulating* comments: directing the thinking of pupils to the question, summarizing a discussion, indicating the beginning or end of a lesson, emphasizing certain features of the course material.
9. *Varying the level* of both cognitive questions and cognitive interaction.

Weeda (1986, p. 69) observes that in the study from which these nine teaching characteristics were drawn, there was much criticism regarding methodology and technique. He divides the later research studies focussed at instructional effectiveness into two areas:

1. pedagogic studies aimed at tracing certain environmental factors and teaching behavior that can influence levels of performance of certain groups of pupils;
2. instructional psychology research aimed at establishing the interaction between teaching variables and pupil characteristics; the *aptitude–treatment–interaction* studies.

A central factor within the first area is that of effective teaching time. The theoretical starting points of this can be traced back to Carroll's teaching-learning model (Carroll, 1963). Chief aspects of this model, as were also cited in the previous chapter, are:

- actual net learning time, which is seen as a result of perseverance and opportunity to learn;
- necessary net learning time, as a result of pupil aptitude, quality of education and pupil ability to understand instruction.

The mastery learning model formulated by Bloom in 1976 was largely inspired from Carroll's model.

The findings of the aptitude–treatment–interaction (ATI) studies were generally judged to be disappointing. Scarcely any interactions were discovered, which was later confirmed by a replication study. De Klerk (1985) regarded the fact that the ATI had failed to reveal any simple interaction between pupil characteristics and instruction method as a challenge to conduct more refined empirical research on more complex interaction patterns.

Stallings (1985) summarized research literature on effective instruction — in as far as it was concerned with primary education —under the headings: effective net learning time, class organization and management, instruction, assessment and teacher expectations.

When studying net learning time it emerged that simply making the school day longer did not necessarily lead to better levels of performance. More important, ultimately, is how effectively time is spent. Stallings and Mohlman (1981) established that effective teachers spent 15% of the school day on organization and management, 50% on interactive teaching and 35% on monitoring pupils' work. Aids for an effective use of instruction time include all types of lesson planning. Under the classification *class organization and management* Stallings discusses streaming and maintaining order. Studies on streaming or working with ability groups as compared to whole class instruction indicate that this type of teaching works more positively with the more gifted pupils, whereas with less able groups — taking the average result of the large numbers of surveys — hardly any effect was found (also according to Kulik & Kulik, 1982; Van Laarhoven & De Vries, 1987, Reezigt, 1993; Slavin, 1987). Moreover, from various types of studies it emerges that in classes where there is disruptive behavior, pupil performance is lower: disruption, naturally enough, is at the cost of effective learning time.

The question of what makes good teaching should be looked at on different levels. For direct question-and-answer type knowledge teaching strategies other than problem solving and acquiring insight are called for. For learning tasks which greatly depend on the memory, a highly ordered and consistent approach is the most effective. For acquiring insight, a clear presentation of the information offered is important, as are questions to check whether pupils have actually absorbed a specific insight. With regard to problem solving, some empirical support is available which shows that it is desirable that pupils take much initiative themselves. Collins and Stevens (1982) mention five teaching strategies to support learning in the sense of problem solving: (a) a systematic variation of examples; (b) counter-examples; (c) entrapment strategies; (d) hypothesis identification strategies; and (e) hypothesis evaluation strategies.

From studies on teacher assessments and expectations of pupils it seems that self-fulfilling prophecies can occur. Once a teacher has formed negative expectations of certain pupils, he or she is likely to give them less attention and expose them less to difficult and challenging tasks. Obviously this is even more of a disadvantage if the initial assessment was a wrong one. Thus it is imperative that teachers should try to avoid negative stereotyping of pupils (van der Hoeven-van Doornum, 1990).

In a review of literature on effective teaching at secondary school level, Doyle (1985) deals broadly with the same category as Stalling's, namely "time on task" and "quality of instruction". Because in secondary education the total teaching spectrum from which a choice must be made is far greater than in primary education, the variable "opportunity to learn" is associated here with the concept of effective net learning time. By "opportunity to learn" it is generally understood that pupils are offered a range of subjects and tasks that cover educational goals. In educational research, opportunity to learn concentrates on the extent to which classroom exercises correspond with the content of the tests for monitoring performance.

As far as the quality of instruction is concerned, there is a stronger emphasis in secondary education on learning higher cognitive processes such as insight, flexibly adopting knowledge and problem solving. Doyle considers the effectiveness of direct teaching, which he defines as follows.

1. Teaching goals are clearly formulated.
2. The course material to be followed is carefully split into learning tasks and placed in sequence.
3. The teacher explains clearly what the pupils must learn.
4. The teacher regularly asks questions to gauge what progress pupils are making and whether they have understood.
5. Pupils have ample time to practice what has been taught, with much use being made of prompts and feedback.
6. Skills are taught until mastery of them is automatic.
7. The teacher regularly tests the pupils and calls on the pupils to be accountable for their work.

The question of whether this type of highly structured teaching works equally well for acquiring complicated cognitive processes in secondary education can be answered in the affirmative (according to Brophy & Good, 1986, p. 367). However, progress through the subject matter can be taken with larger steps, testing need not be so frequent and there should be space left for applying problem-solving strategies flexibly. Doyle also emphasizes the importance of varying the learning tasks and creating intellectually challenging learning situations. For the latter an evaluative climate in the classroom, whereby daring to take risks even with a complicated task is encouraged, is a good means. In addition, Doyle deals with the effect of certain ways of working and grouping, including individual teaching and working together in small groups. A meta-analysis by Bangert *et al.* (1983) revealed that individualized teaching in secondary education rarely led to higher achievement and had no influence whatsoever on factors such as the self-esteem and attitudes of pupils.

Evaluation studies on special programs to stimulate working in small groups reveal that some of these have a positive effect on lower attaining pupils. Generally speaking, from other reviews of research on the effects of cooperative learning it appears that there is no conclusive empirical evidence to support the positive influence of this type of work on performance. Vedder (1985) explained the lack of an unequivocal positive influence of group work by the possibility that, owing to the way pupils work together, there is insufficient cognitive stimulation present.

Making up the balance

Five types of educational effectiveness research were discussed in this section. The fact that various subdisciplines of educational science all address effectiveness questions indicates the centrality of these questions in educational inquiry as a whole. The five effectiveness-related areas of research examined in this chapter were:

1. research on equality of opportunities in education and the significance of the school in this;
2. economic studies on education production functions;
3. the evaluation of compensatory programs;
4. studies on effective schools and the evaluation of school improvement programs;
5. studies on the effectiveness of teachers, classes and instruction procedures.

The first two research traditions focussed on "material" school characteristics (e.g. teacher salaries, building facilities and teacher/pupil ratio). The results were rather disappointing in that no substantial positive correlations of these material investments and educational achievement could be established in a consistent way across individual studies.

In-depth process studies connected with large-scale evaluations of compensatory programs pointed out that programs which used direct, i.e. structured, teaching approaches were superior to more open approaches. The research movement known as research on exemplary effective schools (or briefly, effective schools research) focussed more on the internal functioning of schools than the earlier tradition of input–output studies (those summarized under headings 1 and 2 above).

These studies produced evidence that factors such as strong educational leadership, emphasis on basic skills, an orderly and secure climate, high expectations of pupil achievement and frequent assessment of pupil progress were indicative of unusually effective schools.

Research results in the field of instructional effectiveness are centered around three major factors: effective learning time, structured teaching and opportunity to learn in the sense of a close alignment between items taught and items tested.

Although all kinds of nuances and specificities should be taken into account when interpreting these general results they appear to be fairly robust, as far as educational setting and type of students are concerned. The overall message is that an emphasis on basic subjects, an achievement-oriented focus an orderly school environment and structured teaching, including frequent assessment of progress, is effective in the attainment of learning results in the basic school subjects. In the following chapters more specific information will be given about the research findings and the way in which they should be interpreted.

A Review of Review Studies

With regard to school effectiveness research, it has sometimes been jokingly remarked that there have been more reviews published than actual original study reports. Certainly when the quality of the studies under review is looked at this is not far from the truth: the number of empirical school effectiveness studies from which generalized conclusions can be drawn is still rather modest. English-language review articles include Anderson (1982), Cohen (1982), Dougherty (1981), Edmonds (1979), Murmane (1981), Neufeld *et al.* (1983), Purkey and Smith (1983), Rutter (1983), Good and Brophy (1986), Ralph and Fennessey (1983), Kyle (1985), Sweeney (1982), Borger *et al.* (1984) and Levine and Lezotte (1990). Reviews in Dutch have been produced by, among others, van de Grift (1987), Scheerens and Stoel (1987), Scheerens, 1992, Creemers 1994 and Lugthart *et al.* (1989).

Early studies

The Purkey and Smith review (1983)

Purkey and Smith distinguish four types of school effectiveness research: studies among outliers, i.e. among extremely high or extremely low attaining schools; case studies; evaluations of school improvement programs and other studies.

Outlier studies

When studying outliers a school is first looked at to see whether it has performed better or worse than one would assume on the basis of the characteristics of the pupil population. By means of regression analysis one is able to tell which pupil background characteristics affect achievement, on the basis of which one is then able to predict the achievement level of the school. If a school clearly comes out above this predicted level, one can hypothesize that this is due to specific structural, curriculum or teaching characteristics. Purkey and Smith discuss nine outlier studies all related to primary schools. While the results are varied, many similarities could also be found. The prevailing effectiveness characteristics found to be in common were: good discipline, teachers' high expectations regarding pupils' performance and an emphasis on educational leadership (a school head that is actively involved with stimulating and monitoring activities related to teaching rather than simply acting as an administrator or business manager). The interpretation of these results, however, is undermined by weak research techniques such as very small random samples (between two and 12 schools) and shortcomings in the identification procedure of the outliers (owing to the fact that the effect of important background characteristics such as socioeconomic status were insufficiently neutralized). There is also criticism of the comparison criterion of the outlier studies. By comparing extremely good schools with extremely weak ones the effect of the school differences is exaggerated; in this context it is proposed that researchers compare the more positive outliers with the average schools. In addition, whenever the effectiveness of extreme schools is compared over several points in time, one is almost certain to find minor effects by later measurements as a result of the regression-towards-the-mean phenomenon.[2]

---

[2]The regression-towards the mean phenomenon appears whenever two groups are selected on the basis of unreliable measurements. As far as high- or low-scoring groups are concerned, capitalizing on chance will occur during selection, owing to the unreliable nature of the measurements.

Case studies

The seven case studies used by Purkey and Smith in their review include well-known ones such as those of Weber (1971), Brookover *et al.* (1979) and Rutter *et al.* (1979). (An eighth study discussed by Purkey and Smith is ignored here because this was, in fact, an evaluation study.) It should be remarked that classifying Brookover's research project as a case study is misplaced: no fewer than 159 schools were involved here!

The school factors that emerged from these case studies as positively influencing effectiveness are summarized in Table 5.2. There was little consistency in the standards used to decide whether or not a factor could be considered important: sometimes this only rested on the subjective opinion of the researchers, without any form of quantitative basis being given.

From this review it is apparent that there is remarkable consensus on the importance of school factors such as strong leadership (mentioned four times), orderly climate (three times), high expectations (six times), achievement-oriented policy (four times) and time on task (three times).

Purkey and Smith state that the case studies are open to the same research technical criticism as was applied to the outlier studies. Bearing in mind the earlier remark made on the study of Brookover *et al.*, this study should be seen as a forceful exception to this unfavorable rule.

Table 5.2 The most important school characteristics that emerged from seven case studies, discussed by Purkey and Smith (1983)

Author	Weber (1971)	Venezky & Winfield (1979)	Glenn (1981)	Cal. State (1980)	Brookover & Lezotte (1979)	Brookover *et al.* (1979)	Rutter *et al.* (1979)
No. of schools	4	2	4	–	8	159	12
Strong leadership	x		x	x		x	
Orderly climate	x		x				x
High expectations	x		x	x	x	x	x
Frequent evaluation	x		x	x			
Achievement-oriented policy		x		x		x	x
Cooperative atmosphere		x	x				x
Clear goals for basic skills			x		x		
In-service training/ staff development		x		x			
Time on task					x	x	x
Reinforcement						x	x
Streaming						x	x

Program evaluations

Purkey and Smith discuss six evaluation studies wherein most of the programs to be assessed were compensatory ones. In three studies in Michigan, factors which appeared to be related positively to achievement in earlier studies were implemented in three schools. After one year these three experimental schools appeared to perform slightly better than the control schools. This result confirms once more the five-factor model (mentioned on p. 146).

Van de Grift (1987, pp. 25–27) discusses the evaluation of three American school improvement programs: the New York City School Improvement Project, the Rising to Individual Scholastic Excellence Project in Milwaukee and the New York Local School Development Project. In these projects the effective school characteristics, by now familiar to the reader, comprised the basis of the reforms. From the evaluations it appears that in each of the three project schools much more progress in achievement level was realized than in schools not involved in the project. The evaluation of school improvement programs greatly inspired by school effectiveness research results is looked at more closely later in this chapter.

Other studies

Under the category "other school effectiveness studies", Purkey and Smith deal with two large-scale surveys: the comparative study of Coleman *et al.* (1981) of public and private schools and the Safe School Study of the American National Institute of Education. Coleman *et al.* (1981) concluded that private schools achieve more than public ones, and this was attributed to factors such as less school absence, more homework, more academic subject matter and greater demands made on achieving. From the NIE's Safe School Study it emerged that school characteristics associated with greater safety were also associated with better performance. Clear-cut rules, high staff morale, strong leadership and press for achievement appeared to be the factors that positively influenced both objectives of safety and a high achievement level.

More recent reviews (1990–1995)

In this section reviews by Levine and Lezotte (1990), Sammons *et al.* (1995) and Cotton (1995) are analyzed by focussing on some 10 key characteristics. The way that these categories are labeled in each study is shown in Table 5.3.

Introduction of the three review studies

Levine and Lezotte (1990, pp. 9–38) discuss key correlates of effective schools. Such correlates are generally referred to as those aspects of organizational functioning on which unusually effective schools rank higher than ineffective schools. By using the distinction between unusually effective and ineffective schools the implicit research design on which Levine and Lezotte's analysis depends is the outlier design, in which, for instance, the 25% best performing schools are compared to the 25% least successful schools.

In their discussion of the key correlates the authors make several qualifying comments regarding the strength and equivocality of interpretation of these factors.

Table 5.3 Effectiveness-enhancing conditions of schooling in three review studies (italics in the column of the Cotton study refers to subcategories)

Levine and Lezotte (1990)	Sammons *et al.* (1995a)	Cotton (1995)
Productive climate and culture	Shared vision and goals A learning environment Positive reinforcement	Planning and learning goals Curriculum planning and development
Focus on central learning skills	Concentration on teaching and learning	School-wide emphasis on learning
Appropriate monitoring	Monitoring progress	Assessment (district, school, classroom level)
Practice-oriented staff development	A learning organization	*Professional development* Collegial learning
Outstanding leadership	Professional leadership	School management and organization Leadership and school improvement Leadership and planning
Salient parent involvement	Home–school partnership	Parent community involvement
Effective instructional arrangements	Purposeful teaching	Classroom management and organization Instruction
High expectations	High expectations	Teacher–student interactions
	Pupil rights and responsibilities	
		Distinct–school interactions
		Equity
		Special programs

Table 5.4 Characteristics of unusually effective schools (cited from Levine & Lezotte, 1990)

Productive school climate and culture
- Orderly environment
- Faculty commitment to a shared and articulated mission focussed on achievement
- Problem-solving orientation
- Faculty cohesion, collaboration, consensus, communications, and collegiality
- Faculty input into decision making
- School-wide emphasis on recognizing positive performance

Focus on student acquisition of central learning skills
- Maximum availability and use of time for learning
- Emphasis on mastery of central learning skills

Appropriate monitoring of student progress

Practice-oriented staff development at the school site

Outstanding leadership
- Vigorous selection and replacement of teachers
- "Maverick" orientation and buffering
- Frequent, personal monitoring of school activities, and sense-making
- High expenditure of time and energy for school improvement actions
- Support for teachers
- Acquisition of resources
- Superior instructional leadership
- Availability and effective utilization of instructional support personnel

Salient parent involvement

Effective instructional arrangements and implementation
- Successful grouping and related organizational arrangements
- Appropriate pacing and alignment
- Active/enriched learning
- Effective teaching practices
- Emphasis on higher order learning in assessing instructional outcomes
- Coordination in curriculum and instruction
- Easy availability of abundant, appropriate instructional materials
- Classroom adaptation
- Stealing time for reading, language, and math

High operationalized expectations and requirements for students

Other possible correlates
- Student sense of efficacy/futility
- Multicultural instruction and sensitivity
- Personal development of students
- Rigorous and equitable student promotion policies and practices

First, they say that analysts have usually tried to "identify correlates at a level of generality sufficient to allow for a variety of manifestations in practice while still pointing toward key specific aspects of school effectiveness". This observation corresponds to the analysis of instruments and operational definitions in the preceding chapter.

Second, they point out that most of the research evidence on which their analysis rests relates to elementary and intermediate schools.

Third, when discussing their key correlates, they sometimes refer to the fact that such factors have received empirical support in some studies, but not in others.

A fourth and final qualifying remark is the assumption that a particular correlate in itself may have a low or insignificant association with educational outcomes, but in conjunction with other correlates may operate as an effectiveness-enhancing composite. So, for instance, consensus and cohesion among staff may not be sufficient in itself to enhance effectiveness, but when it stands for a shared achievement-oriented effort, this may be an important conglomerate of effectiveness-enhancing conditions.

The effective schools correlates distinguished by Levine and Lezotte are summarized in Table 5.4.

In their review, Sammons *et al.* (1995a) followed a similar approach to the Levine and Lezotte review. They depend less on case-study evidence, and have also used results from more quantitative school effectiveness studies, such as the British Junior School Project (Mortimore *et al.*, 1988) and teacher effectiveness studies. Next, their review makes considerable use of other reviews, particularly those by Levine and Lezotte (1990) and Scheerens (1992). The further development in school effectiveness research since the Levine and Lezotte review is also evident in the predominant operational definitions of effectiveness in both reviews. Levine and Lezotte's definition depends on the distinction of unusually effective versus ineffective schools and looking back at these schools for discriminating conditions, whereas the basic definition provided by Sammons *et al.* is based on the concept of value added. In Mortimore's terms (Mortimore, 1991, cited by Sammons *et al.*, 1995a, p. 3) "an effective school is one in which students progress further than might be expected from consideration of its intake". Key characteristics of effective schooling, following this definition, are basically correlates of intake-adjusted or value-added outcomes. In their introduction to the review Sammons *et al.* (1995a) pay more attention than Levine and Lezotte to methodological issues, the size of school effects, differential effectiveness and the trustability of school effectiveness findings across contexts and (national) cultures. The key characteristics and components of effective schools distinguished by Sammons *et al.* are summarized in Table 5.5.

The review by Cotton (1995), referred to as a research synthesis by the author, describes "characteristics and practices identified by research as

Table 5.5 Eleven factors for effective schools (cited from Sammons et al., 1995a, p. 8)

1. Professional leadership	Firm and purposeful A participative approach The leading professional
2. Shared vision and goals	Unity of purpose Consistency of practice Collegiality and collaboration
3. A learning environment	An orderly atmosphere An attractive working environment
4. Concentration on teaching and learning	Maximization of learning time Academic emphasis Focus on achievement
5. Purposeful teaching	Efficient organization Clarity of purpose Structured lessons Adaptive practice
6. High expectations	High expectations all round Communicating expectations Providing intellectual challenge
7. Positive reinforcement	Clear and fair discipline Feedback
8. Monitoring progress	Monitoring pupil performance Evaluating school performance
9. Pupil rights and responsibilities	Raising pupil self-esteem Positions of responsibility Control of work
10. Home–school partnership	Parental involvement in their children's learning
11. A learning organization	School-based staff development

associated with improvement in student performance". The types of research on which the results are based are listed in six categories:

- School effects research: studies of whole schools undertaken to identify schoolwide practices that help students learn;
- Teacher effects research: studies of teachers and students in the classroom to discover effective practices;

- Research on instructional leadership: studies of principals and other building leaders to determine what they do to support teaching and learning;
- Curriculum alignment and curriculum integration research: examinations of alternative methods of organizing and managing curriculum to determine effective approaches;
- Program coupling research: inquiries into the interrelationships among practices used at the district, school building and classroom levels;
- Research on educational change: studies to identify conditions and practices that promote significant, durable change in educational programs" (Cotton, 1995, p. 2).

The results are summarized within a more elaborate framework than was used in the two other reviews discussed in this section. Cotton summarizes her findings within three sections: the classroom, the school and the district level.

The resulting category system forms the table of contents of the review (see Table 5.6).

In Cotton's review effectiveness-enhancing conditions supported in empirical research are described in terms of precise recommendations for school practice.

An attempt will be made to capture the core meaning from the elaborate descriptions of the main correlates and key characteristics and subfactors of each of the reviews.

Table 5.6 Categories for listing effective school practices (cited from Cotton, 1995, pp. 7–9)

Classroom characteristics and practices	School characteristics and practices	District characteristics and practices
Planning and learning goals	Planning and learning goals	Leadership and planning
Classroom management and organization	School management and organization	Curriculum
Instruction	Leadership and school improvement	District–school interactions
Teacher–student interactions	Administrator–teacher–student interactions	Assessment
Equity	Equity	
Assessment	Assessment	
	Special programs	
	Parent and community involvement	

Discussion of the main effectiveness-enhancing conditions.

Productive school climate and culture.

From the overview in Table 5.4 it is clear that in Levine and Lezotte's review orderliness, a shared achievement orientation, a problem-solving attitude, cohesion and collaboration, participative decision making and recognition of positive performance are seen as the main components of a "productive school climate and culture".

A "safe and orderly environment" is further qualified by the statement that "discipline derives from 'belonging and participating' rather than 'rules and external control'" (p. 9).
The authors say that this component is probably most relevant in the case of out-of-control schools, whereas in more average schooling situations it will probably not be a discriminating factor between effective and ineffective schools. Manifestations of a safe and orderly environment that are mentioned are the establishment of a "mental health team", plans to reduce absenteeism and truancy, rigorous disciplinary policies and school-wide prevention strategies (p. 11).

From the description of "faculty commitment to a shared and articulated mission focussed on improving achievement" it becomes clear that "goals coordination" among staff is seen as a major focus of working together as a team. Strict control on homework assignment and a close monitoring of student pacing and progress are mentioned as examples of an articulated achievement-oriented mission.

The "problem solving" orientation that is distinguished as a third element of a productive school climate refers to a willingness of all staff to experiment and actively look for solutions to overcome obstacles in student learning, particularly with respect to low achievers.

"Faculty cohesion, collaboration, consensus, communications and collegiality" are seen as particularly important with respect to central organizational goals. Further manifestations are "more and/or better communications". The authors point out that, given the nature of schools as organizations, with many difficult and sometimes conflicting goals, enhanced cohesion and congruence is particularly important. The fact that in several studies this factor could not be shown to discriminate between effective and ineffective schools, may be explained by the limitations of the research methods (i.e. paper and pencil instruments) used to capture these features.

"Faculty input in decision making" refers to a type of school decision making that is participatory rather than centralistic.

"School-wide emphasis on recognizing positive performance" is described as a set of practical measures that can also be seen as manifestations of other general factors such as "high expectations", "leadership" and "achievement orientation". Public honoring of academic achievement is mentioned as an example of such measures.

Although each of these six elements of a "productive climate" adds a distinct and specific feature, some of them seem to be closely interrelated, where an achievement-oriented mission appears to be the main normative principle. The degree to which the type of climate that is meant is permissive versus strict, informal versus formal is rather mixed. In the area of controlling assignments and monitoring progress, formal and strict procedures are mentioned, whereas, as cited before, maintenance of discipline is described in terms of "belonging and participating", rather than "rules and external control". Nevertheless, strict policies on truancy and absenteeism are mentioned as effectiveness-enhancing conditions. From analyses such as these it appears that it is not simple, and probably not useful, to classify the school effectiveness philosophy in terms of "formal versus informal" and "bureaucratic/mechanistic versus organic". It is clearly a mixture of both main strands of organizational structuring.

No fewer than three of Sammons' key characteristics of effective schools (Sammons *et al.*, 1995a) can be more or less subsumed under Levine and Lezotte's category "productive school climate and culture": shared vision and goals; a learning environment; and positive feedback.

Unity of purpose and collegiality and collaboration are the two aspects of a shared vision and goals. Like Levine and Lezotte, Sammons *et al.* stress the unity and coordination of goals in educational organizations. One of the sources of evidence they use to make this point is the research literature about the relatively greater effectiveness of Catholic schools, which is attributed, among other things, to strong institutional norms and shared beliefs. Sammons *et al.* also refer to the more "practical" as compared to the cultural and normative aspects of consistency and collegiality. In this they refer to common and agreed approaches to assessment, the enforcement of rules and policies and the integration of curriculum and instruction.

With respect to collegiality and collaboration, Sammons *et al.* (p. 12) refer to "a strong sense of community among staff and pupils, fostered through reciprocal relationships of support and respect".

The two aspects of the key factor "a learning environment" that Sammons *et al.* distinguish are an orderly atmosphere and an attractive working environment.

With respect to "an orderly atmosphere", the description is strongly in line with Levine and Lezotte's, emphasizing the importance of reinforcement of good practice. The element of an attractive working environment is less explicit in Levine and Lezotte's review. The assumption is that attractive and stimulating working conditions tend to improve morale.

The key factor "positive reinforcement" has two elements: clear and fair discipline and feedback.

The description of clear and fair discipline is similar to Levine and Lezotte's. Feedback is pictured as a school-wide practice in recognition of academic success and other positive behavior, more than an aspect of effective instruction (although Sammons *et al.* do not exclude the operation at the classroom level).

Altogether, the two reviews are very much in line with each other on the general factor of a "productive school climate and culture". In Sammons' review there is less emphasis on pressing for achievement and a problem-solving orientation towards the enhancement of the position of low-achieving pupils. They deal with "focus on achievement" under the heading of "concentration on teaching and learning".

The categories of conditions in Cotton's review that come closest to Levine and Lezotte's "productive climate and culture" and in which a proactive, orderly, well-coordinated pressing for achievement is the main general trend, are: curriculum planning from the district level; planning and learning goals (school level); and planning and learning goals (classroom level).

With respect to district-level curriculum planning the general imperative that Cotton states is as follows: "District leaders and staff conduct careful curriculum planning to ensure continuity". In order to further operationalize "careful" planning, a set of statements follws, which recommends the setting of quality standards, establishment of priority objectives in collaboration with schools, the alignment of learning materials, staff resources, other facilities to objectives or goal areas, the identification of "validated" instructional strategy, "district-wide curriculum and review efforts to ensure high quality of instruction and consistency across schools", an active role in collaborating with schools on curriculum and instruction and "providing support for integration of traditional subject areas" (Cotton, 1995, p. 40); in short, a synoptic planning approach in which the district leaders are explicitly concerned with the curriculum and the instructional design at school level, in order to attain achievement standards.

At the school level five main orientations to planning and learning goals are stated:

1. "everyone in the school community emphasizes the importance of learning", among other things by stating "high expectations" and emphasizing academic learning;
2. "administrators and teachers base curriculum planning on clear goals and objectives", e.g. by clear definition of goals, establishing clear relations among learning goals, instructional activities and student assessments;
3. "administrators and teachers integrate the curriculum, as appropriate", the essence of this category being the message to integrate traditional subject-matter content around broad themes, "where this approach is appropriate";

4. "administrators and teachers provide computer technology for instructional support and workplace simulation"; examples of more concrete activities within this category are training of administrators and teachers to use computer-assisted instruction effectively and the use of computer-assisted instructional activities for chronically misbehaving students;
5. "administrators and teachers include workplace preparation among school goals": this category refers to secondary school students in particular and includes "the development of qualities required for workplace success-dependability, positive attitude towards work, consciousness, cooperation, adaptability, and self-discipline" and "tasks for older students which approximate those performed by people in real work settings" (Cotton, 1995, pp. 23–25).

Evidently in this latter aspect of school planning statements are made about the priority of particular goals, rather than antecedent conditions or means to attain outcome criteria.

At the classroom level the category "planning and learning goals" includes two subcategories:

1. "teachers use a preplanned curriculum to guide instruction": again, a proactive, structured approach is recommended that is close to well-known classical prescriptive models of instructional design, such as the Gage model;
2. "teachers provide instruction that integrates traditional school subjects, as appropriate": again, occasional thematic integration of subject-matter areas is recommended, other aspects being "team teaching" and "performance assessments that allow students to demonstrate knowledge and skills from several traditional subject-matter areas" (Cotton, 1995, p. 11).

To summarize, according to Cotton's review two factors are important in curriculum planning and instructional design at the school and classroom level: detailed rational planning and "when appropriate" thematic integration of traditional subject-matter areas. A striking third element is the recommendation to give a certain priority to particular educational goals, namely the development of workplace-oriented skills.

The research-based recommendations that are subsumed under the categories district level curriculum and planning and learning goals at school and classroom level in Cotton's review also partly cover Levine and Lezotte's category "focus on central learning skills". Aspects such as cohesion and collaboration and efficient use of time are subsumed under other categories in Cotton's review.

Focus on student acquisition of central learning skills.

Within this general orientation two components are distinguished in Levine and Lezotte's review: "maximum availability and use of time for learning" and "emphasis on mastery of central learning skills".

With respect to the first component, measures at school level aimed at maximizing effective use of time are referred to, e.g. ensuring that relatively little time is expended in passing between classes. The authors refer to a "no-nonsense" approach to the use of academic time in effective schools.

The second component, "emphasis on mastery of central learning skills", contains three aspects: concentration on academic content, the use of principles of mastery learning and explicit instruction on central learning skills. The combination of mastery learning and general cognitive skill training is an interesting mixture of two different instructional traditions with a behavioristic versus a more cognitivist, constructivist background.

When one realizes the number of further nuances, questions and possible ways of combining or mixing these two general instructional approaches, which are addressed in current research on instructional technology (e.g. Brophy, 1996), it becomes clear that the effective schools correlates are sometimes very general indeed.

Levine and Lezotte's effective schools correlate "focus on student acquisition of central learning skills" is closely related to the key factor "concentration on teaching and learning" in Sammons *et al.*

In this respect, Sammons *et al.* mention three elements: maximization of learning time, academic emphasis and focus on achievement. With respect to the first of these, Sammons *et al.* draw attention to the issue of single-subject teaching versus sessions in which work on several different curriculum areas is ongoing. In the latter case more time may be needed for administrative interaction.

In stressing academic emphasis Sammons *et al.* do not explicitly raise the issue of time to be spent on basic subjects as compared to other subjects and personal development as do Levine and Lezotte. They draw attention to two additional aspects, however: teachers' subject knowledge and curriculum coverage. They cite Tizard *et al.* (1988, p. 172) in stating that "it is clear that attainment and progress depend crucially on whether children are given particular learning experiences".

The subcategory "focus on achievement" is used by Sammons *et al.* under the heading of "concentration on teaching and learning", while Levine and Lezotte refer to this orientation under the heading of "a productive school culture and climate".

The overall gist of Sammons' key characteristic "concentration on teaching and learning" is that effective schools focus on the maximization of the output of their primary process of teaching and learning, by efficient use of time, targeted curricular choices and a result-oriented attitude.

In Cotton's review the domain "planning and learning goals", which was discussed in reference to the first category of school effectiveness-enhancing conditions, "productive climate and culture", is also associated with the category discussed in this subsection, namely "focus on central learning skills". In particular, the subdomain at the school level, referred to as a "school-wide emphasis on learning" is relevant to a focus on central learning skills. In this subdomain the following specific recommendations to administrators and teachers are made:

- "Have high expectations for student achievement; all students are expected to work hard to attain priority learning goals".
- "Continually express expectations for improvement of the instructional program".
- "Emphasize academic achievement when setting goals and school policies".
- "Develop mission statements, slogans, mottos, and displays that underscore the school's academic goals".
- "Focus on student learning considerations as the most important criteria for making decisions" (Cotton, 1995, p. 23).

As in the reviews by Levine and Lezotte and Sammons *et al.*, in Cotton's review pressing for achievement in academic subjects is emphasized as the school's "core business". In the latter review the overt and publicly visible continuous statement of this ideology is particularly stressed.

Appropriate monitoring of student progress

Levine and Lezotte (1990, p. 15) say that research support for monitoring of student progress as an effective schools correlate is "probably weaker" in terms of providing specific guidance for practice than any other frequently cited correlate. Different authors have referred to different types of evaluative activities. They also warn against an uncritical use of "frequent" or "continuous" monitoring, and imply that "overtesting" of mechanical skills may be counterproductive.

With respect to this particular correlate a further delineation on which forms and types of evaluative activity at school are to be considered effective is not provided.

As do Levine and Lezotte, Sammons *et al.* mention "*monitoring progress*" as an effectiveness-enhancing condition.

In their review Sammons *et al.* distinguish three types of monitoring: (a) monitoring the performance and progress of pupils; (b) monitoring classes and the school as a whole; and (c) the evaluation of improvement programs. They offer an interesting analysis of the relationship of this key characteristic to other school effectiveness correlates. "First it is a mechanism for determining the extent to which the goals of the school are being realized. Second, it focuses the attention of staff, pupils and parents on these goals. Third, it informs planning, teaching methods and assessment. Fourth, it gives a clear message to pupils that teachers are interested in their progress. This last point relates to teachers giving feedback to pupils" (p. 20).

At the school level evaluating school performance can be seen as an essential part of effective leadership. An interesting evaluation-related phenomenon that these authors mentioned, and which received support in the British Junior School Project (Mortimore *et al.*, 1988), is record-keeping and teachers combining the results of objective assessments and their own judgments of pupils.

Where Levine and Lezotte and Sammons *et al.* use the term monitoring, Cotton speaks of assessment in referring to evaluative activities as a general effectiveness-enhancing condition. At the district level Cotton's general imperative is that "district leaders and staff monitor student progress regularly". Concrete activities to do so are: collection and summarizing of information about student performance at a regular basis, coordination of assessment effort, checking the alignment among tests, curriculum and instructional, district-level assessments, specific routines for scoring, storing, reporting and analyzing results and direct support for building- and classroom-level report results (Cotton, 1995, pp. 42, 43).

A second main imperative is that "district leaders and staff support school's development and use of alternative assessments". Alternative assessments are described in terms of "exemplary tasks" and "task templates" (p. 43).

At the school level "administrators and other building leaders" are to "monitor student learning closely", by means of building assessment skills, data collection to identify and treat young children with learning difficulties, use of all kinds of results, reports and records to diagnose and change instructional programs, and the inclusion of assessment of school climate (pp. 34, 35).

At this level too, administrators and building leaders are urged to encourage alternative assessments, such as "performance tasks" and "portfolio assessments".

At the classroom level, specific recommendations are given for teachers to develop and use both routine and alternative assessment procedures. Alternative assessments involving performances and products are seen as closely related to preparation for life outside the school.

Practice-oriented staff development at the school site

According to Levine and Lezotte (1990), the core of the kind of in-service training that is associated with school effectiveness is that it is ongoing, practice-oriented and actively involves teachers in intragrade and cross-grade level meetings.

The authors say little about the contents of such staff development, but on one occasion refer to "selection and assessment of skills emphasized in a mastery framework" (p. 16).

Sammons *et al.* likewise state that development and learning of staff in effective schools is to be seen as a continuous process. They use the term "learning organization" to express that learning is to be school-wide rather than specific to individual teachers. Like Levine and Lezotte they stress the "school-based" nature of staff development, "tailored to the specific needs of staff" (Sammons *et al.*, 1995a, p. 23).

Staff development is a subcategory of "leadership and school improvement" in Cotton's category system of effectiveness-enhancing conditions. Like the other reviewers she mentions that professional development activities should be considered relevant by the participants, and that follow-up activities should be provided to ensure that newly acquired knowledge and skills are applied in the classroom (Cotton, 1995, p. 30). She also mentions that structures should be created for staff members to learn from one another through peer observation/feedback and other collegial learning activities. This is close to what Sammons *et al.* refer to as a "learning organization".

Outstanding leadership

In their review Levine and Lezotte take a broader outlook on leadership than merely discussing "educational" or "instructional" leadership. Although they provide a detailed conceptualization of the latter, they also refer to a broader set of aspects of school leadership, namely:

- vigorous selection and replacement of teachers;
- "maverick orientation" and buffering;
- frequent, personal monitoring of school activities and sense making;
- high expenditure of time and energy for school improvement actions;
- support for teachers;
- acquisition of resources;
- availability and effective utilization of instructional support personnel.

Teacher recruitment is one of the malleable conditions of schooling that may have a great impact on overall quality, although this is sometimes constrained by system-level conditions such as a rather immobile population of teachers owing to oversupply, conditions of labor and other circumstances.

The feature "vigorous selection and replacement of teachers", mentioned by Levine and Lezotte, implies that heads of unusually effective schools seize the opportunity to play an active role in the selection and replacement of teachers. When becoming an effective school is seen as a process over time, it is clear that recruitment policies are particularly important in the early development phases, while later on the stability of experienced faculty may emerge as a correlate of effectiveness.

The active orientation of heads of unusually effective schools is also present in the phenomenon of "maverick orientation and buffering". In this feature head teachers sometimes have to work against the tide of external regulations and other pressures. Head teachers of less effective schools, in contrast, have been shown to have a more externally directed outlook, in which the blame for occurring problems is put on external forces outside the control of the school. This element becomes more relevant when external circumstances are difficult or hostile, as may be the case in U.S. inner city schools.

"Frequent monitoring of school activities" may be the core of educational leadership. It manifests itself in frequent visits to classrooms and conversations with staff. Apart from controlling and/or motivating aspects of this type of leadership orientation, Levine and Lezotte emphasize the cognitive merits of this approach. Close contact with the school's primary process helps the school leader to make sense of complex phenomena and gain insight into which conditions may be relevant for goal attainment and effectiveness in a particular situation.

High expenditure of time and energy for school improvement actions

Although it is not at all certain that simply working hard could be an effectiveness-enhancing condition on its own, it is assumed to be a relevant feature, both as a more direct cause of school improvement and as a way in which to communicate and spread commitment in the organization.

With respect to "support for teachers", the school effectiveness literature refers to "emotional encouragement and practical assistance in acquiring materials and in handling difficult teaching assignments" (Levine & Lezotte, 1990, p. 20).

In line with the effective head teacher as a hard worker, a risk taker and a problem solver, is his or her inventiveness and effort in acquiring additional resources. To the degree that these resources are considered as human resources, this factor can also be seen as another component of good school leadership mentioned by Levine and Lezotte, namely "availability and effective utilization of instructional support personnel". It is important to note that support personnel concerned with the school's primary process of learning and instruction is referred to in this context.

This leaves "instructional leadership" as the remaining main component of Levine and Lezotte's analysis of "outstanding leadership". In defining the elements of instructional leadership, Levine and Lezotte refer to Murphy's administrative functions:

- developing mission and goals and communicating these school-wide;
- managing the educational production function, by supervising and evaluating instruction by means of indirect methods such as informal meetings and observations, by allocating and protecting instructional time (e.g. by reducing interruptions and other threats to instructional time); coordinating the curriculum (e.g. attainment of continuity across grade levels); promoting content coverage (e.g. enforcement of policies that require regular homework assignments) and by monitoring student progress;
- promoting an academic learning climate, emphasizing positive expectations and standards, while maintaining high visibility, providing incentives for teachers and students and promoting professional development;
- developing a supportive work environment, creating a safe and orderly environment, providing opportunities for meaningful student involvement, developing staff collaboration and cohesion, securing outside resources in support of school goals and related action and forging links between the home and the school (Murphy, 1989, cited by Levine & Lezotte, 1990, p. 21).

According to this overview outstanding school leadership involves many factors: an active or even assertive use of all available measures to improve schooling, protection, stimulation and monitoring of the school's primary process and, in the somewhat narrower sense of instructional leadership, the meta-control of most of the other correlates of effective schooling. In this latter sense the statements on factors that work in education obtain a large degree of redundancy or, stated with a somewhat different connotation, "mutual support". Certain key aspects are repeated as features of leadership, school organization and management, classroom processes, the curriculum or teaching materials. "Result orientation", "structure", "high expectations", "opportunities for monitoring" and "efficient use of time and resources" are examples of features that could be borne by any of these modes of schooling.

In their review Sammons *et al.* (1995a) speak of "professional leadership". They distinguish three main aspects: strength of purpose, a participative approach and "the leading professional".

Strength of purpose can appear from a proactive orientation, for instance in the recruitment of staff but also from the ability to "buffer" the school from external pressures. Another element of "strength of purpose" is the "key role of leadership in initiating and maintaining the school improvement process" (Sammons *et al.*, 1995a, p. 9).

Like Levine and Lezotte, Sammons *et al.* mention the importance of a participative approach in leadership and refer to several European empirical studies that have supported this aspect, also with respect to the role of deputy head teachers and heads of department.

In summing up their first two features of effective leadership (strength of purpose and a participative approach) Sammons *et al.* say that "effective leadership requires clarity, avoidance of both autocratic and over-democratic ways of working, careful judgement of when to make an autonomous decision and when to involve others, and recognition of the efficacy of the leadership role at different levels of the school" (Sammons *et al.*, 1995a, p. 10).

The concept of instructional leadership is referred to with the term " the leading professional", in Sammons' review. Elements that are mentioned are:

- the head teacher should know what goes on in the classroom;
- the head teacher should support teachers, by both encouragement and practical assistance;
- projecting a "high profile";
- teacher assessment.

The authors conclude that such leadership is operating indirectly rather than directly and should work in conjunction with all the other key characteristics of effective schools.

The concept of "outstanding school leadership" in Cotton's framework has meaning at the district level as well as the school level. At the district level she uses a general category "leadership and planning", whereas at the school level two general categories relate to leadership: "school management and organization" and "leadership and school improvement".

At the district level district leaders and staff should "hold and communicate high expectations for the entire school system". The instruments that district level officials can use for this general orientation are focussing on goals for improving student performance, support of plans for improving instructional effectiveness, and the review of recruitment, selection or promotion policies.

Next, they are expected to hold and communicate the conviction that all children can be successful learners. Concrete measurable goals and monitoring of these are important instruments in making this strongly achievement and equity-oriented ideology happen.
School-based management and shared decision making among district and school administrators are also recommended.

At the school level Cotton (1995, p. 25) discusses five main aspects of school management and organization:

1. school-based management; this involves the gradual development of school level discussion *vis-à-vis* district level officials and teacher participation in decision making;
2. the grouping of students in ways that promote effective instruction; i.e. heterogeneous grouping, use of instructional aids, in class instruction in

small groups of low achievers (avoid the stigma of pull-out classes); ability groups, when used, should be of short duration, exploring of the possibilities of multiage grouping;
3. optimal use of school time for learning; e.g. avoidance of disruptions, provision of extra learning time for students who need or want it, firm policies with respect to "tardies", absenteeism and appropriate classroom behavior (Cotton, 1995, p. 26);
4. clear, consistent discipline policies, e.g. a written code of conduct, "create a warm, supportive school environment" (p. 26), deliverance of sanctions and positive reinforcements, agreements with parents about ways to reinforce school disciplinary procedures at home;
5. provision of a pleasant physical environment for teaching and learning.

With respect to "leadership and school improvement" four main components are distinguished by Cotton (1995, pp. 28–30):

1. school restructuring efforts as needed to attain agreed goals for students; the main message being that restructuring should only take place if it will contribute to the improvement of student learning;
2. strong instructional leadership, including high expectations, achievement orientation, a clear instructional focus in the school's mission, recruitment of staff that will support the school's mission, validated teaching and learning principals in modeling effective teaching practices for staff, checking of the alignment of curriculum with instruction and assessment, frequent checking of student progress, relying on explicit performance data, working with staff to set standards and address discrepancies, supervision of instructional improvement at classroom level, high expectations of success of instructional improvement programs, involvement of the full staff in the planning of implementation strategies;
3. continuous striving for the improvement of instructional effectiveness: a rational, participative research-based approach to instructional improvement is recommended;
4. engagement of staff in professional development activities (this category of conditions was discussed under "staff development").

The type of management and leadership that is stressed in Cotton's elaborate research-based recommendations is strongly result-oriented, rational and with clear top–down responsibilities, although participative aspects are mentioned frequently. It appeals both to strong normative and ideological aspects (high achievement for pupils at all ability levels) and to structured, technological approaches. Her prescriptions are fully in line with the concept of instructional leadership.

Salient parent involvement

Levine and Lezotte state as a general observation that research generally supports the conclusion that it is desirable and important to enhance parents' "involvement in education and to have high levels of school and home cooperation" (Levine & Lezotte, 1990, p. 22). However, they also refer to the fact that in school effectiveness studies some studies "either have failed to find support for a relationship between involvement and unusual effectiveness or have concluded that less effective schools may have more involvement of some kinds than more effective schools".

Difficulties in detecting the influence of this factor on achievement are the interaction with socioeconomic status of the parents and the fact that there are many forms and varieties of involvement, therefore making it difficult to measure (Levine & Lezotte, 1990, p. 22). As to the first point, parent involvement appears to be particularly relevant in American inner city schools. With respect to the different types of parent involvement the following main categories can be distinguished:

- good communication between home and school;
- involving parents politically on behalf of the school, e.g. by exerting pressure on public officials;
- facilitation of parents' involvement in their children's learning;
- sharing in school governance (Stedman, 1987, p. 219, cited by Levine & Lezotte, 1990, p. 23).

Sammons *et al.* (1995a) also refer to parental involvement as a key factor, and like Levine and Lezotte, admit that "the actual mechanisms by which parental involvement influences school effectiveness are not entirely clear". They cite more recent studies by McBeath (1994) and Coleman *et al.* (1993) which stressed aspects such as an active approach by schools for parental support and the importance of the relationship between teacher and parent(s).

Considering "parent involvement" Cotton uses the category "parent community involvement" as one of the relevant school-level characteristics. In this area the following more concrete practices are recommended:

- written policies legitimizing parent involvement;
- clear communication to parents of the procedures for involvement;
- participation of parents and community members in school-based management teams;
- special efforts to involve the parents of disadvantaged and minority students;
- working with cultural minority parents;
- involvement of parents and community members in decision making regarding school governance;

- monitoring of parent/community involvement;
- publishing of indicators of school quality in order to provide them to parents and community members;
- involvement of business, industry and labor in helping to identify important learning outcomes and in providing opportunities to apply school learning in workplace settings (Cotton, 1995, pp. 37, 38).

Effective instructional arrangements and implementation

Under this heading Levine and Lezotte (1990, pp. 24–34) discuss the following components:

- successful grouping and related organizational arrangements;
- appropriate pacing and alignment;
- active/enriched learning;
- effective teaching practices;
- emphasis on higher order learning in assessing instructional outcomes;
- coordination in curriculum and instruction;
- easy availability of abundant appropriate instructional materials;
- classroom adaptation;
- stealing time for reading, language and math.

In short, if instruction is to be effective, each and every aspect of the teaching learning environment should function well: grouping, structuring of tasks, teaching and instructional materials, while sufficient time should be given to basic subjects. In addition, learning should be active and curricula should be coordinated and adapted to the actual learning and teaching situation.

The component "emphasis on higher order learning in assessing instructional outcomes" seems surprising in this context. It presupposes that "higher order learning" would always be the core educational objective in schools that were studied for their effectiveness. Further, such assessment would refer to a particular *criterion* to measure effectiveness and not a *correlate*. One may wonder to which criterion assessment of higher order learning should be correlated.

The authors appear to be caught in a mixture of normative and empirical statements. They say that much effective schools research has defined effectiveness "largely or entirely" (p. 31) in terms of low-level learning, so it follows that their correlates will mostly depend on exactly such "low-level learning" assessments. A plea for more higher order learning may be sympathetic but falls out of line in a review of correlates of school effectiveness when in most studies school effectiveness is expressed in terms of fewer higher level learning outcomes.

In their discussion of "successful grouping and related organizational arrangements" Levine and Lezotte (1990, p. 24) say that in empirical studies all kinds of grouping arrangements, ranging from whole class instruction to individualized instruction, have sometimes been associated with effectiveness, and sometimes not. In their further discussion they point out that "workable arrangements" for helping low achievers may be the core effectiveness-enhancing element. Methods for bringing about such arrangements include a careful matching of talented teachers to groups of slow learners, other types of targeted intensified arrangements for low achievers such as daily one-to-one tutoring, having more than one teacher in the classroom, a reduction in class size, and extra school time for groups of slow learners. The authors further discuss practices such as pull-out approaches and leveling. They emphasize that such grouping arrangements may only be effective if supported by fitting instructional approaches and skillful teaching.

Levine and Lezotte therefore basically see grouping as an effectiveness correlate, to the degree that it singles out subgroups of low-achieving students and provides these groups with extra resources and adaptive, targeted instruction, at the same time actively working against potential negative effects of such grouping (e.g. stigmatizing, lowering of standards and expectations).

With respect to "appropriate pacing and alignment", Levine and Lezotte refer to the structure and tempo of instruction. Regarding structuring they refer to practices such as direct instruction and mastery learning. Considering the pace of instruction they warn against the inclination to have "slow" instruction for low-achieving groups; instead there should be "relatively rapid pacing".

"Active/enriched learning" refers to emphasizing higher order learning, in enriched environments (e.g. library facilities, media) involving considerable interaction with teachers and other students. The authors stress that "emphasis on active and enriched learning does not necessarily conflict with other effective schools correlates ... or with ... research that supports the utility of teacher-centered, whole-class instruction (i.e. "direct instruction") as a strategy for improving achievement when used appropriately to teach key learning skills, build high expectations for student performance, expedite classroom management, and/or accelerate the pacing of instruction" (p. 30). They say that active and direct instructional approaches should not be seen as opposite ends on a single bipolar continuum, and admit that the process of "interweaving" different approaches is very demanding to teachers. The distinction, "competition" or even "reconciliation" of direct teaching and constructivist orientations to independent learning will be further discussed in other chapters.

When referring to "*effective teaching practices*", Levine and Lezotte (1990, p. 30) refer to features such as time-on-task, recognition for academic success, appropriate reinforcement, lesson sequence, "wait time" after questions and teacher–student interaction guidelines. They feel that introduction and development of effective teaching practices should be part of effective schools projects, but only as one component in an effort to address the larger set of correlates discussed in their monograph.

Levine and Lezotte have made no attempt to conduct a comprehensive review of the research literature on instructional and teacher effectiveness. In the section that is referred to here, they appear to shift from a descriptive review of "what works" to prescriptive notions about school improvement and normative orientations with respect to active learning *vis-à-vis* direct instruction.

"Coordination in curriculum and instruction" refers to consistency in the continual development of competencies across grades and teachers and in the approaches to teach these skills. Coordination in low socioeconomic status schools may take the form of coordination between regular instruction and special services. As a result of either resourceful active educational leaders or engaged parents, unusually effective schools in disadvantaged schooling areas have distinguished themselves from ineffective schools by more "abundant and appropriate instructional materials".

Adaptation of curriculum materials and instructional methods by active and sophisticated teachers has emerged as a school effectiveness correlate in several studies.

The school effectiveness correlate "stealing time for reading, language and math" is a rephrasing of Edmonds' factor emphasizing basic skills. There appears to be some self-evidence in the notion that when achievement in basic reading, language and math skills is the main effectiveness criterion, increased lesson time in these subjects will increase effectiveness. The verb "stealing" points to the difficult choices that are likely to be at stake when emphasis, in terms of lesson time, on subjects like art or science, is sacrificed in order to increase time for basic reading, language and mathematics skills.

Sammons *et al.* use the term "purposeful teaching" to refer to high-quality teaching, thus emphasizing a result-oriented, proactive attitude to teaching.

With respect to instructional technology Sammons *et al.* mention "structured lessons" and, amongst others, cite Joyce and Showers (1988) on characteristics of effective teachers:

- "teach the classroom as a whole;
- present information or skills clearly and animatedly;
- keep the teaching sessions task-oriented;
- are non-evaluative and keep instruction relaxed;
- have high expectations for achievement (give more homework, pace lessons faster, create alertness);
- relate comfortably to the students, with the consequence that they have fewer behavior problems" (Sammons *et al.,* 1995a, p. 16).

Sammons *et al.*, unlike Levine and Lezotte, do not discuss active learning *vis-à-vis* direct instruction, but they do refer to the question whether or not a structured approach is also fitting in secondary rather than "inner city" primary schools and for basic as for higher order cognitive skills.

In the final aspect of "purposeful teaching", Sammons *et al.* use "adaptive practice" not so much in the sense of individualized instruction but in the sense of teachers modifying and adapting their teaching styles (Sammons *et al.*, 1995a, p. 17).

Effective instructional arrangements are described by Cotton under the headings "classroom management and organization" and "instruction".

In classroom management the following components are distinguished:

- grouping: whole group instruction is recommended for introducing new concepts; ability groups are only to be used for short periods, grouping arrangements should be adjusted regularly and heterogeneous grouping should be used as much as possible; cooperative learning and peer-tutoring are also mentioned;
- efficient use of time, e.g. by minimizing disruption, maintaining a brisk pace, presentation of learning activities that is neither too easy nor too difficult, providing short homework assignments, etc.;
- the use of smooth, efficient classroom routines, e.g. smooth rapid transitions between activities, well-prepared lessons with materials and assignments ready when the students arrive;
- clear standards for classroom behavior, e.g. use of written behavior standards, involvement of older student in helping to establish standards and sanctions, application of consistent, equitable discipline for all students, positive reinforcement of positive behaviors and skills.
(Cotton, 1995, pp. 12, 13).

Under the heading "instruction" there are seven main components:

- teachers carefully orient students to lessons, e.g. explanation of lesson objectives, relation of current lessons to key concepts or skills previously encountered, arousal of students' interest and curiosity by relating the lesson to things of personal relevance to them, use of "advance organizers", "study questions and prediction";
- teachers provide clear and focussed instruction, e.g. clear directions, focussed lectures, opportunity for independent practice, instruction in strategies for learning as well as test-taking skills, use of validated strategies to develop students' higher level thinking skills, such as instruction in study and problem-solving skills;
- providing feedback and reinforcement regarding students' learning progress, immediate feedback, feedback related to unit and course goals, praise for correct answers ("avoid the use of unmerited or random praise"), corrected homework, peer evaluation techniques;

- reviewing and reteaching in order to help all students master learning material, e.g. provision of review of key concepts and skills throughout the year;
- use of validated strategies to help build students' critical and creative thinking skills, e.g. instruction in thinking skills and strategies for problem solving, development of meta-cognitive skills (with older students); asking of higher order questions, "incorporate computer-assisted instructional activities into building thinking skills such as verbal analogy, logical reasoning, induction/deduction, elaboration and integration" (p. 16); use of specific thinking skills development programs, maintenance of a supportive classroom environment in which students feel safe experimenting with new ideas and approaches;
- teachers' use of effective questioning techniques to build basic and higher level skills, e.g. classroom-questioning, lower cognitive (fact and recall) and higher cognitive (open-ended and interpretative) questions, allowing generous wait-time, particularly for higher order questions, support for pupils who do not initially respond correctly to higher order questions;
- integration of workplace readiness skills into context area instruction, e.g. decision-making skills, learning strategies, and creative thinking and attitudes such as consciousness, cooperation adaptability and self-discipline; means to attain these skills are the use of work-based learning experiences and the organization of the secondary curriculum around broad occupational themes or categories (Cotton, 1995, pp. 15–17).

The outlook on effective instruction provided by Cotton places a stronger emphasis on the attainment of higher order cognitive skills in relationship to general skills and attitudes relevant for the "after school world" than do the other two reviews. Despite the reference to higher order cognitive thinking goals, the teaching approach that is recommended is quite structured, with emphasis on careful planning and close monitoring of the learning process. It is striking in a way that there is less emphasis on active, independent and discovery learning.

High operationalized expectations and requirements for students

This factor is closely related to other factors such as "a productive climate", "educational leadership" and "monitoring of students' progress". Levine and Lezotte (1990, p. 34) call this factor a crucial characteristic of virtually all unusually effective schools described in case studies. Aspects of the concept of "high expectations" are:

- the fact that demands and expectations are concrete rather than abstract and short term rather than long term;
- keeping students strictly accountable for their performance.

Despite referring to such concrete and operationalized practices, the authors also say that it is important that *something* is done systematically and vigorously to communicate and ensure a strong academic press and a climate conducive to learning. The underlying "mind set" or "climate" is essential, although its manifestations may vary.

Finally, they warn that rigorous behavioral requirements, which are often the overt manifestations of "high expectations", should also be accompanied by providing students with relevant support and guidance.

Sammons *et al.* (1995a) draw attention to the fact that "high expectations" may be a side-effect rather than a cause of good school performance. An active challenging approach, however, as opposed to lack of control over pupils' difficulties and a passive approach to teaching, is seen as an active rather than a reactive condition. "High expectations correspond to a more active role for teachers in helping pupils to learn and a strong sense of efficacy" (Sammons *et al.*, 1995a, p. 18). These high expectations include communicating expectation, e.g. by means of positive reinforcement and providing intellectual challenge, for instance, by encouraging pupils to use their imagination and problem-solving capacities.

The "high expectations" category has already been referred to in Cotton's treatment of leadership at the district and school level. At the school level high expectations are again referred to as part of "administrator–teacher–student interactions". A firm belief that all students can be successful, professional growth norms for teachers and a positive warm atmosphere in which excellence is explicitly recognized are the key features of attitudes and practices relating to high expectations (Cotton, 1995, pp. 30–32).

Other characteristics.

Levine and Lezotte (1990, pp. 36–38) refer to several other possible correlates of effective schools:

- student sense of efficacy or futility (which they conclude is perhaps more of an outcome than a correlate of effectiveness);
- multicultural instruction and sensitivity;
- personal development of students (several studies are mentioned in which achievement gains were associated with increased faculty attentiveness to students' personal development and problems);
- rigorous and equitable student promotion policies and practices;
- student responsibility for learning.

A key characteristic mentioned by Sammons *et al.* (1995a) that is not explicitly represented in Levine and Lezotte's review of effective schools correlates is "pupil rights and responsibilities". Aspects of this general characteristic are: raising pupil self-esteem, providing pupils with positions of responsibility in the school system and providing pupils with opportunities to control their work independently, at least over "short periods of time".

This last "key characteristic" of effective schooling once again refers to a central bipolarity in the conceptualization of effectiveness-enhancing conditions, namely structure and control on the one hand versus autonomy and independence on the other. On the whole, the outlook taken by Sammons *et al.* leans slightly more towards a predominance of the control and structuring pole than Levine and Lezotte's. They refer to independent learning less frequently than Levine and Lezotte, and make less of a problem of mechanistic versus higher order cognitive skills development. An explanation for the emphasis of control and structure in both reviews may be the fact that most empirical school effectiveness research has been carried out in primary schools in lower socioeconomic status neighborhoods and that structured approaches may work particularly well for this type of school, but not necessarily for others.

"District school interactions" is an area of effectiveness-enhancing conditions that is discussed by Cotton, but not by the other reviewers. Relevant aspects of district school interactions are a considerable delegation of decision-making authority to schools, the encouragement, support and monitoring of school improvement efforts by district leaders and staff, the explicit recognition and rewarding of excellence by district leaders and the assistance of district leaders of schools in carrying out prevention activities and the support of high-needs students.

Other areas dealt with in the Cotton review that are not presented in the other two reviews are "equity" and "special programs". Multicultural activities are prominent as an emphasis in equity-oriented schooling. Next, giving prominence to an equitable distribution of achievement is to be given sufficient priority as a goal area, whereas the more technical recommendations on how to teach high-needs students are quite similar to the general recommendations for effective instruction. Special programs that are discussed are aimed at drop-out prevention, tobacco, alcohol and drugs prevention and support for families with urgent health and/or social service needs (Cotton, 1995, pp. 35–37).

Final comments on the three reviews

Levine and Lezotte's review presents rich descriptions of the most salient antecedent conditions in the empirical literature on effective schooling. As stated earlier, their evidence is mostly based on case studies in which unusually effective schools were compared with ineffective schools. This is definitely a defendable source of empirical evidence on school effectiveness, its strong points being rich descriptions with ecological validity, whereas lack of precision in quantifying strengths of associations and complex interactions are its weaker aspects.

A stronger emphasis is placed on factors at the school, rather than the classroom level. With respect to instructional conditions their review suffers from occasional insertions of normative statements in the issue of active learning versus direct instruction. The treatment of this distinction is not completely satisfactory. To a degree the impression is given that practices favored to ensure basic skills development in core subject matter areas and practices aimed at active, independent learning can be combined to give favorable results in both lower and higher order cognitive skills learning.

Given the nature of the majority of studies that they have reviewed, it is perhaps not surprising that there is little questioning about the size of school effects and the size of correlations of antecedent conditions and outcomes.

The Sammons *et al.* (1995a) review depends on a wider scope of effectiveness studies and reviews, encompassing comprehensive, quantitative studies rather than mainly case studies of unusually effective or ineffective schools, and including European studies. A strong point of this review is the incidental effort to relate several key characteristics to each other, as was done for "monitoring progress" and "professional leadership".

A strong point of the Cotton review is its comprehensiveness; in particular, the fact that district-level, school-level and classroom-level interventions are jointly considered is useful in showing mutually enforcing strategies that operate at several levels. This is particularly evident in attitudinal and normative positions on "high expectations" and "result orientation" at each level and also in the area of assessment.

Although the author is explicit in warning against the use of her recommendations as recipes, the way that they are presented makes them quite amenable to such interpretation. Nevertheless, specific as they are, it is also clear that an educational professional who is knowledgeable about the particularities of the school's setting should judge for herself or himself which degree of ability grouping is "appropriate" in a given situation and to what extent a particular group of pupils can handle a certain level of higher order cognitive skills.

A disadvantage of the wide scope of the review, from a scientific point of view, is the difficulty in relating the conclusive statements to the research evidence. Since the body of research studies on which the Cotton review is based is so large, and references to specific conclusions and interpretative issues of the various studies are not included, there is no way to verify the conclusions other than to turn to the original studies. It seems almost inevitable that certain schemata will have been used in order to summarize the results; in other words specific lenses used by the reviewer in looking at the evidence. This is not to say that the summary would be biased, but rather that using certain frames is almost unavoidable in creating the neat picture of the results .

Another aspect of this is that there is hardly any reference to either the methodological difficulties of individual studies or the technical difficulties of uniting the evidence from large bodies of studies, of which the majority will have addressed only a very partial issue.

A common characteristic of these qualitative reviews is that given their nature these reviews lack quantitative estimates on the size of school effects and the size of associations of key correlates with achievement measures. Such estimates, together with counts of significant effects across individual studies, are contained in the research syntheses referred to in the next chapter.

Review of Five Illustrative Individual Studies

Brandsma (The Netherlands, 1993): Characteristics of primary schools and the quality of education

Research questions and general focus

The central research questions in this Dutch study were as follows.

- To what degree are there differences in effectiveness between schools?
- Which school characteristics are related to these expected differences?

The study not only focussed on differences in attainment levels (i.e. adjusted mean achievement test scores), but also looked into differences in the "compensatory potential of schools", defined with respect to differences in pupil background characteristics such as socioeconomic status. Schools with a steeper regression of achievement on such background characteristics were seen as having a lower degree of compensatory potential than schools with a flatter regression slope.

Methods

The study had a sample of 252 schools, representative for the population of primary schools in The Netherlands. All pupils in the seventh grade of these schools were selected as secondary sampling units.

The design of the study used pretests and posttests in (Dutch) language and arithmetic at the stage of the end of grade 7 and the end of grade 8, respectively. Therefore, the progress that was measured should be interpreted as the gain in these subjects during the year that the pupils were in grade 8 (i.e. the final grade in Dutch primary schools).

The following instruments and data collection procedures were used to measure variables at the pupil level:

- standardized tests in language and arithmetic;
- socioeconomic status of the parents of each pupil, as judged by the teachers;
- non-verbal intelligence (three subscales of the factor "spatial intelligence");
- ethnicity and gender of pupils, school career in terms of whether or not (if yes, how often) pupils had to repeat a grade.

At the school level, questionnaires administered to the head teacher and teachers were used to measure variables in the domains of the school context, school leader functioning and the school organization. Furthermore, questionnaires were administered to the teachers to measure classroom organization and teaching practice variables.

The following school context variables were measured:

- school size, i.e. total number of pupils;
- time for principal available for management
- number of ordinary classroom teachers;
- age composition of the school team;
- gender composition of the school team;
- team stability (number of changes during the last five years);
- the number of school sites;
- the age of the school building;
- whether or not the school had a multifunction room;
- special teaching aids for pupils with learning problems;
- denomination (i.e. public, Roman Catholic, Protestant or neutral);
- average socioeconomic status of the pupils;
- whether the total number of pupils in the school was rising or falling over the years;
- average advice given to pupils at the end of the primary school period concerning the type of secondary education;
- percentage of pupils that repeated grades;
- percentage of teachers with specialized training in remedial teaching;
- percentage of teachers qualified to teach in a special school.

The following school leader variables were measured:

- instructional leadership;
- the proportion of time head teachers spend on administrative versus educational tasks;
- achievement orientation of the school leader;
- amount of task delegation from the head master to the teachers;
- staff participation in decision making;
- degree of job satisfaction of the school leader;

- the attitude towards innovation of the school leader;
- emphasis on basic skills (also measured among teachers, aggregated to the school level).

School organization variables included in the study were:

- the degree to which staff meetings were prepared and planned systematically;
- frequency of staff meetings;
- parental involvement;
- evaluation orientation of the school;
- orientation towards innovation and innovation history;
- openness to the local community (e.g. to other schools, to parents);
- use of the School Counseling Services;
- team homogeneity (similarity in opinion and attitudes of staff on organizational and educational issues);
- emphasis on order and discipline;
- clarity of rules.

At the classroom level variables were measured in the domains of classroom context, teacher characteristics and classroom organization, by means of questionnaires administered to teachers. With respect to classroom context the following variables were measured:

- class size;
- whether or not the teacher had part-time employment;
- multigraded classes;
- the percentage of girls in the class;
- the percentage of pupils of foreign origin;
- the average socioeconomic status of the class;
- the percentage of pupils repeating classes.

The following variables were used to measure teacher perceptions and attitudes:

- cooperation with other teachers;
- participation in decision making;
- general satisfaction with the school;
- attitude towards innovation;
- curricular emphasis;
- readiness to receive in-service training.

In the domain of classroom organization the following variables were measured:

- planning behavior of teachers;
- pressing for achievement ;
- time on task;

- feedback on the basis of pupils' achievement;
- grouping practices;
- individualization;
- classroom rules;
- pupil responsibility and autonomy.

The data collection among head teachers and teachers took place several weeks before the achievement data were collected on pupils at the grade 7 level and again one year later. Data were analyzed by testing a two-level random coefficient model with random intercepts and random slopes. The VARCL program developed by Aitkin and Longford (1986) was used to carry through these analyses. The two levels used in this study are the school/classroom level and the pupils level. Since only a few schools had more than one class at the grade level of interest, the school and classroom level were combined, i.e. school level variables were disaggregated to the classroom level.

Results

The study design and the multilevel modeling technique employed allowed for the adjustment of the posttest scores, in terms of (a) adjustment for prior achievement and (b) adjustment for prior achievement *and* other pupil background characteristics. Using the latter adjustment procedure the between-school variance with respect to the posttest arithmetic score was 11.6% of the total variance. When applying the same adjustment to the data collected on the basis of the language posttest score the net between-school variance was 8%.

An overview of the school- and classroom-level variables that significantly reduced the between-school variance on the adjusted language and arithmetic test scores is given in Table 5.7.

It was concluded that only a relatively small number of variables explained the net variance between schools and pupils. School context and school organization variables explained more variance than classroom characteristics, teacher characteristics and school leadership variables in language, while in the domain of arithmetic classroom organization variables also had a relatively important impact.

In terms of variables that are commonly thought to be relevant in the context of school effectiveness, only variables in the area of frequent evaluation and frequency of cooperation (i.e. number of meetings) proved to be relevant in this study. Variables such as educational leadership, an orderly climate, emphasis on basics and achievement orientation had no significant effect.

At the instructional level (classroom organization) relatively little variance was explained, with a few variables, in line with a structured approach (planning, feedback, homework) showing significant associations.

Table 5.7 Results of the Brandsma study (source: Brandsma, 1993)

Type of variables	Between-school variance explained (language)	Total variance explained (language)	Between-school variance explained (arithmetic)	Total variance explained (arithmetic)
Context variables	20%	2%	27%	3.1%
Denomination Catholic[a]+				
Education priority policy[b]+				
% Female teachers[a]–				
Referrals to special education[a]–				
Average socioeconomic status[c]+				
School leader	2.5%	0.2%	1%	0.3%
Positive attitude[a]+				
Satisfaction[c]+				
School organization	43%	4%	42%	5%
Use standardized test[a]+				
Frequency of team meetings[a]+				
Number of implemented innovations[a]+				
Frequent evaluation[c]+				
Time school counseling service[c]–				
Classroom context	3.5%	0.5%	1%	0.3%
Foreign students[b]–				
Part-time teacher[b]–				
Multigraded class[c]–				
Teacher characteristics	3.5%	0.3%	1%	0.4%
Female teacher[a]–				
Innovation attitude[c]–				
Classroom organization	3.5%	0.3%	23%	3%
Planning[a]+				
Teaching method[c]				
Strict rules[c]–				
Differentiation[c]–				
Feedback[c]+				
Homework[c]+				

[a]Significant for language and arithmetic; [b]significant for language only; [c]significant for arithmetic only.
+: Positive association; – : negative association.

Results with respect to differences in "compensatory potential" between schools will not be summarized in detail. The overall conclusion was that these differences were largely negligible.

Concluding comments

The study, given the usual limitations of correlational research, and the limitation that no direct observation methods could be employed for budgetary reasons, was properly conducted in a methodological sense (value-added measures and appropriate analysis techniques, instruments that met standards of reliability — usually internal consistency). Another limitation of the study was that school and classroom level could not be separated, so that a three-level analysis was not feasible.

Hill et al. (Australia, 1995a): The Victorian Quality of Schools Project

Research questions and general focus

After providing an account of the state of the art of school effectiveness research, in which the authors conclude that recent comprehensive studies have indicated relatively small school effects and more substantial class-level effects, the authors state the aims of their central research questions in the following terms:

1. "What are the characteristics of schools in which students make rapid and sustained progress in English and mathematics, after adjusting for their initial levels of achievement?"
2. "What are the characteristics of schools in which there are positive student attitudes and behaviors, positive perceptions by teachers of their work environment and high levels of parent participation in and satisfaction with their child's schooling?" (Hill *et al.*, 1995a, p. 5)

In addition to these research objectives the study aimed to contribute to school improvement processes and to system-level quality assurance, policy development and planning.

Design and methods

The study employed a fixed occasion, repeated-measures design, involving a probability proportional-to-size cluster sample consisting of five entire year-level cohorts of 13,909 students, including their parents and teachers, drawn from 90 government, Catholic and independent elementary and secondary schools (Hill *et al.*, 1995a, p. 2; Rowe *et al.*, 1994). The sample consisted of 59 primary and 31 secondary schools, including 365 and 538 teachers, respectively (where 65 teachers were in combined primary and secondary schools). At the primary level there were 6678 pupils and at the secondary level 7231 pupils.

Pupil data were collected in three waves (1992, 1993 and 1994). In the study reported here, only 1992 and 1993 data were used. Over a period of three years, pupil data were collected for each of the compulsory years of schooling (from preparatory up to year 11).

In the study a remarkable type of outcome measures was used to measure performance in English language (reading, writing and spoken language), and mathematics (space and number). Because there were serious reservations about the validity of available standardized achievement tests with respect to the curriculum, a deliberate choice was made to use "systematic teacher assessment of student progress using specially developed rating scales or subject profiles, specifically designed to reflect the curriculum in Victorian schools — these subject profiles are ordered sets of indicators describing observable learning behaviors that have been empirically calibrated on to common measurement scales..." (Hill *et al.*, 1995a, p. 8).

The scales for reading, writing and spoken language were integrated in a weighted composite, labeled "English achievement" and the same was done with the two mathematics subscales to create a composite score for "Mathematics achievement". The reliability of these subscales, as assessed through various methods (internal consistency, test–retest reliability and inter-rater reliability), appeared to be sufficiently high.

Apart from these cognitive outcome measures the following non-cognitive outcome measures were used:

- student attentive/inattentive behavior as measured by means of teacher ratings.
- student attitudes to learning (composite score for attitudes to reading, writing and maths) based on students' responses and students' perceptions of the quality of school life and curriculum/usefulness.

The study used the following student background variables:

- student ability, as measured by means of standardized reading comprehension and mathematics tests;
- socioeducational level, based on parents' occupation, level of education and income;
- non-English-speaking background;
- rural/non-rural;
- gender.

Instructional characteristics used in the study were:

- English homework (students' self-report);
- mathematics homework (students' self-report);
- composite classes;
- teachers' participation in structured Literacy In-Service Programs;
- teacher affect, a combination of two scales (self-perceptions) measuring energy/enthusiasm and warmth.

The last category of variables used in the study is subsumed under the heading "teachers' perceptions of their work environment".

On the basis of 12 subscales of the School Organizational Health Questionnaire, two factors labeled "professional culture" (containing professional development, peer support, goal congruence, role clarity and discipline policy) and "teacher–student interaction" (containing morale, curriculum coordination, student orientation, feedback and discipline policy) could be extracted.

Finally, a measure of teachers' perceptions of leadership support was included in the study.

The analyses consisted first of two- and three-level variance components models to obtain estimates of explained variance among students, among classes and between schools. Second, a three-level regression model was used to investigate which variables explained progress in English and math at primary and secondary level. Third, a procedure to conduct multilevel path analysis (Bosker & Scheerens, 1994, p. 169) was used to investigate more structural relationships among the explanatory and outcome variables.

Results

When two- and three-level variance components models were used to estimate the proportion of total variance in student progress (adjusted for year level and prior achievement) in English and math tied by either between school differences or among-class differences, the results as summarized in Table 5.8 were obtained. When differences between classes are taken into account, differences between schools appear to be relatively small.

The results of fitting a three-level regression model to detect significant relationships of school- and classroom-level explanatory variables with progress in English and math indicated that the effects of particular

Table 5.8 Percentage of variance in student progress (with adjustments for year level and prior achievement) accounted for by among-classes and between-school differences (source: Hill *et al.*, 1995a)

	Two-level model	Three-level model	
	Between schools (%)	Among classes (%)	Between schools (%)
English			
Primary	17.0	45.4	8.6
Secondary	18.2	37.8	7.4
Mathematics			
Primary	16.4	54.7	4.1
Secondary	18.9	52.7	8.4

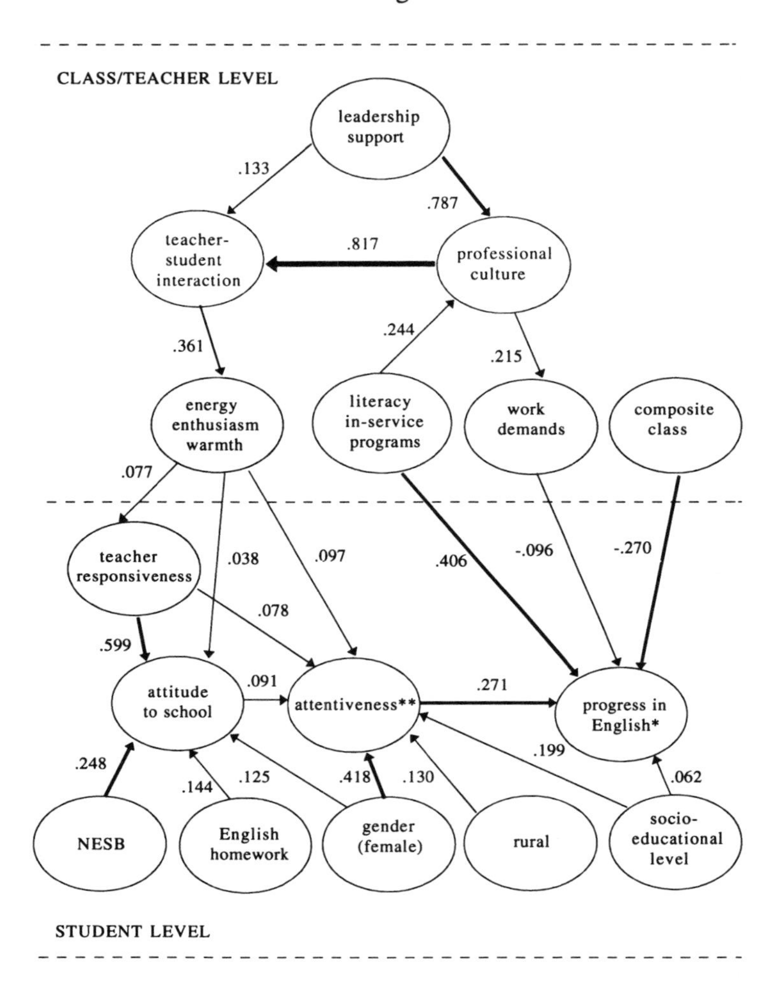

Figure 5.2 Multilevel path model of primary student progress in English. *Progress in English was measured as English achievement adjusted for Year Level and prior achievement, with prior achievement specified as random at levels 2 and 3. **As a predictor of progress in English, attentiveness was specified as random at level 2.

explanatory variables were not always the same across subjects and across primary and secondary schools (Hill *et al.*, 1995 p. 18). A second interesting result was that no variables measured at the school level were found to be significant in explaining student progress.

Classroom-level variables that were significant for more than one subject–school-level combination were: participation in Literacy In-Service Programs (+); composite class (–); and work demands (–).

The results of the multilevel structural equation modeling, with regard to primary student progress in English, are summarized in Fig. 5.2 (cited with the permission of the authors; Hill *et al.*, 1995a). Apart from the participation of teachers in Literacy In-Service Programs, and whether or not the student was placed in a composite class, student attentiveness had the largest association with progress in English.

Although relatively high coefficients were found between some of the school-level variables (leadership and professional culture, and professional culture and work demands) these patterns of relationships were not found to be linked to progress in English (see the small negative effect of "work demands").

Concluding comments

The study yielded some important falsifications of central hypotheses in school effectiveness thinking:

- factors that work are subject and context (school level; primary/secondary) specific, rather than generalizable over school subjects and contexts;
- school differences explain relatively little variance, after differences between classes have been taken into account;
- classroom-level explanatory variables are associated with learning progress more frequently than school-level explanatory variables;
- no support could be found for indirect effects of school-level variables, such as leadership support, mediated by class/teacher-level variables on adjusted outcomes.

Slavin (U.S.A., 1996): Success for All

Background

A specific type of research bearing on educational effectiveness issues is the evaluation of innovatory programs that are based on the educational effectiveness knowledge base.

From a research perspective there are advantages and disadvantages to such evaluations. Advantages are the ecological validity of the research, the fact that educational conditions are systematically manipulated (so that field-experimental designs become feasible), and the possibility of observing the cumulative, holistic, or synergetic effect of severable effectiveness-enhancing conditions that may support each other. This last possibility can also be seen as a disadvantage since, in most cases, it will be impossible to establish which exact features have determined program success (or failure). Finally, there are also specific research-methodological problems associated with this approach, such as the Hawthorne effect and the whole range of factors that threaten the internal validity of quasi-experimental research.

Apart from rigorous application of the methodology of evaluation research, the usefulness of such studies depends on the explicitness of the program rationale and clear guidelines, standardized procedures and special materials (e.g. curricula, teaching methods, computer hardware and software, standardized assessment methods) to control program implementation.

A recent example that appears to meet these criteria is the evaluation of the Success for All program in the U.S.A. (Slavin, 1996).

Program description

Success for All is a program for inner-city schools where the general aim is not only to raise achievement levels, but also that all students succeed, at least in reading (the subject-matter area on which the program is focussed).

According to the scale of its dissemination (presently more than 400 schools in 26 U.S. states and three countries, reaching about 200,000 children) the program appears to be quite successful. Success for All concentrates at the lower grades of primary schools, kindergarten and even prekindergarten.

The core mission of Success for All is concerned with "prevention and early intervention rather than remediation" (Slavin, 1996, p. 85).

The prevention aim is tackled by providing structured curricula, classroom management and assessment procedures from age four onward. Intensive early intervention is accomplished by means of one-to-one tutoring by certified teachers or para-professionals, and close cooperation with social services and medical services as well as active attempts to work with parents.

The instructional principles that the reading program uses are:

- providing structure and gradually giving pupils more opportunities for independent reading (scaffolding);
- cooperative learning "built around story structure, prediction, summarization, vocabulary building, decoding practice, and story-related writing" (p. 87);
- direct instruction in reading comprehension skills;
- "a relentless focus on the success of every child" (p. 91);
- daily homework assignments (20 minutes reading);
- regrouping according to reading performance levels;
- increase in time for direct instruction by means of the regrouping arrangements and tutoring;

- eight-week, curriculum-based, reading assessment (formal methods and teacher observations and judgments);
- teaching of meta-cognitive skills by tutors.

The support context of the instructional program consists of a family support team, a program facilitator per school, teacher training, and an advisory committee composed of the building principal, the program facilitator, teacher representatives, parent representatives, and family support staff.

Evaluation research results

Slavin (1996) reports the results of evaluations of 19 Success for All schools, each of which was compared with a matched control school, at grade levels 1 through 4. The unit of analysis used in the comparison of program and control schools were grade-level cohorts (students in all classes in that grade in a given year); the number of cohorts in the experimental schools matched the number of cohorts in the experimental schools.

Results, adjusted for pretest performance, on four reading tests were used to compare program and control schools. Results were expressed in terms of the proportion of the standard deviation for the control group cohorts. The adjusted means for the program cohorts differed from the adjusted means for the control cohorts. The program cohorts significantly outperformed the control cohorts on all 16 comparisons (4 grade levels × 4 reading tests). The average effect size was about half a standard deviation.

Another relevant evaluation finding was that effect sizes increase with the years of implementation of the program, in other words the gap between program and control schools became larger over time.

Interpretation of results

The success of Success for All can be seen as just one more important piece of evidence indicating that the instructional level determines the difference between effective and less effective schools. It is clear, however, that actions and structures at the above-classroom level were required to make implementation at the classroom level possible.

Given the substance of the program, particularly the nature of the instruction and classroom management principles that were used, the evaluation results support the characteristics of effective instruction that also have been found in other types of study and the research syntheses of these studies.

The author states in his conclusion that the program, which is essentially an instructional program, has been demonstrated to be replicable across different contexts, and over schools that are not always exceptional. However, "systemic changes in assessments, accountability, standards and legislation can facilitate the implementation of Success for All Schools"

(Slavin, 1996, p. 109). As stated more explicitly by Wang *et al.* (1993), the message seems to be that systematic innovation and restructuring of school administration and organization should be seen as facilitative to educational reform rather than as its essence. The core of educational reform should be the primary process of learning and instruction.

The Success for All evaluation also has an important message with respect to innovation strategy. The program is externally developed, with specific materials, manuals and structures and, contrary to the view that comprehensive school reform must be invented by school staff themselves, does not support the view that "every school staff should reinvent the wheel". Similar conclusions are drawn by Stringfield *et al.* (1995) in their assessment of another highly structured program, the Calvert program.

Sammons et al. (U.K., 1995b): The Differential Secondary School Effectiveness Project

Aims and scope

This study, conducted in the U.K., had three major aims:

1. to obtain knowledge about the size, extent and stability over time of differences between secondary schools in their overall effectiveness in promoting pupils' GCSE attainments;
2. to explore the extent of internal variations in school effectiveness with respect to departments and different groups of pupils (classified by gender, eligibility for free school meals, ethnic group and prior attainment;
3. to investigate whether variations in school and departmental processes are important in accounting for better performance, both overall performance in GCSE and performance in specific subjects (English and mathematics).

Considering these aims, it should be noted that the study addressed three major themes in school effectiveness research: assessment of school effects, the consistency of school effects across time and school organizational subunits (departments and subgroups of pupils), and the search for explanatory process conditions of effective schooling.

The study was of a longitudinal nature, using assessment results over a five-year period, while gathering information on school processes both concurrently with the final wave of outcome measurement (1994/1995) and retrospectively (1990–1992).

The study was conducted among 94 secondary schools in eight inner London Local Education Authorities (LEAs), involving GCSE candidate cohorts of approximately 7000 pupils in any one year.

Design and methods

The project was structured into three phases. In the first phase, school- and department-level value-added effects were analyzed to obtain a proper picture of differences between schools. The final outcome measures were total GCSE scores and GCSE scores in six subjects: English, English literature, mathematics, French, history and science. Final scores were adjusted for prior achievement and other student background factors such as age, gender, ethnic background and eligibility for free school meals. School and departmental residuals were calculated and used as the basis for selecting schools for detailed case studies.

During the second phase detailed qualitative case studies were carried out. Three types of schools were selected for these case studies:

1. academically ineffective–broadly negative effects;
2. academically more effective–broadly positive effects;
3. highly mixed effects (e.g. significantly positive effects in both years for mathematics and significantly negative effects in both years for English and history.

The case studies were conducted for in-depth qualitative analysis of the processes and characteristics in these three different categories of schools.

During the third phase the results of the case studies were used to develop instruments for the collection of information about school and departmental processes from schools in the complete sample.

These instruments (questionnaires to head teachers and heads of departments) were administered, and the results were related to value-added outcome data. Essentially, the questionnaires asked the respondents to express their opinion on:

- principal goals that the school tries to achieve;
- factors to be taken into account when judging effectiveness (i.e. which effectiveness criteria ought to be applied);
- factors which are expected to contribute most to the effectiveness of the school;
- factors which are seen as barriers to the effectiveness of the school.

The association between process characteristics and value-added outcomes was investigated both by means of correlations and *t*-test comparisons of high- and low-scoring schools of each separate variable, and by means of simultaneous analyses of the most promising factors employing a three-level multilevel analysis (school, year and student level).

Results

The results of the first two phases will not be summarized in any detail in this review. The results of the first phase indicated that only a small number of schools could be viewed as outliers (consistently highly academically effective or ineffective or ineffective over several years) and that, for most schools, the results were far less clear cut "and some schools had highly mixed effects at the departmental level (highly effective and highly ineffective departments coexisting in the same school)" (Sammons *et al*., 1995b, p. 45).

These results have important practical implications within a national context where there is a policy to publish league-tables of schools, based on "raw" student outcomes, the overall message being that only providing value-added outcomes does not solve the problems with ranking schools; first, because confident discrimination can be made only at the extremes, and second, because of the doubtful relevance of overall achievement scores (such as total GCSE), given the observed phenomenon of inconsistency of effectiveness across subject-matter areas.

The second phase of the study, case studies of three types of schools, yielded eight factors which were found to be important in explaining the (in)effectiveness of schools and departments:

- the importance of school and departmental histories and the impact of change;
- high expectations;
- academic emphasis — including examination entry policy and monitoring;
- shared vision/goals;
- an effective School Management Team;
- the quality of teaching;
- parental involvement.

When comparing the schools that were consistently effective across departments with schools that were mixed in this respect, the authors point to the relevance of facilitative conditions at the school level.

> "Whilst the departmental level was undoubtedly very important, in some schools it was apparently "easier" for all departments to function effectively, due to a more supportive context, shared whole school emphasis on the importance of student learning and achievement, and the apparently mutually beneficial impact of successful departments supporting each others' efforts and sparring each other on to further success. Conversely, in other schools it was "harder" for departments to be effective due to lack of overall leadership, shared goals and vision, poor expectations and inconsistent approaches."

The results of the third phase consist of separate associations of each of the process variables with value-added outcomes and a simultaneous analysis, using a three-level multilevel analysis.

Table 5.9 Features associated with greater academic effectiveness at GCSE (cited from Sammons *et al.*, 1995b)

1. High expectations	
2. Strong academic emphasis	
3. Shared vision/goals	School-wide With individual departments
4. Clear leadership	School Departmental
5. Effective school management team	Team work Staff morale
6. Consistency in approach	School policies and practice Departmental policies and practice
7. Quality of teaching	For all ability groups Work focus Effective control Enthusiasm Student feedback Pupil grouping Staff absence/shortages
8. Student-centered approach	Pastoral environment Staff/student relationships
9. Parental support/involvement	

A large number of questionnaire items was found to be significantly associated with value-added measures of school effectiveness. The authors summarize these results by mentioning 11 interdependent factors (summarized in Table 5.5, cited from Sammons *et al.*, 1995a; see Table 5.9).

The multilevel analyses resulted in seven different types of process–outcome relationships:

1. total GSCE performance scores and head teacher variables;
2. total GSCE performance scores and English head of department variables;
3. total GSCE performance scores and mathematics head of department variables;
4. English GSCE scores and head teacher variables;
5. English GSCE scores and English head of department variables;

6. mathematics GSCE scores and head teacher variables;
7. mathematics GSCE scores and mathematics head of department variables.

In all of these different arrangements the process characteristics accounted for a large degree or even for all of the between-school variance, after pupil background variables had been fitted. For example, in the case of association 1 (total GSCE-score and head teacher variables) the total school level variance was 7.21% of the total variance in student GSCE score, which was reduced to 1.82% after controlling for pupil background factors and prior achievement. When the process variables were added this percentage dropped to 0.58, indicating that the six most relevant process characteristics accounted for 68% of the relatively small school level variance.

In each of the seven types of process–outcome associations a somewhat different set of most relevant (i.e. significant) process variables was identified. Only three head teacher variables occurred more than once:

- an effective School Management Team;
- academic emphasis;
- parental involvement.

The authors conclude that eight factors were found to be of particular importance in relation to the effectiveness of schools as institutions and individual departments:

- the importance of school and departmental histories and the impact of change;
- high expectations;
- academic emphasis, including examination entry policy and monitoring;
- shared vision/goals;
- head teachers' and heads of departments' roles and leadership.
- an effective School Management Team;
- the quality of teaching;
- parental involvement and support.

It should be noted that this list is identical to the one cited earlier, as a result of the second phase, the case-study analysis.

Commentary

The differential school effectiveness project should be considered as a very rich source of information considering foundational and fundamental issues in school effectiveness; in terms of:

- value-added adjustments;
- consistency, stability and differential effects for subgroups of pupils;

- cognitive and non-cognitive outcomes;
- combination of quantitative and qualitative approaches;
- close alignment with educational practitioners (particularly in the way that process characteristics were selected);
- application of multilevel analysis.

As a whole, the results confirm the picture that also emerges from other recent studies, with respect to the subject and subgroup specificity of educational effectiveness.

With respect to the identification of key process factors, the study gives a far more optimistic impression than some of the other studies that were reviewed:

- many process variables are significantly associated with value-added outcomes;
- the factors are interpretable in a way that confirms the evidence from qualitative research reviews;
- process variables account for a very large amount of the between-school variance.

The fact that the identified process variables account for most of the between-school variance, should be seen against the background of a relatively low proportion of variance that is between schools. This implies that the amount of total variance accounted for by the identified process variables is very small indeed (in the case of the association of head teacher variables and total GSCE performance, for instance, no more than 0.6 of 1%).

Grisay (France, 1996): Evolution of cognitive and affective development in lower secondary education

Aims and general design

This study, conducted in France by A. Grisay of the Université de Liège in Belgium, took place in the four lower grades of secondary education (the collège). The study had two main goals: an instrumental goal, to develop and use a battery of research instruments to measure knowledge and skills in the socioaffective domain, applicable to adolescents; and an explanatory goal, to establish those characteristics of the educational environment (comprising both family and school) which could be associated with a "good" or "less good" development in cognitive and socioaffective performance.

In the background to the study, reference is made to the school effectiveness research tradition. The aim is expressed not to frame questions on the effective school in too narrow a perspective, and merely to investigate cognitive outcomes. In addition, the concept of value added, and techniques to establish the net effect of schooling was to be further explored.

The study was conducted in a sample of 100 collèges in the school districts of Lille, Amiens, Dijon, Grenoble, Montpellier and Toulouse. The sample could not be considered representative; it was composed with stratification to obtain a balance on school background characteristics such as rural/urban, privileged, average and disadvantaged school populations, and level of performance of the student populations at school entrance.

In each of the secondary schools 80 pupils in the first grade (sixième) were selected at random, as well as 10 teachers of the first grade. The following waves of data collection took place.

- In Autumn 1990 a number of tests and scales measuring socioaffective characteristics of pupils was administered, as well as the tests in French and mathematics which are part of the national entrance evaluation among pupils of the first grade (sixième).
- In Spring 1991, school principals, the selected pupils and teachers completed questionnaires dedicated to collecting information on the educational environment of the pupils.
- The operation described in the first two steps (including administering the questionnaires on the educational environment) was repeated in May/June 1992, when the majority of the 1990 cohort had reached the end of the second grade of secondary education (cinquième). This time the teacher questionnaires were administered to 10 teachers who taught in the second grade.
- The last wave of data collection took place in May/June 1994, when the majority of the cohort had reached the end of the fourth grade and thus were near termination of their career at the collège. Again, teacher questionnaires were administered to teachers of the grade in question, in this case, grade four.

Problems of sample attrition that occurred during the four years of the study are discussed in detail in the report, but will not be reproduced here. In fact, around one-quarter of the pupils from whom test results were obtained in 1990 were retested when the final data collection wave took place in 1994. In general, the "survivors" had a more favorable starting profile than the pupils who disappeared over the years.

Input, process and output measures

The variables in the study can be categorized according to an input–process–output framework. As mentioned before, the output category contained both cognitive and non-cognitive measures. Factor analyses were conducted to reduce the large set of process variables measuring the

educational environment. In the overview presented below, only the factors will be mentioned. Input variables were pretests on the output variables on the one hand and other pupil background variables on the other.
Output measures. Cognitive: French; mathematics; general knowledge (history, economy, legislation, labor, civics); *non-cognitive*: opinions concerning citizenship; study skills; motivation to learn at school; locus of control; self-image; perspective on the future; relationships with others (cooperation, social skills, aggression).
Input measures. Pretests on the output measures; socioeconomic background; gender; age; grade petition; score on a social desirability scale; educational climate in the family; nationality.

Process measures (indicators based on factor analyses).

Indicators based on pupil questionnaires.

1. Positive judgment on the quality of education and the school climate (based on variables such as judgment on the competence of teachers, education perceived as structured, experience of formative evaluation, warm relationships with teachers, sense of being treated with justice and respect).
2. Lack of discipline and loss of time (frequent deviant actions, absenteeism, (poor) time management).
3. Students' rights and responsibilities (rights and responsibilities and frequency of external contacts).
4. Structure (dispositifs d'encadrement) (remedial courses, frequency of use of school library, rules used to combat absenteeism, frequent presence of the principal in the classroom).
5. Opportunity to learn (the degree to which pupils are familiar with the test items).
6. Clarity of rules (the degree to which pupils know the school regulation).

Indicators based on teacher questionnaires.

1. Positive judgment on the role of school management (perception of the role of the principal, participation of teachers in decision making, contacts with families and positive judgment on the development of the school).
2. Positive judgment on the reputation of the school.(opinion on pupils' motivation, reputation, prognoses on the proportion of pupils that will go to the second cycle of general secondary education, peaceful atmosphere, good relationships among pupils, good discipline).
3. Innovatory, pupil-centered instruction (pupil-centered instruction, stimulating cooperation among students, structured teaching, participation in innovations, teachers' locus of control, knowledge on pupils).

4. Age and experience of the teacher (this factor, apart from age and number of years on the job, also contains a loading on the variable "knowledge on pupils").
5. Integration and alignment (feeling of being well integrated in the school team, alignment and coherence of practice, participation in decision making, teachers' perception of positive atmosphere among pupils, professional satisfaction).
6. Discipline (judgment on the efficacy of discipline policies, time needed for keeping order, professional satisfaction): the positive pole corresponds to teachers being satisfied about discipline in their school, the negative pole to teachers who lose much time in keeping order and see many acts of deviant behavior by pupils.
7. Material conditions (state of classrooms and availability of resources).

Indicators based on principal questionnaires

1. Success expectations (the indicator loads positive on positive expectations of pupils' careers and on "school policy oriented towards high standards", as opposed to "school policy trying to avoid failures").
2. Discipline (few acts of deviance and lack of discipline, absence of serious problems like drugs, violence and thefts).
3. Material conditions (principals' opinion on state of classrooms and availability of resources).
4. School management showing an active interest in teacher training, goal coordination and the work of pupils (aspects of "educational leadership").
5. Coordination measures (actual measures to enhance cooperation and alignment).
6. Active support policy (policy aimed at the success of all pupils, maintaining support structures at school, e.g. remedial courses, courses in study skills, a proactive role of the school leader).

The report gives plenty of details on the reliability of the instruments and the longitudinal use of the output measures. IRT analyses were used to solve problems of vertical equating tests. These aspects will not be further discussed here.

Analyses

An important part of the report discusses the development of the attainment by pupils on the cognitive and non-cognitive output measures of the four-year period of the study. Although very interesting and rich in description and interpretation, that part of the report will be left unmentioned here.

This summary will concentrate on the part of the report in which the explanatory power of schools and classes as such and process indicators is investigated with respect to the outcome variables.

The analyses conducted to answer this type of question used a three-level analysis, in which variance between schools, between classes within schools and between pupils within classrooms could be established separately.

Three types of explanatory variables were used to fit alternative models:

A pretests
B other pupil background variables
C characteristics of the educational environment.

It should be noted that in category C no distinction was made between variables defined at the school level and variables defined at the classroom level. Therefore the set of alternative models that was tested does not allow for a comparison between the explanatory power of the total set of school level and the total set of classroom level process indicators. However, the significance of each of the individual process indicators in explaining between-pupil variance was established in an additional analysis.

Apart from the usual sequence in the fitting of alternative models, starting with a null model containing no explanatory variables, followed by a model fitting all pupil level covariables and finally a model in which covariables *and* other explanatory variables are fitted, the study fitted additional models. The sequence of fitted models was as follows:

O the empty model;
A fitting of pretests;
B fitting of other covariables (discarding pretests);
C fitting of process indicators (discarding A and B);
AB pretests and other background characteristics;
AC pretests and process indicators;
BC after background variables and process indicators;
ABC pretests, other background characteristics and process indicators.

Usually, in these terms, the sequence of fitted models would be O, A, AB and ABC. The other models (B, C, AC and BC) were included in order to achieve a grasp on the supposed interactive nature of educational environment and pupil background characteristics.

Results

The results on the evolution of cognitive and non-cognitive attainment show significant gains in the cognitive areas (French, mathematics and general knowledge — connaissance civique), but a decline of scores in areas such as attitudes towards school, motivation, locus of control work methods, self-image and a stagnation in study skills. The results in the non-cognitive domain

Table 5.10 Variance components for math and French (source: Grisay, 1996)

	Math			French		
	Between schools	Between classes	Between pupils	Between schools	Between classes	Between pupils
O model	18.7%	21.6%	59.7%	16.7%	15.4%	67.4%
Pretest + covariables (AB)	5.4%	4.7%	48.6%	2.6%	1.9%	45.4%
Pretests, covariables, process indicators (ABC)	2.3%	3.1%	46.1%	0.5%	1.7%	42.6%

Table 5.11 Significant process indicators (cited from Grisay, 1996)

(+)	Positive judgment on the quality of education and the climate[ab]
(–)	Lack of discipline and loss of time[ab]
(+)	Rights and responsibilities of pupils[a]
(+)	Structured support (tutoring, remedial courses)[ab]
(+)	Opportunity to learn[ab]
(+)	Positive judgment on the reputation of the school[a]
(+)	Integration and alignment[a]
(+)	Age and experience of teachers[b]

[a]Significant in math; [b]significant in French; [ab]significant in math and French

show relatively little variation between schools and between classes within schools (about 5–6% of the total variance between pupils).

Given a school system that still bears traces of both a strong centralistic tradition and an overt policy of creating equal chances in education for all pupils, the proportion of variance between schools and between classes is relatively large for mathematics and French (about 40% of the total variance; i.e. the gross between-school variance added to the gross between-class variance is about 40% of the total variance that lies between pupils).

The amplitude of the various variance components when applying several of the above model variants for mathematics and French is summarized in Table 5.10. These results indicate that for math the net proportion of variance determined by attending a particular school and a particular classroom within that school is reduced to about 10% (5.4 + 4.7) of the total variance ("net" in this case means after fitting the pretest and the other covariates). For French this net proportion of variance is 4.5%, whereas around half of this net proportion of variance for math and French is explained by the process indicators (environnement scolaire).

Individual process indicators that were significant in explaining variance are summarized in Table 5.11. It should be noted that practically all significant process indicators are based on the variables measured by means of the pupil questionnaires.

Final remarks

First of all, the results of this study confirm the picture drawn by other technically sophisticated and recent school effectiveness studies, namely that the net between-school variance in the domain of cognitive achievement is relatively small (5–10%).

An important finding was that between-school and (especially) between-classroom variance seemed to increase significantly in math during the four years covered by the study, and that this increase can probably be attributed to the fact that the schools and classrooms which receive high socioeconomic status and/or initially high-achieving students offer them a much better environment compared with other classrooms or schools.

The study does not confirm the conclusions of other studies in indicating larger variations between classrooms within schools than between schools. Although "teachers" were not analyzed as a separate level, the researcher states that there is evidence that schools as institutions do not differ as much as do teachers among themselves (Grisay, 1996, p. 281). Since process indicators were not distinguished in terms of classroom-level and school-level factors, the question of whether school-level factors explain less than classroom-level factors (e.g. Hill *et al.*, 1995a; Scheerens *et al.*, 1989) is not answered in this study. Most of the significant process indicators in this study have a meaning at both the school and the classroom level.

The "classical" factors associated with school effectiveness are only partially supported by the results of this study. Elements such as achievement orientation and educational leadership are not supported. Other factors such as favorable climate, discipline, structured teaching and formative evaluation are supported.

The fact that pupil judgments about likely effectiveness-enhancing conditions showed most of the significant process–output associations raises an interesting methodological point. Pupil judgments and observations on school organizational functioning and teaching can be seen as having two advantages: avoiding self-reports by teachers and head teachers, and measuring these process conditions at the level where the impact on actual performance is likely to be the greatest. Perhaps this practice should be followed more often in school effectiveness studies.

Since the aim of this review of individual studies was to concentrate on aspects that are comparable across studies, relatively little attention has been paid to the non-cognitive part of the study. It should be noted, however, that this part is very innovative and interesting. As was evident from the list of output factors cited earlier, a wide range of non-cognitive outcomes has been measured in this study.

Some general conclusions concerning the non-cognitive domain are: there was less variance among schools in this area as compared to scores in the cognitive domain; there was a strong influence of pupil background characteristics on attitudes towards schooling; and roughly the same process indicators are significantly associated with attitudes towards schooling as compared to achievement in French and mathematics.

Summary and Conclusions

In the introductory part of this chapter five research traditions, each concentrating on a different aspect of educational effectiveness, were discerned:

- school factors in research on equality of educational opportunity;
- research on education production functions (also labeled input–output studies);
- the evaluation of compensatory programs;
- effective schools research;
- research on instructional effectiveness.

The first two types of studies yielded results which played down the importance of school difference as such and the relevance of monetary or resource inputs in particular. The next two types of studies shed more light on process characteristics inside the black box of schooling.

Factors that received empirical support in case studies and program evaluations were:

- strong educational leadership;
- emphasis on the acquiring of basic skills;
- an orderly and secure environment;
- high expectations of pupil progress;
- frequent assessment of pupil progress.

Studies on instructional effectiveness revealed the importance of time on task and structured approaches to learning and instruction, such as direct teaching and mastery learning.

In an influential review of early school effectiveness studies, mostly based on case studies of exceptually well-performing as compared to low-performing schools, Purkey and Smith (1983) indicated that several studies provided support for the following factors, seen as effectiveness-enhancing conditions:

- strong leadership;
- orderly climate;

- high expectations;
- frequent evaluation;
- achievement-oriented policy;
- cooperative atmosphere;
- clear goals for basic skills;
- in-service training/staff development;
- time on task;
- reinforcement;
- streaming.

Three extensive more recent reviews structured their conclusions by distinguishing the following set of general factors:

- productive climate and culture;
- focus on central learning skills;
- appropriate monitoring;
- practice-oriented staff development;
- professional leadership;
- parental involvement;
- effective instructional arrangements;
- high expectations.

The ways in which the various authors have given meaning to each of these general factors is rendered in detail in the main body of the chapter. Taken together, the explanation of the various factors provides an elaborated view on effective schooling and instruction. Although different reviewers have slightly different focusses, overall there appears to be consensus on the general outlook of what makes schooling effective. It is also striking to note the agreement with the factors that were generated from the early school effectiveness studies. This agreement among reviewers from different countries, who have sometimes concentrated on different kinds of effectiveness-oriented studies, points to the existence of an international agreed-on "educated" common sense on "what works" in education. This condition in itself, together with the fact that these factors are also recognized by educational professionals other than researchers, provides support for the strength of this knowledge base.

Although efforts to reduce further the set of relevant factors to a more limited set of essential ingredients for enhancing effectiveness does not do justice to the level of detail that is yielded by the studies, the following structure depicted in Fig. 5.3 comes to mind. Effective schooling is seen to be a product of *vision*, supported by an achievement-oriented policy, production or result-oriented management, and which is shared by a common climate of quantity and targetedness of *exposure*, in terms of time on task and test–curriculum overlap, and appropriate *technology*, in which close guidance, monitoring, feedback and reinforcements are key elements.

As compared to the modes of schooling (see Chapter One) the areas that have received most emphasis in empirical school effectiveness are:

- various aspects of management, notably production management, including monitoring and control, and consensus building;
- culture and climate, shared vision, achievement orientation, orderly arrangements;
- curriculum, curriculum alignment, structured planning;
- instruction, structured approaches with a growing interest in meta-cognition and self-directed learning;
- parents as the major external constituency;
- human resource development, e.g. practice-oriented in-service training.

Despite the strength of the qualitative reviews, given the impressive international consensus, there are also weaknesses concerning how precise and convincing the studies are.

First, the qualitative reviews provide no insight into effect sizes, neither regarding the impact of school differences as a whole, nor with respect to the strength of association between the key factors and (value-added) output.

Second, the issue of aggregation levels within the school organization, particularly the distinction between school level and classroom level conditions, becomes blurred in the reviews. It is often unclear whether factors are supposed to work at the school level, at the classroom level or both. As shown in the

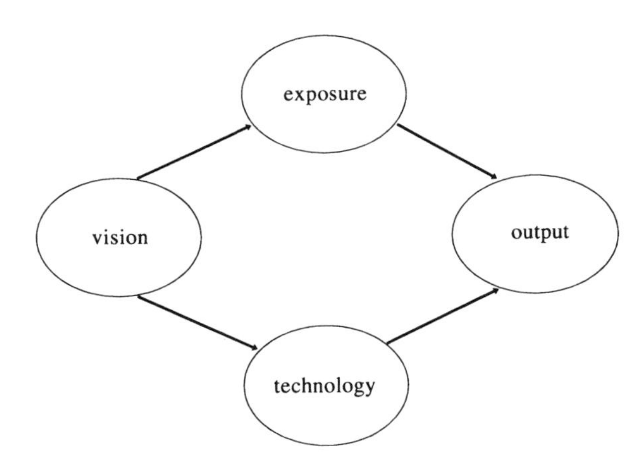

Figure 5.3 Essential ingredients of effective schooling.

chapter on models and clearly illustrated by the Australian study by Hill *et al.* (1995a), the relative impact of school- and classroom-level factors, as well as the way in which cross-level operation is considered to work, is a very important issue in a further structuring of the school effectiveness knowledge base.

Third, to some extent, the picture on effectiveness-enhancing factors of schooling appears to be almost too broad, in giving the impression that effective schooling implies optimizing each and every aspect of school functioning. Critics of the field (e.g. van der Velden, 1996) have pointed to the tautological nature of statements like a good school is the sum total of good teaching, strong leadership, good teachers, ample resources, a favorable environment, etc. Although, strictly speaking, the term tautology does not apply when comparing the qualification of a whole to qualifications of the sum total of its parts, the point underscores the image of a set of conditions that is rather broad, and perhaps not sufficiently focussed.

In subsequent chapters a review of quantitative research syntheses and an exploration of theoretically embedded principles will be used to compensate for the second and third weaknesses of the qualitative reviews, by offering more precision and detail and an attempt to penetrate into the mechanisms behind the factors which are believed to work.

CHAPTER SIX

The Knowledge Base On Effectiveness-Enhancing Conditions, Part 2: Quantitative Research Syntheses

Introduction

There is a vague transgression from qualitative reviews to quantitative meta-analyses or research syntheses. Scheerens (1992) summarized his review of some 40 school effectiveness studies, carried out before 1989, in terms of the "amount of empirical support" for particular school effectiveness correlates (see Table 6.1). According to this overview instructional conditions such as "structured teaching" and aspects of classroom management such as "effective learning time" have received the most convincing empirical support. Other factors that have a meaning at both the school and the classroom level and that were considered to have "a reasonable empirical basis" are "opportunity to learn", "pressure to achieve", "high expectations" and "parent involvement". Most factors that are predominantly defined at the level of school organization and management have only a "doubtful empirical confirmation" or are just "hypothetical".
Creemers (1994, p. 9) provided a similar type of summary of results, focussed at classroom level instruction; his overview of empirical evidence for the characteristics of effective instruction is presented in Table 6.2.

In this overview curriculum, grouping procedures and teacher behavior are seen as basic categories. Characteristics that are considered as having strong empirical support sometimes have a meaning for more than one of the three basic categories. Examples are evaluation, feedback and advance organizers.

Table 6.1 The degree to which the most important school and instruction characteristics relevant to effectiveness have been confirmed by empirical research (from Scheerens, 1992, p. 84)

Characteristics	Multiple empirical research confirmation	Reasonable empirical basis	Doubtful empirical confirmation	Hypothetical
Structured teaching	x			
Effective learning time	x			
Opportunity to learn		x		
Pressure to achieve		x		
High expectations		x		
Pedagogic leadership			x	
Assessment ability			x	
School climate			x	
Recruiting staff				x
Organizational/structural preconditions			x	
Physical/material school characteristics		o		
Descriptive context characteristics			x	
External stimuli to make schools effective				x
Parental involvement		x		

x: Meaningful influence; o: more marginal influence.

Table 6.2 Overview of empirical evidence for the characteristics of effective instruction (cited from Creemers, 1994, p. 9)

Characteristics	Strong empirical evidence	Moderate empirical evidence	Plausible
Curriculum		x	
Grouping procedures	x		
Teacher behavior	x		
Curriculum			
Explicitness and ordering of goals and content	x		
Structure and clarity of content		x	
Advance organizers	x		
Evaluation	x		
Feedback	x		
Corrective instruction			x
Grouping procedures			
Mastery learning	x		
Ability grouping		x	
Cooperative learning			x
– Differentiated material			x
– Evaluation	x		
– Feedback		x	
– Corrective instruction		x	
Teacher behavior			
Management/orderly and quiet atmosphere	x		
Homework	x		
High expectations		x	
Clear goal setting		x	
– Restricted set of goals		x	
– Emphasis on basic skills		x	
– Emphasis on cognitive learning and transfer			x
Structuring the content		x	
– Ordering of goals and content		x	
– Advance organizers	x		
– Prior knowledge		x	
Clarity of presentation		x	
Questioning	x		
Immediate exercise		x	
Evaluation	x		
Feedback		x	
Corrective instruction			x

The next step from qualitative review to quantitative meta-analysis is the "vote-counting procedure". According to this procedure the number of significant positive associations between hypothetical effectiveness-enhancing conditions and educational output are counted and compared to the total number of possible positive associations (i.e. the number of studies in which a particular effectiveness-enhancing condition was measured). An example of a vote-counting procedure is the table cited from Borger *et al.* (1984) in a previous chapter.

Another example, cited from Scheerens and Creemers (1996), is shown in Table 6.3, which gives an overview of 42 Dutch effectiveness studies. It is clear from the table that only in a small fraction of studies were significant positive associations between hypothetical effectiveness-enhancing conditions and outcome (adjusted for pupil's background characteristics) found.

A more developed form of meta-analysis considers a set of studies selected for further analysis as a sample from a universe of studies and establishes statistical estimates on average effects across studies. The magnitude of the estimates can then be expressed by means of significance levels and confidence intervals.

There are several conditions that should be met in order to make such meta-analysis possible (see Cooper & Hedges, 1994). First of all there needs

Table 6.3 Number of Dutch studies in which certain school and instructional conditions correlated significantly with outcome variables, after controlling for pupils' background characteristics (cited from Scheerens & Creemers, 1996, p. 187)

	Primary level		Secondary level	
	Positive association	Negative association	Positive association	Negative association
Structured teaching/ feedback	5		1	
Teacher experience	3	1		1
Instructional leadership		2	1	
Orderly climate	2		3	1
Student evaluation	5		0	
Differentiation		2	0	
Whole class teaching	3		0	
Achievement orientation	4		4	
Team stability/cooperation		3		3
Time/homework	4		4	
Other variables	16		8	
Average between-school variance	9		13.5	
Number of studies	29		13	

Not all variables mentioned in the rows were measured in each and every study.

to be sufficiently detailed information on the individual studies, in terms of operational variables, the way in which outcomes are measured and adjusted, the number of cases, the reliability of measures and the statistical analyses used.

Second, there needs to be — at least — a common core in the explanatory variables (in our case hypothetical effectiveness-enhancing conditions), in order to make comparisons feasible. From the divergence in the operationalization of key factors that was shown from our analyses in Chapter Four, it is clear that this condition may be quite problematic in this field. Bosker and Witziers (1996) provide some practical solutions to this type of problem.

Third, it should be clear what type of "raw" or adjusted outcomes will be used in determining effects of each study that will be entered in the meta-analysis. In the chapter on modeling educational effectiveness four different types of outcomes were discussed (raw, unadjusted outcomes, outcomes adjusted for pupil background factors such as intelligence or socioeconomic status, progress or gain scores, and gain scores adjusted for pupil background characteristics).

Fourth, a choice of effect measure will have to be made. Examples are: the average correlation of a particular process–outcome association, the standardized difference between the average achievement in the treated and untreated experimental conditions, e.g. in terms of tenths of one standard deviation (1 s.d.) of the untreated condition or the ratio of the residual between school variance and the original total variance (cf. Bosker & Witziers, 1996).

It is clear that the first three conditions pose problems with respect to the comparability of individual studies. The fourth condition gives rise to the fact that the comparability of meta-analyses, among themselves, may become difficult. Nevertheless, examples do exist of such more complete meta-analyses on educational effectiveness (e.g. Wang *et al.*, 1993; Bosker & Witziers, 1996).

The discussion on meta-analysis of educational effectiveness studies will be centered under three headings: reconsideration of "economic" input variables of schooling, "instructional effectiveness" and "overall school effectiveness".

When comparing effect sizes from different studies using different effect size indices, the reader might use as a rule of thumb: twice the correlation is equal to Cohen's standardized effect size measure.

Resource Inputs

In the previous chapter the rather disappointing results of research on education production functions, focussing on "technical efficiency", to use Cheng's (1993) term, were mentioned. To illustrate this further, a table is

Table 6.4 Review of 147 input–output studies

Input	Statistically significant			Not statistically significant			
	Number of studies	+	–	Number of studies	Total +	–	Unknown signs
Teacher/pupil ratio	112	9	14	89	25	43	21
Teacher training	106	6	5	95	26	32	37
Teacher experience	209	33	7	69	32	22	125
Teachers' salaries	60	9	1	50	15	11	24
Expenditure per pupil	65	13	3	49	25	13	11

+: Positive correlation; –: negative correlation between input and output variables. (Based on Hanushek, 1986, p. 1161.)

reproduced here from Hanushek's review study (1986) summarizing the findings of 147 studies on education production functions. From Table 6.4 it appears that only the factor "teacher experience" shows some consistency across studies: 30% of the estimated coefficients are statistically significant. Hanushek's general conclusion is, however, that there is currently no correlation between educational expenditure and attainment. According to Hanushek, only if the differences between schools in these sort of inputs were considerably greater than they are at present, could a clear effect of these factors be expected. Thus teacher salaries now vary somewhat but an effect may only show up when widening the difference between the maximum and minimum salary by a factor around two. In most countries these salaries are linked to strict regulations governing training and years of experience. By implementing a system of merit pay there would no doubt be a wider discrepancy and thus a significant correlation with attainment might appear.

Hanushek's overall conclusion was that there is no systematic relation between resource inputs and school outcomes, when controlling for student characteristics such as socioeconomic status. When considering teacher–pupil ratio, teacher training, teacher experience, teacher salaries and overall expenditure per pupil, the vote-count procedure showed positive significant effects in a small minority of cases.

In their reanalysis of Hanushek's data, Hedges *et al.* (1994) point to some of the limitations of the vote-count procedure; for example, the fact that vote counts cannot provide an indication of the magnitude of a relation.

When treating effects of individual studies as replications in a statistical inference procedure, a different type of reasoning is applied. In order to draw general conclusions by combining evidence across different studies, combined significance tests and effect magnitude analyses can be conducted. When Hedges *et al.* (1994) applied these two approaches to a cleaned version of

Hanushek's data set, they concluded that the null hypothesis of no positive effect on the outcome variable had to be rejected for the variables per pupil expenditure, teacher experience, and teacher–pupil ratio, while for these variables the "negative case" null hypothesis of no negative effect on the outcome variables could *not* be rejected. "These results suggest that since there are positive relations between outcome and per pupil expenditure (PPE), and teacher experience, and teacher/pupil ratio but no negative relations between outcome and these resource inputs, the typical relation is positive" (Hedges *et al.* 1994, p. 10).

When examining effect sizes in terms of the number of standard deviations of change in output with a one-unit change in input, Hedges *et al.* conclude that there is an effect of per pupil expenditure of "considerable practical importance" (an increase of PPE by $510 would be associated with a 0.7 s.d. increase in student outcome).

For teacher experience the effects are also generally positive, but with a magnitude too small to be considered of practical importance. For teacher education the median effects are negative, while the evidence with respect to teacher salary and class size is "confounded by the use of both starting and average salaries" or "mixed", respectively.

Apart from a few more technical points of criticism, Hanushek's rejoinder comes down to the statement that Hedges and colleagues' statistical interpretation of his original findings is unrealistic ("absurd" is the word that he uses). He says that the inconsistency in findings across studies is what one would expect. "Some districts use resources well while others make very poor use of resources". In their reply to this rejoinder Hedges *et al.* (1994) conclude that the exchange has led to at least one common interest for future study, namely the question "of *how* money matters".

Even in this relatively clearly delineated subarea of educational effectiveness research, the following general issues in the interpretation of findings may be illustrative of the whole field.

1. Important differences that depend on the methodology and statistical analyses that are employed.
2. The issue of the "practical" meaning of (usually small) effect sizes.
3. The question of the use of general categories of effectiveness enhancing conditions compared with more specific "educational treatments". In the case of the variables analyzed by Hanushek, *per pupil expenditure* is the most general variable. One of Hanushek's criticism of the Hedges analyses is the fact that they find a relatively important effect of the overall PPE measure, which does not coincide with substantive positive effects of the main variables that would contribute to PPE, such as pupil–teacher ratio. In the case of the basic question "does money matter", the next, deeper probing

question is "how" does money matter. Similar reasoning is found with respect to the interpretation of "time" as an effectiveness-enhancing condition. Not only is overall "gross" or "net" time for instruction considered important, but also the question of whether the available time is used for good-quality teaching.

The Quantitative Evidence on Instructional Effectiveness

In the 1980s, several influential research syntheses were carried out by Walberg (1984) and Fraser *et al.* (1987). The teaching conditions for which Walberg found the highest effects were:

- reinforcement (reward and punishment);
- special programs for the educationally gifted;
- structured learning of reading;
- cues and feedback;
- mastery learning of physics;
- working together in small groups.

Fraser *et al.* (1987) even provided a synthesis of 134 meta-analyses which together comprised 7827 individual studies. Part of their findings is summarized in Table 6.5. Specific variables, included in the main categories in Table 6.5 which correlate highly with achievement, are: quality of teaching, $r = 0.47$; amount of instruction, $r = 0.38$; cognitive background characteristics, $r = 0.49$ and feedback, $r = 0.30$.

A remarkable conclusion that Walberg attaches to his research syntheses is the statement that the findings apply for all types of schools and all types of pupils. Walberg expresses this in the saying "What's good for the goose is good for the gander". He adds that this especially applies to the more powerful factors (i.e. the factors that correlate the highest with levels of performance).

Table 6.5 Effects of teaching and pupil characteristics on performance tests (based on Fraser *et al.*, 1987)

Factor	Result (correlation)
School characteristics	0.19
Social background characteristics of pupil	0.21
Teacher characteristics	0.22
Teaching characteristics	0.24
Pupil characteristics	0.14
Instruction method	0.28
Learning strategies	0.12

When one looks at these powerful factors, it seems that highly structured learning or direct teaching, which emphasizes testing and feedback, again emerges as the most effective teaching form. Yet, in Walberg's research syntheses there are also forms of individual teaching and teaching adapted to fit the specific needs of pupils as well as working together in small groups that come quite strongly to the fore. He even values "open teaching" in which cooperation, critical thinking, self-confidence and a positive attitude are important objectives. Walberg's and other meta-analyses reveal that open teaching has no adverse consequences for cognitive achievement, while there is a positive influence on creativity, social behavior and independence. In the meta-meta-analysis of Fraser *et al.*, individualizing emerged as a less powerful factor ($r = 0.07$).

No matter how impressive the huge data files may appear on which the research syntheses are based, there are some limitations attached to the findings. In all cases, simple correlations are presented whereby it cannot be ruled out that a particular correlation is carried for the most part by a third variable, which in these simple analyses cannot be made visible. This problem exists partly because it can be assumed that many of the individual effectiveness predictors are correlated among themselves; and where this problem applies to the general analyses it can by no means be ruled out that this is also the case with many of the individual studies on which the syntheses are based.

Finally, with regard to this survey of instructional effectiveness it must again be pointed out that within the scope of this book only a broad summing up of the most important research findings on school effectiveness is possible. Even if the conclusion is that a few prominent characteristics of effective teaching can be distinguished — the amount of instruction and a structured approach — that apply to any given teaching situation, it should certainly not be forgotten that with a less general treatment all types of nuances exist that are linked to differences in subjects taught, pupil characteristics, school type and educational goals. For a review in which these nuances are well expressed reference is made to Brophy and Good (1986).

In a more recent synthesis of meta-analyses and reviews (maybe the term *mega*-analysis would be appropriate for this work), Wang *et al.* (1993) summarize the current knowledge with respect to the influence of educational, psychological and social factors on learning. Although the evidence they present is comprehensive in the sense that school-context factors and school-level factors are included, the majority of studies concern "design and delivery of curriculum and instruction" (36%), "student characteristics" (24%) and "classroom practices" (18%).

A first main outcome of the Wang *et al.* research synthesis is a rank ordering of the relative importance of "distal" versus "proximal" factors in influencing achievement. Distal factors are less directly associated with the primary process of learning and instruction, for example, "state and district governance and organization" and "school demographics, culture, climate, policies and practices". Student characteristics and classroom practices are considered as proximal factors, close to the instructional process. The results of the syntheses show that the more proximal factors have a stronger positive association with educational achievement than the more distal factors. The rank ordering presented by Wang *et al.* is (ordered from high to low):

- student characteristics;
- classroom practices;
- home and community educational contexts;
- design and delivery of curriculum and instruction;
- school demographics, culture, climate, policies and practices;
- state and district governance and organization.

Leaving aside student characteristics, more specific factors that have the strongest association with achievement are "classroom management" and "student and teacher social interaction" (both aspects of the more general factor "classroom processes") and "home environment" (aspects of the more general factor "home and community educational contexts"). An illustrative variable within the "classroom management category" is "group alerting" (teacher uses questioning/recitation strategies that maintain active participation by all students). Other variables that are relatively influential within the classroom practice factor are: classroom climate, classroom assessment, quantity of instruction (e.g. time on task) and "student and teacher interaction" (e.g. "students respond positively to questions from other students and from the teacher").

In their interpretation of these effective classroom practices the authors emphasize the following points:

- academic student–teacher interactions should "make students aware of subject-specific knowledge structures", for example, by an appropriate use of questioning by the teacher;
- social teacher and student interactions should dissuade students from disruptive behavior and "establish a classroom atmosphere conducive to learning"; besides, the use of praise and corrective feedback is mentioned.

In summary, it appears that, as far as classroom instruction is concerned, this recent meta-analysis enforces the importance of general instruction approaches that are quite structured, such as mastery learning and direct instruction. At the same time, the interpretation that the authors give to "academic teacher and student interactions", namely in terms of laying bare "knowledge structures" and the importance of "meta-cognition" as a student background factor, provide support to an emerging, more cognitivist–constructivist view on learning and instruction (also see Chapter Two of this book).

The effects of instructional conditions reported in meta-analyses are generally larger than the sizes of the effect of resource input variables (in terms of proportion of 1 s.d. the latter are in the order of 0.20–1.17 and in terms of correlations in the order of 0.20). In most instances, however, the associations reported reflect relationships unadjusted for student background characteristics.

One of the strong points of these mega-meta-analyses is that information with a wide scope is provided, while using global categories of effectiveness-enhancing conditions. At the same time, however, this characteristic may also be taken as one of the limitations of this approach: the results do lack a certain degree of specificity. Another point of criticism is that the variance in research quality of the studies used as a basis for meta-analysis is not always taken into account.

In order to overcome some of these disadvantages of meta-analysis Slavin (1996) introduced an approach which he calls "best-evidence synthesis". He describes this approach as a combination of the methods of systematic literature review and meta-analysis. In "best-evidence synthesis" the quantification of study outcomes is combined with the discussion of substantive and methodological issues of narrative reviews (p. 3). In his recent book *Education for all* (Slavin, 1996), best-evidence syntheses on several instructional practices are presented: cooperative learning, ability grouping in elementary and secondary schools and mastery learning.

With respect to cooperative learning, Slavin analyzed a total of 99 studies in which different types of cooperative learning were investigated. Of these 99 studies 64% showed a significant advantage of the experimental group over the control group. In only 5% of the study were control groups favored. The median effect sizes for different types of cooperative learning range from 0.04 s.d. to 0.86 s.d The mean effect size over all studies was 0.26 s.d., which can be seen as an educationally significant effect.

Factors that make group work "work" are the use of group rewards based on the individual learning of all group members and the direct teaching of structured methods to students to work together or teaching them learning strategies closely related to the instructional objectives (Slavin, 1996, p. 57).

Applying the same method of "best-evidence synthesis", Slavin (1996) also analyzed ability grouping, by analyzing studies in which a form of ability grouping was compared with a control condition of heterogeneously grouped classes. He concluded that ability grouping between classes in order to create groups that are homogeneous in ability level or, to use his term, "ability-grouped class assignment" is generally ineffective (effect sizes are either negative or close to zero, with only a few exceptions of studies reporting small positive effects).

Grouping for reading across grade lines (the Joplin Plan) has a consistently positive effect (median effect size = +0.44 s.d.); the same conclusion is drawn with respect to within-class ability grouping in mathematics in the upper elementary school (median effect size = +0.34 s.d.).

In his explanation as to *why* certain types of grouping (e.g. within classes) work and others (ability-grouped class assignment) do not, Slavin applies three criteria:

1. grouping must reduce heterogeneity in the specific skill being taught;
2. the grouping plan must be flexible enough to correct for misassignments and changes in student performance;
3. "Teachers must actually vary their pace and level of instruction to correspond to students' levels of readiness and learning rates" (p. 158).

He concludes that ability-grouped class assignment fails the first and probably also the second criterion.

In addition to ability-grouped class assignment being generally ineffective, it is disadvantageous in the sense that it is likely to cause segregation and institutionalize low teacher expectations.

Relevant explanatory background factors considered by Slavin regarding the question of why certain types of grouping are effective and others ineffective, are instructional time and the use of assessment-based adaptive instruction. The latter characteristic is likely to be conducive to instruction that is closely tailored to students' levels of performance. With respect to instructional time, within-class ability grouping will usually require some loss of time for transition of the teacher from one group to the next, and will also imply that groups will have to work for a considerable amount of time without direct teacher instruction. Slavin's results indicate that within-class ability grouping is more effective than whole class teaching to heterogeneous classes, despite a certain loss of direct instructional time. To suppress the negative impact of loss of instructional time there should not be too many ability groups within the class.

On the basis of these findings Slavin proposes the following tentative recommendations (1996, p. 164):

1. Leave students in heterogeneous classes most of the time and regroup by ability only in subjects (reading, mathematics) in which reducing heterogeneity is particularly important.
2. Grouping plans should reduce heterogeneity in the specific skill being taught.
3. Grouping plans should be flexible and allow for easy reassignment.

4. Teachers should actually vary their level and pace of instruction to correspond to student performance level.
5. In within-class ability grouping, the number of groups should be small.

The lack of effect of ability grouping between classes (streaming or tracking) was reproduced in a "best-evidence synthesis" in which studies on secondary schools were analyzed (Slavin, 1996, pp. 167–188). The median effect size that was found was –0.02.

Mastery learning ranked high in the meta-analyses by Walberg (1984) and Fraser *et al.* (1987) cited in a previous section. Slavin (1996) conducted a best-evidence synthesis and found moderately positive effects on experimenter-made achievement measures (about 0.25 s.d.), closely tied to the objectives taught in the mastery learning classes, but practically no effects when standardized achievement tests were used as the dependent variable.

He concludes that his findings do *not* support the claim that mastery learning is more effective than traditional instruction given equal time and achievement measures "that assess coverage as well as mastery" (Slavin, 1996, p. 253). The explanation that Slavin gives for the fact that his outcomes are much more modest than those of other reviewers and meta-analysts is that he selected only studies that met the criterion of a study duration of at least four weeks, whereas the other reviewers included many short-term studies, and studies in which the effect of mastery learning was inflated because of more instruction time — even one-to-one tutoring — in the experimental group.

Therefore, the selection of studies used by Slavin was higher in ecological validity than many of the more laboratory-based studies included in the other meta-analyses and reviews. In real-life classroom situations the major principles of mastery learning, and particularly the amount of corrective instruction, may not be applied optimally ("too little and too late"; Slavin, 1996, p. 256). The key element in mastery learning, according to Slavin, is the frequency of testing and feedback. Other related elements are well-specified educational objectives and basing teaching decisions on the results of these assessments. Finally, the time factor plays a central role in the success of mastery-learning approaches (although this factor has sometimes confounded the outcomes of experimental studies).

The basic factors that emerge from the meta-analyses and best-evidence synthesis cited in this chapter are:

- time on task;
- closeness of content covered to assessment instrument;
- a structured approach: specific objectives, frequent assessment and corrective feedback;
- types of adaptive instruction that can be managed by teachers (e.g. no more than two within-class ability groups per classroom).

Again, the effects of these various instructional conditions are larger than the effect sizes noted for resource input factors

The Impact of School-Level Organization and Management Conditions

As cited in the Introduction to this chapter, Fraser *et al.* (1987) estimated the average impact of school characteristics, in terms of a correlation with achievement, at 0.12. This is a value that is lower than the estimated correlation for teaching conditions and pupil characteristics (approximately 0.20). It should also be noted that these are raw rather than partial correlations, in other words this concerns unadjusted rather than value-added achievement measures.

Wang *et al.* (1993) elaborated on this issue by considering the relative impact of broad categories of antecedent conditions of learning (i.e. achievement). They too found that the impact of school characteristics such as school demographics, culture, climate policies and practices is lower than the impact of factors that are closer to the actual learning process. At the same time factors that are even further removed from the primary process of learning and instruction subsumed under the heading "state and district governance and organization" have an even lower impact than school characteristics.

From these findings the authors draw an important conclusion for educational innovation by stating that schools should begin solving problems by addressing "proximal variables like curriculum, instruction and assessment which emphasize student outcomes" (Wang *et al.*, 1993, p. 276). They also state that the relatively small impact of school-level conditions is inconsistent with "current conventional wisdom" which emphasizes policy-driven solutions like school restructuring.

The number of studies in which school characteristics were included in these meta-analyses was relatively small compared with the number of studies that looked into the effects of teaching conditions (e.g. in the Wang *et al.* synthesis 8% of all reviews are syntheses considering school-level factors). In fact, we know of no meta-analyses that were targeted on the syntheses of school effectiveness studies. This was one of the reasons for including meta-analyses of school effectiveness studies in our own research program.

Meta-Analyses on Presumed School Effectiveness-Enhancing Factors

In one of the previous chapters the question of what the "true" size of school effects amounts to was answered by applying a statistical meta-analysis. The second step consists of determining, via a meta-analytical approach, whether variables mentioned in the school effectiveness literature have a positive relation with relevant output measures and what the estimated effect size of these variables might be. By this new approach to the field of school effectiveness inquiry, a more thorough contribution to the knowledge base can be made. The research reported here is drawn from studies by Bosker and Witziers (1996) and Witziers and Bosker (1997).

Design and Model for the Meta-Analyses

For the meta-analyses on school-level effectiveness-enhancing factors the multilevel model suggested by Raudenbush (1994) was applied. The selected studies are considered as a sample from the population of studies on school effects. Nested under each study are the secondary units: the schools. Each study then can be viewed as an independent replication. This idea could be applied, but it does not solve the problem of multiple results from one study, e.g. effects are reported for mathematics and language achievement separately while using the same sample of schools and students in a study. The ideal solution would be the application of a multivariate multilevel analysis, but in that case each study should report results on both subject domains. Since this condition could not be met, the two-level model for meta-analysis was generalized to a three-level model for meta-analysis, in which the highest level of the studies is referred to as the across-replication level, and the multiple results within a study as the within-replication level. The lowest level is the level of the schools and, as is the case in a variance-known situation, only the variance of the estimate (the square root of which is the standard error, s.e.) was modeled on this level. In many cases studies did not report the standard error of the estimate, in which case it was simply derived from the sample size of the schools. The main advantage of the statistical meta-analysis over simple procedures like vote-counting is that the information from each study is weighted by the reliability of the information, in this case the sample size. Moreover, differences in reported effect sizes can be modeled as a function of study characteristics.

The multilevel model then, starting with the within-replications model, is (cf. Bryk & Raudenbush, 1992, pp. 158–161):

$$d_{rs} = \delta_{rs} + e_{rs}. \tag{6.1}$$

The effect size in replication r in study s (d_{rs}) is an estimate of the population parameter (δ_{rs}) and the associated sampling error is e_{rs} (since in each replication only a sample of schools is studied).

The between-replications model is:

$$\delta_{rs} = \delta_s + u_{rs}. \tag{6.2}$$

In this model the true replication effect size is a function of the effect size in study s and sampling error u_{rs}. Finally, the between-studies model is formulated as follows:

$$\delta_s = \delta + v_s. \tag{6.3}$$

The true unknown-effect size as estimated in study s (δ_s) is a function of the effect size across studies (δ) with random sampling error v_s (since the studes are sampled from a population of studies).

In assessing effects of subject domain, model (6.2) is extended to:

$$\delta_{rs} = \delta_s + \gamma_1 \text{subject}_{rs} + u_{rs}. \tag{6.4}$$

Furthermore, effects of country, sector, study design (with or without adjustments for covariates; referred to as "value added") and statistical modeling technique (multilevel or not) can be modeled by extending model (6.3):

$$\delta_s = \delta_0 + \gamma_2 \text{sector}_s + \gamma_3 \text{country}_s + \gamma_4 \text{design}_s + \gamma_5 \text{model}_s + v_s. \tag{6.5}$$

Subject and country are polytomous variables, represented in the analyses by dummies. Thus, the variables to be used in the meta-analyses are:

country–USA	0 = else, 1 = U.S.A.
country–NL	0 = else, 1 = The Netherlands
design	0 = gross, 1 = value added (correction for prior achievement and/or background variables)
subject–math	0 = composite score for math and language, or a special domain
	1 = math score only
subject–lang	0 = composite score for math and language, or a special domain
	1 = language score only
sector	0 = primary education, 1 = secondary education
model	0 = multilevel
	1 = otherwise
respondent	0 = teacher
	1 = school leader

This last predictor was only used in the meta-analysis on the effect of school leadership, in which analysis the variable model was dropped. Therefore in equation (6.5) δ_0 is the estimated effect size for studies where all predictors have value 0. The variables for which the effect was assessed are:

- cooperation;
- school climate;
- monitoring measured at: (a) school level; (b) class level;

- opportunity to learn, which concept is indicated by: (a) content coverage; (b) homework; (c) time;
- parental involvement;
- pressure to achieve;
- school leadership.

Excluded from the studies for the meta-analyses are international comparative assessments made by the International Association for the Evaluation of Educational Achievement (IEA). A re-analysis of one of these studies will be presented and reanalyzed in Chapter Seven of this book. Including them in the meta-analyses would lead to doubling of the research evidence. Meta-analysis with the IEA studies included are presented in Witziers and Bosker (1997).

The correlation coefficient was used as an index for the effect of the factors. Not all studies have presented their results in terms of correlations, so all coefficients were transformed using formulae presented by Rosenthal (1994). The most well-known effect size coefficient, Cohen's *d*, is related to the correlation coefficient in the following way:

$$d = 2r/\sqrt{(1 - r^2)}. \quad (6.6)$$

As will be seen by the results later on, a good rule of thumb is that the effect size is twice the correlation coefficient when the latter is lower than, say, 0.15. In order to be able to synthesize the results the correlation coefficient is transformed to Fisher's Z:

$$Z_r = \tfrac{1}{2}\ln((1 + r)/(1 - r)). \quad (6.7)$$

Again, for small values of the correlation coefficient Z_r and r do not differ much, but the reader should bear in mind that all of the following tables refer to Z_r.

Cooperation

Table 6.6 contains the results of the meta-analysis on the effects of cooperation on pupil achievement. In total, 20 studies (N_a) are included, some of which produced multiple results so that the total number of within and across replications (N_w) is 41.

The estimated mean effect size of cooperation across all studies, is expressed in Z_r, is 0.0292, which is significant at the 10% level (one-tailed). The estimated variance across all studies (both within and across replications) is 0.0023, which indicates that the 95% prediction interval around the mean effect size runs from $Z_r = -0.0648$ to $Z_r = 0.1232$. The prediction interval, unlike the confidence interval, describes the distribution based on the

Table 6.6 Estimated effect size for cooperation and variance across and within replications (N_a=20, N_w=41)

	Effect	S.E.	*p*-Value
Mean effect size	0.0292	0.0185	0.06
Variance across replications	0.0023	0.0018	0.03
Variance within replications	0.0000	0.0000	0.50

estimates. The confidence interval only gives information on the amount of precision with which the mean of that distribution is estimated.

The results of the analyses trying to predict differences between effect sizes with study characteristics (or moderators) such as subject matter, sector, country, study design and statistical model employed show that none of the predictors have a significant relationship with the effect size (see Table 6.7).

Table 6.7 Predicting differences in cooperation effect sizes

	Effect	S.E.	*p*-Value
Intercept	–0.0046	0.1277	0.97
Secondary	–0.0530	0.0869	0.55
Arithmetic/math	–0.0282	0.0361	0.44
Language	–0.0195	0.0373	0.60
U.S.A.	0.0798	0.0579	0.19
The Netherlands	–0.0658	0.0781	0.41
Value added	0.0860	0.0900	0.36
Not multilevel	0.0209	0.0998	0.84
Variance across replications	0.0000	0.0000	0.50
Variance within replications	0.0000	0.0000	0.50

School climate

The estimated mean effect size of school climate across all studies is Z_r = 0.1090, which is significant at the 1% level (one-tailed) (see Table 6.8). The estimated variance across all studies (both within and across replications) is 0.0145, which indicates that the 95% prediction interval around the mean effect-size runs from Z_r= –0.1270 to Z_r= +0.3450.

The results of the analyses trying to predict differences between effect sizes with study characteristics (or moderators) such as subject matter, sector, country, study design and statistical model employed show that two predictors have a significant relationship with the effect size (see Table 6.9). Both predictors are concerned with the research design. The stricter the design (with adjustment for intake differences between schools and employing a multilevel statistical model) the less pronounced the effects of school climate. For studies that included both design features the estimated effect size equals zero.

Table 6.8 Estimated effect size for school climate and variance across and within replications (N_a=22, N_w=62)

	Effect	S.E.	*p*-Value
Mean effect size	0.1090	0.0333	0.00
Variance across replications	0.0145	0.0069	0.00
Variance within replications	0.0000	0.0000	0.50

Table 6.9 Predicting differences in school climate effect sizes

	Effect	S.E.	*p*-Value
Intercept	0.0626	0.1048	0.55
Secondary	0.0416	0.0742	0.58
Arithmetic/math	–0.0091	0.0537	0.86
Language	0.0122	0.0549	0.82
U.S.A.	–0.1442	0.0918	0.13
The Netherlands	–0.0023	0.0942	0.98
Value added	–0.0958	0.0483	0.06
Not multilevel	0.2704	0.0981	0.02
Variance across replications	0.0103	0.0052	0.03
Variance within replications	0.0000	0.0000	0.40

Monitoring

For monitoring there are two indicators, the first at school level and the second at the class level.

Table 6.10 Estimated effect size for monitoring at school level and variance across and within replications (N_a=24, N_w=38)

	Effect	S.E.	*p*-Value
Mean effect-size	0.1481	0.0370	0.00
Variance across replications	0.0160	0.0088	0.04
Variance within replications	0.0016	0.0031	0.30

Table 6.11 Predicting differences in school level monitoring effect sizes

	Effect	S.E.	*p*-Value
Intercept	0.1036	0.3143	0.74
Secondary	0.0317	0.2365	0.89
Arithmetic/math	–0.0868	0.0626	0.18
Language	–0.0065	0.0636	0.92
U.S.A.	0.0462	0.3218	0.89
The Netherlands	0.0318	0.2956	0.92
Value added	N.A.		
Not multilevel	0.0736	0.1467	0.62
Variance across replications	0.0250	0.0113	0.01
Variance within replications	0.0000	0.0000	0.45

Monitoring at school level

The estimated mean effect size of monitoring across all studies is Z_r=0.1481, which is significant at the 1% level (one-tailed) (see Table 6.10). The estimated variance across all studies (both within and across replications) is 0.0176, which indicates that the 95% prediction interval around the mean effect size runs from Z_r= –0.1119 to Z_r= +0.4081.

The results of the analyses trying to predict differences between effect sizes with study characteristics (or moderators) such as subject matter, sector, country, study design and statistical model employed show that no predictor has a significant relationship with the effect size (see Table 6.11).

Monitoring at class level

The estimated mean effect size of monitoring at class level across all studies is Z_r = 0.1147, which is not significant at the 10% level (one-tailed) (see Table 6.12). The estimated variance across all studies (both within and across replications) is 0.1207, which indicates that the 95% prediction interval around the mean effect size runs from Z_r= –0.5662 to Z_r= +0.7956.

The results of the analyses trying to predict differences between effect sizes with study characteristics (or moderators) such as subject matter, sector, country, study design and statistical model employed show that two predictors have a significant relationship with the effect size (see Table 6.13).

Both for arithmetic/mathematics and for language (as compared to composite scores or other domains) there are moderate but significant effects of monitoring at class level.

Table 6.12 Estimated effect size for monitoring at class level and variance across and within replications (N_a=43, N_w=70)

	Effect	S.E.	*p*-Value
Mean effect size	0.1147	0.0822	0.16
Variance across replications	0.1119	0.0426	0.00
Variance within replications	0.0088	0.0070	0.00

Table 6.13 Predicting differences in monitoring at class-level effect sizes

	Effect	S.E.	*p*-Value
Intercept	–0.1091	0.2815	0.70
Secondary	–0.0191	0.2678	0.94
Arithmetic/math	0.2908	0.0720	0.00
Language	0.2633	0.0721	0.00
U.S.A.	–0.3745	0.4543	0.42
The Netherlands	–0.1592	0.3521	0.57
Value added	N.A.		
Not multilevel	0.2854	0.2718	0.43
Variance across replications	0.1609	0.0573	0.00
Variance within replications	0.0000	0.0000	0.33

Table 6.14 Estimated effect size for content coverage and variance across and within replications (N_a=N_w=19)

	Effect	S.E.	*p*-Value
Mean effect size	0.0886	0.0386	0.02
Variance across replications	0.0169	0.0088	0.00

Opportunity to learn

This concept contains three variables: content coverage, homework, and time.

Content coverage

The estimated mean effect size of content coverage across all studies is Z_r= 0.0886, which is significant at the 5% level (one-tailed) (see Table 6.14). The estimated variance across all studies is 0.0169, which indicates that the 95% prediction interval around the mean effect size runs from Z_r= –0.1662 to Z_r= +0.3434.

The results of the analyses trying to predict differences between effect sizes with study characteristics (or moderators) such as subject matter, sector, country, study design and statistical model employed show that three predictors have a significant relationship with the effect size (see Table 6.15).

The results show that the effects of content coverage are less clear if math is the criterion variable, and even less so in the case of language achievement. For both subject domains the estimated effect sizes are clearly negative. The only other effect has a negative connotation: inadequate statistical techniques produce more positive effect size estimates for content coverage.

Table 6.15 Predicting differences in content coverage effect sizes

	Effect	S.E.	*p*-Value
Intercept	–0.0517	0.1897	0.78
Secondary	–0.0230	0.0735	0.76
Arithmetic/math	–0.1962	0.0874	0.05
Language	–0.3362	0.1010	0.01
U.S.A.	0.1638	0.1126	0.17
The Netherlands	0.0527	0.0884	0.67
Value added	0.2505	0.1992	0.24
Not multilevel	0.1622	0.0857	0.08
Variance across replications	0.0016	0.0026	0.11

Table 6.16 Estimated effect size for homework and variance across and within replications (N_a=13, N_w=41)

	Effect	S.E.	*p*-Value
Mean effect size	0.0574	0.0260	0.03
Variance across replications	0.0041	0.0029	0.01
Variance within replications	0.0000	0.0000	0.49

Homework

The estimated mean effect size of homework across all studies is Z_r= 0.0574, which is significant at the 5% level (one-tailed) (see Table 6.16). The estimated variance across all studies (both within and across replications) is 0.0041, which indicates that the 95% prediction interval around the mean effect size runs from Z_r= –0.0681 to Z_r= +0.1829.

The results of the analyses trying to predict differences between effect sizes with study characteristics (or moderators) such as subject matter, sector, country, study design and statistical model employed show that no predictor has a significant relationship with the effect size (see Table 6.17).

Time

The estimated mean effect size of time and variance across all studies is Z_r = 0.1931, which is significant at the 1% level (one-tailed) (see Table 6.18). The estimated variance across all studies (both within and across replications) is 0.0364, which indicates that the 95% prediction interval around the mean effect size runs from Z_r= –0.1808 to Z_r= +0.5670.

Table 6.17 Predicting differences in homework effect sizes

	Effect	S.E.	*p*-Value
Intercept	0.0617	0.2720	0.82
Secondary	–0.2007	0.1626	0.26
Arithmetic/math	0.1741	0.1613	0.29
Language	0.1594	0.1610	0.33
U.S.A.	0.1100	0.1218	0.40
The Netherlands	–0.1328	0.1878	0.50
Value added	–0.0408	0.1089	0.72
Not multilevel	0.0798	0.1264	0.56
Variance across replications	0.0033	0.0026	0.05
Variance within replications	0.0000	0.0000	0.42

Table 6.18 Estimated effect size for time and variance across and within replications (N_a=21, N_w=56)

	Effect	S.E.	*p*-Value
Mean effect size	0.1931	0.0491	0.00
Variance across replications	0.0354	0.0151	0.00
Variance within replications	0.0010	0.0031	0.36

The results of the analyses trying to predict differences between effect sizes with study characteristics (or moderators) such as subject matter, sector, country, study design and statistical model employed show that one predictor has a significant relationship with the effect size (see Table 6.19). Without adjustments for intake the effects of time allotment tend to be more positive.

Table 6.19 Predicting differences in time effect sizes

	Effect	S.E.	p-Value
Intercept	0.1067	0.1514	0.48
Secondary	0.1525	0.0992	0.14
Arithmetic/math	–0.0659	0.1133	0.56
Language	–0.1276	0.1114	0.26
US.A.	0.2163	0.1306	0.11
The Netherlands	0.1413	0.1164	0.24
Value added	–0.0995	0.0562	0.10
Not multilevel	0.0905	0.0898	0.33
Variance across replications	0.0262	0.0125	0.21
Variance within replications	0.0000	0.0000	0.29

Table 6.20 Estimated effect size for parental involvement and variance across and within replications (N_a = 14, N_w = 29)

	Effect	S.E.	p-Value
Mean effect size	0.1269	0.0596	0.03
Variance across replications	0.0386	0.0195	0.00
Variance within replications	0.0000	0.0000	0.47

Parental involvement

The estimated mean effect size of parental involvement in all studies is Z_r = 0.1269, which is significant at the 5% level (one-tailed) (see Table 6.20). The estimated variance across all studies (both within and across replications) is 0.0386, which indicates that the 95% prediction interval around the mean effect size runs from Z_r= –0.2670 to Z_r= +0.5032.

The results of the analyses trying to predict differences between effect sizes with study characteristics (or moderators) such as subject matter, sector, country, study design and statistical model employed show that no predictor has a significant relationship with the effect size (see Table 6.21).

Table 6.21 Predicting differences in parental involvement effect sizes

	Effect	S.E.	*p*-Value
Intercept	0.1608	0.4165	0.70
Secondary	0.1049	0.1712	0.56
Arithmetic/math	–0.0261	0.0544	0.64
Language	–0.0485	0.0544	0.38
U.S.A.	0.2634	0.2696	0.36
The Netherlands	0.1074	0.3179	0.66
Value added	–0.0441	0.0691	0.54
Not multilevel	–0.0222	0.3198	0.51
Variance across replications	0.0395	0.0180	0.00
Variance within replications	0.0000	0.0000	0.33

Table 6.22 Estimated effect size for pressure to achieve and variance across and within replications (N_a=26, N_w=74)

	Effect	S.E.	*p*-Value
Mean effect size	0.1327	0.0441	0.00
Variance across replications	0.0337	0.0136	0.00
Variance within replications	0.0139	0.0049	0.00

Pressure to achieve

The estimated mean effect size of pressure to achieve across all studies is Z_r = 0.1327, which is significant at the 1% level (one-tailed) (see Table 6.22). The estimated variance across all studies (both within and across replications) is 0.0476, which indicates that the 95% prediction interval around the mean effect size runs from Z_r = –0.2949 to Z_r = 0.5603.

The results of the analyses trying to predict differences between effect sizes with study characteristics (or moderators) such as subject matter, sector, country, study design and statistical model employed show that only one predictor has a significant relationship with the effect size (see Table 6.23).

The results show that the strongest effects of achievement pressure can be found in the U.S.A. For other countries the effects of this variable are almost non-existent.

Table 6.23 Predicting differences in pressure to achieve effect sizes

	Effect	S.E.	*p*-Value
Intercept	0.0233	0.1187	0.84
Secondary	0.0235	0.0635	0.72
Arithmetic/math	0.0332	0.0728	0.32
Language	–0.0008	0.0707	0.99
U.S.A.	0.2499	0.1097	0.03
The Netherlands	0.0355	0.0975	0.64
Value added	–0.0452	0.0684	0.52
Not multilevel	0.0427	0.0721	0.56
Variance across replications	0.0224	0.0107	0.02
Variance within replications	0.0175	0.0059	0.03

Table 6.24 Estimated effect size for school leadership and variance across and within replications (N_a=38, N_w=108)

	Effect	S.E.	*p*-Value
Mean effect size	0.0499	0.0225	0.03
Variance across replications	0.0072	0.0043	0.08
Bariance within replications	0.0056	0.0035	0.01

School leadership

The estimated mean effect size of school leadership across all studies is Z_r = 0.0499, which is significant at the 5% level (one-tailed) (see Table 6.24). The estimated variance across all studies (both within and across replications) is 0.0128, which indicates that the 95% prediction interval around the mean effect size runs fromZ_r = –0.1719 to Z_r = 0.2717.

The results of the analyses trying to predict differences between effect sizes with study characteristics (or moderators) such as subject matter, sector, country, study design and statistical model employed show that two predictors have a significant relationship with the effect size (see Table 6.25).

It can be seen that the research design of the study affects the effect sizes found on school leadership. Adjustments for intake differences produce higher effect sizes. However, this positive effect was not seen at all for The Netherlands.

Table 6.25 Predicting differences in school leadership effect sizes

	Effect	S.E.	*p*-Value
Intercept	0.0905	0.0805	0.26
Secondary	–0.0380	0.0492	0.49
Arithmetic/math	0.0281	0.0431	0.52
Language	0.0129	0.0437	0.77
U.S.A.	0.0169	0.0699	0.81
The Netherlands	–0.1669	0.0797	0.04
Value added	0.0784	0.0375	0.04
School leader is respondent	0.0037	0.0417	0.93
Variance across replications	0.0046	0.0033	0.10
Variance within replications	0.0056	0.0035	0.02

Conclusions

The meta-analyses presented in the first part of this chapter indicate that per pupil expenditure, teacher experience and pupil–teacher ratio are potential achievement-enhancing factors, but that this interpretation is still contested. Teacher salary and teacher training appear to have a negative association with pupil achievement. All effects are smaller than what Cohen calls a small effect size, with the exception of per pupil expenditure. Hedges *et al.* (1994, p. 11) interpret this effect as, "It suggests that an increase of PPE by $500 (approximately 10% of the national average) would be associated with a 0.7 s.d. increase in student outcome. By the standards of educational treatment interventions, this would be considered a large effect". The rigorous approach of a statistical meta-analysis on school effectiveness studies (or studies that produce results relevant for this field) show consistent positive but moderate effects for most of the factors studied. On average these effects are equal to a correlation of 0.10. In Cohen's terminology this would be a small effect. For time allotment, the effect is substantially higher (0.19), which is between a small and medium effect; for homework, cooperation and school leadership it is substantially lower. For monitoring at class level it is generally absent, although more refined analyses show that its effects are present for math and language as subject domains. On average, the correlation of 0.10 would be considered a weak effect, but it should be borne in mind that schools on average only account for 10% of the variation in achievement. In that case, the effectiveness-enhancing factors considered here each accounts uniquely for 10% of the school-level differences in achievement. An even more optimistic picture might be derived if corrections for attenuation could be made, thus adjusting for errors of measurement in both the dependent and independent variables, i.e. in the achievement variable and the school effectiveness-enhancing factors. The correlations reported here might then increase substantively. As Cohen (1988, p. 79) writes about correlations of 0.10:

> ... many relationships pursued in "soft" behavioral science are of this order of magnitude. Thurstone once said that in psychology we measure men by their shadows. As the behavioral scientist moves from his theoretical constructs, among which there are hypothetically strong relationships, to their operational realizations in measurement and subject manipulation, very much "noise" (measurement unreliability, lack of fidelity to the construct) is likely to accompany the variables. This, in turn, will attenuate the correlation in the population between the constructs as *measured*. Thus, if two constructs in theory (hence perfectly measured) can be expected to correlate .25, and the actual measurement of each is correlated .63 with its respective pure construct, the observed correlation between the two *fallible* measures of the construct would be reduced to .25 (.63) (.63) = .10. Since the above values are not unrealistic, it follows that often (perhaps more often than we expect), we are indeed seeking to reject null hypotheses about r_s when r is some value near .10.

One should not forget, however, that generally in meta-analysis effect sizes are slightly larger because of, amongst other things, publication bias (the tendency not to report non-results). Moreover, features of more adequate research designs and statistical models generally lead to less pronounced results. In some instances (for content coverage and monitoring) effects appear to be domain specific. School sector differences are also present, showing more positive effects of cooperation in secondary schools, and country differences could also be demonstrated showing the absence of school leadership effects in The Netherlands and clear-cut effects of achievement pressure and cooperation in the U.S.A.

The most powerful factors, however, are not surprisingly located at the classroom level: a structured approach to teaching, time on task, and grouping across grade lines appear to be the most promising factors. For time on task the effect is between small and medium, as is the case for differentiation and grouping across grade lines. For aspects of structured teaching, feedback and reinforcement appear to have large effects, whereas cooperative learning has a medium effect.

CHAPTER SEVEN

AN INTERNATIONAL COMPARATIVE SCHOOL EFFECTIVENESS STUDY USING READING LITERACY DATA

Introduction

Although many models of educational effectiveness were presented in the first chapters of this book, and subsequently questioned in the following research review chapters, their cultural specificity has been almost unquestioned so far. The results of the meta-analyses indicate that the size of school effects differs across countries, and so do the effects of their antecedent conditions that are seen as their cause. Although the Edmonds' five-factor model of (1) strong educational leadership, (2) emphasis on basic skills achievement, (3) safe and orderly climate, (4) high expectations of pupils' achievement, and (5) frequent evaluation of pupils' progress was advocated in the 1980s as the factors that make schools work, as did extensions and refinements of this model such as those presented by, for example, Scheerens (1992) and Creemers (1994) (see Chapter Two), it might as well be that some of these factors work in some educational systems but not in others. In other words: to what extent is the educational effectiveness model generalizible across countries? In this chapter previous research into this area will be reviewed, and a reanalysis of the

Reading Literacy Study on 9-year-old children by the International Association for the Evaluation of Educational Achievement (IEA) will be presented.

The main advantages of using IEA data over meta-analysis are that a common research design was employed in all countries, and that all variables have been operationalized in English and then have been translated to the national language using strict translation procedures. Moreover, one can be sure that, when creating a value-added-based school effect, this effect is constructed statistically in the same way in all countries. The disadvantage of the IEA data is that it was a multipurpose study, not specifically aimed at detecting educational effectiveness-enhancing factors.

Previous Research

There have been earlier attempts to assess the context specificity of educational effectiveness models. These research projects will be described briefly.

A reanalysis of the SIMMS data

Scheerens *et al.* (1989) report a reanalysis of the Second International Mathematics and Science study of 14-year-old pupils by the IEA. They included in their analysis the mathematics scores of students from Belgium (Flemish and French), Canada (Ontario and British Columbia), Finland, France, Hong Kong, Hungary, Israel, Japan, Luxembourg, The Netherlands, New Zealand, Scotland, Sweden, Thailand and the U.S.A.

To begin their analysis the authors first investigated within-country differences between schools and classes. Stated otherwise: they looked at the size of unadjusted class effects and unadjusted school effects, as well as effects adjusted for the occupational level of the parents of the pupils, and compared these across the countries selected. The results are summarized in Table 7.1.

Since in eight countries only one class per school was selected, classroom variance could not be separated from school variance in these cases. Four groups of countries can be distinguished from the results. First, some countries (Belgium Flemish, Belgium French and The Netherlands) showed vast differences in the mean achievement of students across schools: this situation involves vertically organized, strongly differentiated school systems. Second, several countries (U.S.A., Sweden, New Zealand and Finland) showed relatively small differences between schools but large differences between classes within schools: this pattern indicates homogeneous grouping of pupils within a horizontally organized, integrated system of secondary schools. Third, there is a group of countries (Canada, France and Israel) where differences both between schools and between classes within schools are relatively small, probably because of (partially) mixed ability grouping within an integrated schooling system. Fourth, some countries do not have a tracked, vertically organized system, but *de facto* there are large quality differences between schools (most notably in Hong Kong and Thailand).

Using multilevel statistical models, and correcting for the occupational and educational level of the student's father and for future aspirations of the student (thus having a value-added-based school effect outcome), the authors analyzed the effects of 13 potentially successfully predictors. The predictors were selected on the basis of raw correlations, which should be at least 0.10 in at least five countries.

Table 7.1 Estimates of the variance explained by schools and classes (cited from Scheerens *et al.*, 1989)

Country	Classroom variance component	School variance component
15 Belgium (Flemish)		0.50
16 Belgium (French)		0.64
22 Canada (British Columbia)		0.27
25 Canada (Ontario)	0.18	0.09
39 Finland	0.45	0.002
40 France	0.17	0.06
43 Hong Kong		0.51
44 Hungary		0.30
50 Israel	0.22	0.10
54 Japan		0.08
59 Luxembourg	0.29	0.15
62 Netherlands		0.67
63 New Zealand	0.45	0.01
72 Scotland	0.34	0.12
76 Sweden	0.45	0.00
79 Thailand		0.39
81 U.S.A.	0.46	0.10

Estimates of the variances expressed in terms of the intraclass correlation coefficient, for all countries, assuming that schools are sampled at random within countries and classrooms are sampled at random within schools. Intraclass coefficients after adjusting for the occupational level of the parents of the pupils only differ marginally.

The antecedent conditions were as follows (Scheerens *et al.*, 1989, p. 792).

1. Teacher characteristics

- experience as a mathematics teacher (in years)
- time spent on keeping order (in min per week)
- time spent on teaching (in min per week)

2. Opportunity to learn

- items to test covered in tuition

3. Expectations

- estimate by teacher of the number of pupils who belong to the top band in mathematics

4. Instructional characteristics

- total time (hours) spent on homework
- the use of published tests
- the use of teacher-made tests

5. School characteristics

- the number of women teachers in mathematics
- the number of men teachers that teach only mathematics
- the number of meetings of mathematics teachers

6. Contextual characteristics

- degree of urbanization of the school area
- class size

From these 13 variables only two showed clear and consistent positive effects. Opportunity to learn had a positive effect in 9 out of 17 countries, and teacher expectations in 13 out of 17 countries. The way that both variables were operationalized in the IEA study, however, makes it clear that contamination with the achievement measure cannot be altogether excluded.

Class size shows a positive effect in 47% of the cases (i.e. relatively large classes did better) which, by the way, is contrary to meta-analytic studies on class size (Glass *et al.*, 1982).

Indications for factors that work in education could furthermore be found for homework, teacher experience, time spent on teaching, and time spent on keeping order (which had the expected negative effect).

When looking into these results, the picture appears to be confusing. One of the reasons for this, when looking at the factors that were studied, is that they are only partially connected to the educational effectiveness model. Conceptually interesting conditions at school level, e.g. educational leadership, were not included because of limitations in the dataset.

A further problem arises from the fact that the countries studied have different educational systems at the secondary level: some are integrated horizontally, whereas others are tracked vertically. If then, for instance, high-aptitude students in a tracked system are separated from the low-aptitude students, factors such as time spent on teaching are contaminated with the (self-)selection of students to schools in the system.

An outlier study using data from the IEA Reading Literacy Study

The second attempt at gaining insight into the cross-cultural generalizability of effectiveness-enhancing educational factors was made by Postlethwaite and Ross (1992). These authors used the Reading Literacy Study (RLS) on 9-year-old students of the IEA (as we will do in the sequel of this chapter). Using a "Home Composite" index (based on the variables: using the language tested at home, number of possessions at home — television, refrigerator, etc., depending on country — number of meals per week, and number of books at home).

Subsequently, this indicator was then used as a proxy for socioeconomic status from which mathematics achievement was predicted. Using the school residuals (the average amount of overachievement of the students in that school, with a correction for sampling error) the 20 most and the 20 least effective schools were then selected within each country.

A selection from the 500 indicators in the RLS was carried out in the following way.

1. Teacher and school variables had to correlate 0.18 or more with school mean achievement.
2. Only those indicators were retained that passed criterion 1 in at least 10 (out of 27) countries.
3. Constructs were formed of the 150 remaining indicators, of which 24 were teacher and 27 were school indicators.

Next, in this exploratory study, the 20 effective schools were compared with the 20 ineffective schools on the 51 indicators. These then were ranked from 1 to 51, with rank 1 assigned to an indicator that discriminated best between the effective and ineffective schools. The results are summarized in Table 7.2., in which the 15 best discriminating teacher and school indicators are presented, with their rank number, and the number of countries in which the indicator successfully discriminated between effective and ineffective schools.

From among the contextual factors, urban–rural, school size and community resources appeared to be associated relatively frequently with achievement. Of the school factors, degree of parental cooperation, no problems in school, percentage of female teachers, reading materials in school, school resources, sponsoring reading initiatives, and the student–teacher ratio are amongst the top 15.

Teacher–classroom indicators that are associated with reading achievement are: reading in class, comprehension instruction, classroom library, teaching experiences, and literature emphasis.

In general, the relations take the expected direction, except for school size and student–teacher ratio. The former indicates that the larger the size of the school the more likely it is to be an effective school, and the latter indicates that the more students per teacher the higher the achievement. The authors themselves prudently state that this analysis had a very exploratory character, and that readers should not deduce causal relationships from it.

In the sequel, results using the same dataset will be presented, applying multilevel models on all schools, teachers and students, without restricting the study to outliers.

Table 7.2 Teacher and school indicators discriminating between effective and ineffective schools (top 15; source: Postlethwaite & Ross, 1992)

Rank	Indicator	No. of countries
1	Degree of parental cooperation	16
2	Reading in class	17
3	No serious problems	18
4	Urban–rural	14
5	School size	12
6	Community resources	14
7	Reading materials in schools	13
8	Comprehension instruction	11
9	Percentage female teachers	14
10	Classroom library	10
11	Total teaching experience	11
12	School resources	13
13	Student–teacher ratio	12
14	Sponsor reading initiatives	13
15	Literature emphasis	9

The Generalizability of the Educational Effectiveness Model Using RLS Data

Dataset and selection of variables

In this part the RLS data of the IEA will also be used, but first multilevel analyses to assess the adequacy of one educational effectiveness model will be applied. As a basis for the selection of predictor variables the multilevel model of educational effectiveness (Scheerens, 1989b) was chosen. The dependent variable was the reading literacy score of the student. Student-level predictors used to create a value-added measure of school effectiveness were: age (in months), gender (0: boys; 1: girls), grade retention (0: no; 1: yes), and a composite index of socioeconomic status (based on the following variables: is the mother tongue of the family at home the same as the language tested? Number of home possessions; number of meals per week; number of books at home; for details see Postlethwaite & Ross, 1992, pp. 59–60). Using these variables as covariates in the analysis, one obtains school effects adjusted for intake differences when assessing effects of input, school and class variables. Furthermore, potential covariates at the classroom level were included: teacher language (was the mother tongue of the teacher the same as the language tested? 0: no; 1: yes) and class size (number of pupils in the class). With respect to the latter variable it should be noted that in many countries this is a proxy of extra resources for schools for helping to overcome inequality of educational opportunities, which implies that the higher the class size the more privileged the student population of the school might be!).

The following potentially effectiveness-enhancing variables were included:

Context:

1. Public versus private.
2. Community type (rural coded as 1, to city with over a million inhabitants 4).

Inputs:

3. Teacher training: did the teacher follow in-service or on-service training with respect to reading in the last three years (1: none; 2: once; 3: twice; 4: three times; 5: four or more times).
4. Class size: number of students in the class that are tested.
5. Parental involvement: the degree of parent cooperation with the school in terms of support for the schools' educational principles or goals (compared with other schools you know) (1: much below average; 2: below average; 3: average; 4: above average; 5: much above average).
6. Resources (composite index based on availability of school library, reading room for students, student/school newspaper or magazine, teacher library).

School processes:

Achievement pressure indicated by three variables:

7. Focus on higher order problem-solving skills (factor score for one of the dimensions found in 28 items on reading activities).

8. Focus on reading (factor score for one of the dimensions found in 28 items on reading activities).
9. Improvement orientation (sum of two items: "does the school sponsor informal initiatives to encourage reading" and "do you have a program for improving reading instruction?").

Educational leadership indicated by four variables:

10. Principal involvement (scale constructed on six items asking teachers whether the principal discusses with them standard setting, asks for results, discusses with them the choice of instructional methods, encourages contacts amongst teachers, stimulates professional development, makes suggestions about content).

11a. Evaluation of teachers (is the work of a teacher evaluated by the school principal, according to the teacher?).

11b. Evaluation of teachers (how often is the work of teachers evaluated, according to the principal?).

12. Leadership (scale based on the importance ranking by the principal of four items out of eight: evaluation of staff, discussing educational objectives with staff, using records of pupils' progress, developing activities aimed at the professional development of teachers).

Consensus and cooperation indicated by:

13. Staff discussions (a combination of two series of questions, one on how often there are staff meeting (never, once a year, ..., weekly) and one on whether the subjects of discussion are mainly on educational topics).

Curriculum quality indicated by

14. Teaching hours (derived from number of hours and minutes of total instruction time and how many school weeks there are in a year, corrected for days of instruction lost due to accidents, floods, strikes, festivals, staff meetings, etc.).

Safe and orderly climate:

15. Problems in school (whether or not the school experiences serious problems in providing for the teaching and learning of reading).

Evaluative potential:

16. Evaluative procedures (how many of the following procedures does the school use to gather information for evaluation: interviews, written or oral self-reports by teachers, observational data on teachers' classroom

work, student ratings of teachers' performance, other forms of systematic evaluation).

Classroom processes:

Time indicated by three variables:

17. Time for instruction (total number of hours and minutes available for instruction in a typical week).
18. Time for reading (total number of hours and minutes available for the teaching and practice of reading in a typical week).
19. Homework (a combination of two questions, one on how often the teacher asks children to read something at home as part of the reading/language program, and the other on the amount of time children are expected to spend on their reading homework).

Direct instruction indicated by three variables:

20. Instructional strategies (a scale score based on items with respect to the frequency in which nine different instructional strategies are applied).
21. Needs assessment by using written material (factor score for one of the dimensions found in nine items on methods to assess students' needs in reading).
22. Needs assessment by using observations (factor score for one of the dimensions found in nine items on methods to assess students' needs in reading).

Opportunity to learn (i.e. curriculum coverage): no indicators available

High expectations: no indicators available

Student monitoring indicated by two variables:

23. Progress assessment using oral performance (score based on three items on how often the teacher assesses a student's progress).
24. Progress assessment using written performance (score based on three items on how often the teacher assesses a student's progress).

Reinforcement: no indicators available

Three categories thus remain uncovered, opportunity to learn, high expectations and reinforcement. For high expectations, this is not so much of a problem because this measure would have been contaminated with the output measure. The results will be examined in three stages: first the differences between countries with respect to the process indicators, then the effects of these process variables within each country, and finally a cross-country educational effectiveness model will be fitted to the data.

Data analysis strategy

The strategy for analyzing the dataset is as follows.

1. First, country means on the variables described above will be presented, to see whether countries differ with respect to the process indicators.
2. Then, 27 multilevel models will be fitted, one for each country. In this only one classroom per school is selected at random, since that it is the most common situation. First, the amount of between-school variation per country will be assessed; second, a model using the student level covariates will be fitted; and then process variables will be introduced. This strategy enables estimation of the effects of the latter and their unique explanatory power.
3. Finally, in a multilevel analysis with country as the superordinate third level next to schools (second level) and students (first level), a cross-country model using Z-scores per country will be fitted. The process variables cannot explain differences between countries in this analysis. What we are searching for is the communality in the explanatory power of process variables across countries.

Results Part 1: Differences Between Countries With Respect to the Process Indicators

First a look will be taken at the between-country differences in the context and input indicators (see Table 7.3). No surprising differences occur with respect to public/private schools. The high figures for The Netherlands derive from denominational public schools, which are subsumed under the heading "private". With respect to community type and teacher training there are no relevant differences. Class size shows remarkable differences, with some countries (Hong Kong, Singapore, Venezuela) having classes containing twice as many pupils as those in other countries (Italy, Iceland). Small class size may be an indicator of pupils at risk. In some countries equal opportunity programs provide schools with extra personnel for underprivileged children, and in other countries (e.g. Italy, Denmark) special education and regular education are integrated. Large differences also occur with respect to school resources, both between industrialized and developing countries, and within these categories.

Finally, some differences in parental involvement can be discerned between countries. However, some caution is warranted: this variable has been operationalized as the perceived difference in the intensity of parental involvement as compared to other schools in the vicinity of the school. One would expect only marginal differences between countries.

Table 7.3 Country averages with respect to input and output variables

	Private	Community type	Teacher training	Class size	Parental involvement	Resources
Belgium	1.00	2.01	1.98	20.29	3.14	1.78
Canada	1.09	2.53	1.90	23.10	3.64	2.15
Cyprus	1.00	2.36	1.99	25.42	3.94	0.03
Denmark	1.08	2.19	2.00	17.29	3.41	3.28
Finland	1.00	1.99	1.99	24.57	3.39	2.49
France	1.00	1.46	1.94	21.47	3.02	1.51
Germany (East)	1.00	1.88	1.99	20.49	3.04	1.21
Germany (West)	1.01	1.95	1.99	21.93	3.17	1.82
Greece	1.06	2.45	1.98	23.33	3.78	1.71
Hong Kong	1.19	3.41	1.97	36.29	4.10	1.63
Hungary	1.13	2.12	2.00	23.44	3.18	2.68
Iceland	1.01	2.09	1.99	16.72	2.86	2.25
Indonesia	1.06	1.42	1.22	32.49	2.89	2.13
Ireland	1.41	2.08	1.94	29.85	3.43	1.26
Italy	1.00	1.77	1.99	16.09	2.90	2.10
Netherlands	1.60	1.49	1.93	24.27	3.28	3.46
New Zealand	1.04	2.84	1.96	29.75	3.43	2.06
Norway	1.01	1.71	1.99	15.20	3.36	2.33
Portugal	1.07	1.51	2.00	20.60	3.29	1.34
Singapore	1.00	4.00	1.16	36.83	3.49	3.31
Slovenia	1.01	2.17	1.99	24.64	3.10	3.21
Spain	1.34	2.31	1.84	27.67	3.20	2.36
Sweden	1.00	1.81	1.99	20.03	3.03	2.40
Switzerland	1.01	1.83	1.98	18.43	2.97	1.88
Trinidad/Tobago	1.12	1.55	1.93	28.46	2.87	1.01
United States	1.15	2.57	1.99	24.20	3.57	2.47
Venezuela	1.19	2.90	1.98	32.43	2.72	1.05
Ranges of coding	1–2	1–4	1–5		1–5	0–4

Table 7.4 contains the means for each country on the school process indicators. The three indicators for achievement pressure do not all point in the same direction because the first two indicators were derived as orthogonal factors from one common scale. The countries where schools focus on higher order problem-solving skills are thus generally not the countries with high scores for focus on reading. With respect to improvement orientation (range from 0 to 2) almost the whole range of possible scores can be seen: from 0.20 and 0.22 for Germany (East) and Switzerland up to 1.65 and 1.77 for the U.S.A. and New Zealand.

Table 7.4 Country averages with respect to school process variables

	Focus on higher order problem solving skills	Focus on reading	Improvement orientation	Principal involvement	Evaluation of teachers (principal)	Evaluation of teachers	Leadership	Staff discussions	Teaching hours	Problems in school	Evaluative procedures
Belgium	−1.03	−0.28	0.86	3.53	3.19	1.60	4.44	4.78	865.22	1.26	1.99
Canada	0.13	1.22	1.37	3.01	2.88	1.92	4.15	5.64	916.08	1.34	2.53
Cyprus	1.16	−0.25	1.42	5.20	3.73	1.96	4.72	7.37	693.32	158	2.41
Denmark	−0.46	−0.83	0.87	1.38	2.45	1.13	5.03	3.52	656.91	1.37	1.84
Finland	−0.53	0.05	1.03	2.64	2.81	1.60	4.58	6.66	705.21	1.54	1.90
France	−1.04	−0.17	0.63	1.90	1.39	1.06	4.81	5.52	868.57	1.24	0.33
Germany (East)	0.07	−0.79	0.45	2.89	2.65	1.81	4.07	5.49	671.63	1.51	2.13
Germany (West)	−0.32	−0.70	0.20	1.95	2.10	1.29	4.51	4.35	648.27	1.49	1.29
Greece	1.38	−0.97	0.68	2.92	1.48	1.12	5.02	6.10	652.76	1.49	0.36
Hong Kong	−0.20	−0.74	1.54	3.59	3.19	1.87	4.52	5.50	994.81	1.18	0.91
Hungary	1.19	−0.39	0.95	3.85	3.44	1.99	3.09	6.06	658.26	1.15	2.37
Iceland	−1.18	0.00	0.41	2.68	2.55	1.07	5.10	5.94	557.68	1.31	1.89
Indonesia	−0.09	0.92	1.15	5.44	3.92	1.99	3.94	10.84	1013.1	1.11	2.52
Ireland	−0.20	−0.06	0.91	2.89	2.42	1.38	4.46	4.19	698.81	1.13	1.14
Italy	1.10	−0.90	0.42	3.52	3.33	1.60	4.03	6.15	894.95	1.24	1.51

(cont'd)

Table 7.4 (continued)

	Focus on higher order problem solving skills	Focus on reading	Improve-ment orien-tation	Principal involve-ment	Evaluation of teachers (principal)	Evaluation of teachers	Leader-ship	Staff discussions	Teaching hours	Problems in school	Evaluative procedures
Netherlands	–0.77	0.13	1.19	2.83	2.20	1.34	4.75	6.66	1002.3	1.57	0.99
New Zealand	–0.32	1.40	1.77	3.86	3.41	1.83	3.98	7.62	957.33	1.29	2.32
Norway	–0.32	–0.19	1.16	3.70	2.59	1.29	4.66	5.64	604.90	1.29	1.27
Portugal	1.03	–0.03	0.53	3.17	1.69	1.51	4.26	6.25	909.47	1.09	0.52
Singapore	–0.16	0.57	1.46	4.41	3.50	1.99	3.38	6.56	966.73	1.19	2.42
Slovenia	0.41	0.04	1.45	4.19	3.24	1.93	3.53	6.04	594.71	1.28	2.85
Spain	0.02	0.04	0.92	3.24	3.10	1.54	4.49	6.60	906.34	1.22	1.51
Sweden	–0.61	0.53	1.26	1.98	2.91	1.14	4.08	6.19	776.49	1.34	1.98
Switzerland	–0.35	–0.61	0.22	2.01		1.90		3.41	879.63	1.40	0.00
Trinidad/Tobago	0.15	0.29	1.21	356	3.83	1.89	3.76	5.35	870.75	1.08	2.11
United States	0.13	0.66	1.65	4.11	3.44	1.99	4.15	5.90	1001.3	1.63	2.83
Venezuela	0.38	0.24	0.87	3.72	3.50	1.83	4.26	5.04	729.65	1.00	2.13
Range of coding			0–2	0–6	1–6	1–2	25–65	1–20		1–2	0–5

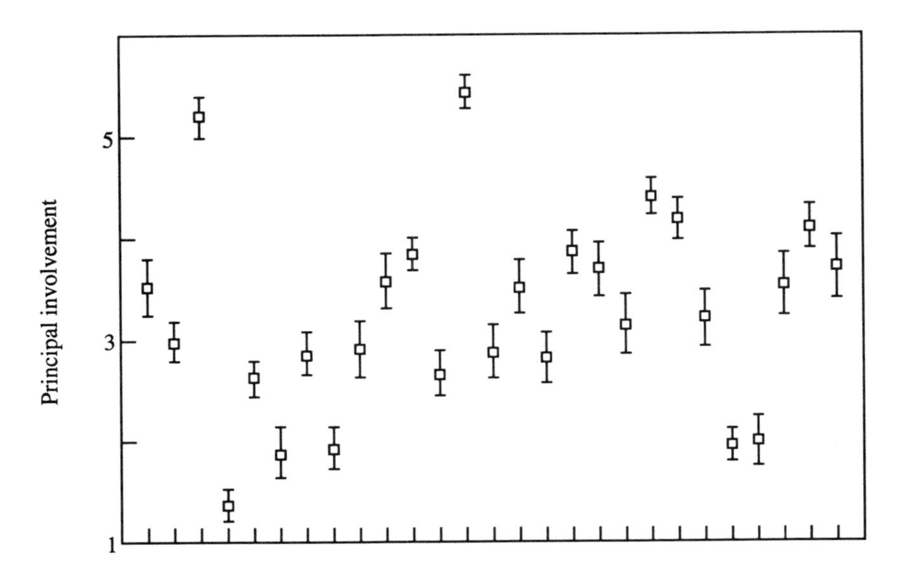

From left to right: Belgium, Canada, Cyprus, Denmark, Finland, France, Germany (East), Germany (West), Greece, Hong Kong, Hungary, Iceland, Indonesia, Ireland, Italy, Netherlands, New Zealand, Norway, Portugal, Singapore, Slovenia, Spain, Sweden, Switzerland, Trinidad/Tobago, United States, Venezuela

Figure 7.1 Level of principal involvement in 27 countries (IEA Reading Literacy data), on a scale from 0 to 6.

Looking at the four indicators for educational leadership a pattern emerges showing Cyprus, Indonesia and the U.S.A. scoring high on all four indicators, whereas France, Germany (West), Sweden and Switzerland score rather low. The differences in principal involvement with teacher classroom practices are presented in Fig. 7.1, which contains the means for each of the 27 countries with the associated 95% confidence interval. On the horizontal axis the 27 countries are depicted. On the vertical axis the level of principal involvement is drawn, with ranges from 1 (no involvement) to 5 (maximal involvement). For each country the box indicates the mean level of principal involvement, whereas the bars represent the confidence intervals. For this indicator there is a group of five countries lagging far behind: Denmark, France, Germany (West), Sweden and Switzerland. On average, these countries score only 2 (out of 6) on the principal involvement scale. Cyprus and Indonesia, and to a lesser extent Singapore, Slovenia and the U.S.A., have clear above-average principal involvement.

Table 7.5 Country averages with respect to the classroom process variables

	Time for instruction	Time for reading	Homework	Instructional strategies	Needs assessment written	Needs assessment observation	Progress assessment oral	Progress assessment written
Belgium	24.13	2.40	13.60	–0.66	0.24	–0.65	–0.48	0.75
Canada	24.30	6.80	14.77	0.35	-1.64	0.61	0.34	–0.44
Cyprus	20.41	2.81	88.62	0.72	0.77	–0.08	0.32	–0.21
Denmark	16.50	2.61	35.08	–0.47	–0.66	0.57	0.04	–0.26
Finland	18.32	1.86	26.75	–0.23	0.00	0.55	–0.44	–0.26
France	24.52	2.87	27.63	–0.47	0.09	–0.83	–0.54	0.99
Germany (East)	18.50	2.81	33.91	–0.03	–0.20	0.40	0.09	–0.62
Germany (West)	17.66	1.60	23.75	–0.76	–0.79	0.14	0.06	–0.97
Greece	19.74	3.19	61.79	0.75	0.96	–0.33	–0.09	0.18
Hong Kong	23.93	1.90	15.61	–0.72	0.56	–0.38	–0.79	–0.48
Hungary	19.08	2.95	79.22	0.43	–0.06	0.91	0.36	0.42
Iceland	16.80	3.08	104.25	–0.81	–0.52	–0.42	–0.36	–0.10
Indonesia	25.22	4.24	43.10	0.03	0.85	0.35	–0.23	0.55
Ireland	21.50	2.84	33.05	0.21	0.18	–0.41	0.36	–0.26
Italy	25.69	3.96	76.52	0.69	0.58	0.21	0.27	0.85
Netherlands	25.21	2.73	0.77	–0.90	–0.75	–0.48	–0.15	–0.60
New Zealand	23.42	5.12	25.50	0.07	–0.97	0.21	–0.09	–0.50
Norway	16.06	1.86	39.33	–0.46	–0.19	0.51	–0.14	0.14
Portugal	22.90	4.82	65.50	0.91	0.83	–0.20	0.30	0.59
Singapore	24.18	2.79	20.93	–0.08	0.49	–0.50	–0.11	0.26
Slovenia	15.90	1.68	32.73	0.04	0.51	0.49	0.25	0.40
Spain	24.01	2.61	34.55	0.30	–0.23	–0.02	0.06	0.27
Sweden	20.00	3.72	31.26	–0.31	-1.02	0.27	–0.08	–0.83
Switzerland	22.83	2.32	19.23	–0.60	–0.39	–0.60	–0.26	–0.53
Trinidad/Tobago	21.17	4.86	53.03	0.33	0.89	0.15	0.37	0.35
United States	26.40	6.15	95.40	0.43	0.09	0.16	0.38	0.16
Venezuela	18.01	4.03	41.88	0.59	0.43	–0.02	0.06	0.24

Consensus and cooperation, as indicated by staff discussions, show a similar pattern as the principal involvement indicator, which may point to the fact that in the most wealthy European countries teacher autonomy is high on the agenda.

Teaching hours show differences amounting up to almost 500 hours, with Slovenia, Denmark and Cyprus at the lower end and Indonesia, The Netherlands and the U.S.A. at the upper end of the distribution.

Problems in school as an indicator for a safe and orderly climate shows some small differences. Evaluative procedures are developed poorly in France, Greece and Switzerland and to some extent Slovenia, Canada, the U.S.A., Indonesia and Singapore.

Table 7.5 contains the means for the teacher–classroom process variables. The first three indicators relate to time allotment. Some countries have longer working days in school, resulting in 24 hours a week or more (Belgium, Canada, France, Hong Kong, Indonesia, Italy, The Netherlands, Singapore, Spain and U.S.A.), whereas others stay under 20 hours (Denmark, Finland, East and West Germany, Greece, Hungary, Iceland, Norway, Slovenia and Venezuela).

At the country level there is a high correlation (0.47) between time for instruction and time for reading, indicating that the necessary condition for much reading instruction (i.e. enough instruction time) is indeed put to use for reading instruction in many cases. The third indicator (homework) shows that in some countries the school day is prolonged over and above much time that is already spent on reading instruction (Italy, Portugal, U.S.A.), whereas others, most notably The Netherlands, do not follow this practice.

For direct instruction there are three variables available: consistently scoring rather low on all three are Belgium, France, Germany (West) and The Netherlands, whereas Cyprus, Greece, Hungary, Indonesia, Portugal and Trinidad/Tobago score consistently high.

The last indicators deal with monitoring pupil progress. The highest scores are found for Portugal, Trinidad/Tobago and Hungary, whereas Finland, The Netherlands and Switzerland score rather low.

Results Part 2: The Factors that Work in Each Country

The rather complex information provided in Table 7.6 is used to derive conclusions on the generalizability of the educational effectiveness model across countries.

First, a look is taken at the column containing the percentage of between-schools variation in reading achievement ("% schools var."). Adjusted quality differences between schools with respect to reading achievement are rather modest in all Scandinavian countries (including Iceland) and Cyprus. The largest differences occur (in order of percentage) in Trinidad/Tobago, Indonesia, Venezuela, Greece, Hong Kong and Italy.

The first six columns contain the estimates for the intercept, and the effects of age, gender, grade retention, home circumstances and teacher language. All effects are standardized and multiplied by 100 (as in the rest of the table). Since the first part of the table is the intercept and covariate part these results will not be discussed in more detail. This part is only listed in the table as a reminder that we are concerned with adjusted school effects when estimating the effects of input, context and process indicators.

The next two columns contain the effects of public versus private and community type. Private schools only outperform public schools in Venezuela. Urban schools (with the exception of the U.S.A. where the relation is reverse) outperform rural schools in 10 out of 27 countries, with estimated effects ranging from 0.08 to 0.29. The estimated effects for the input variables are contained in the next four columns. Teacher training never has an association with value-added school effects, and class size has a positive effect in three countries.

Parental involvement shows a clear consistent positive relationship in many countries, as was the case in the outlier study of Postlethwaite and Ross (1992). Only in the case of Portugal is a reversed association shown. With one positive and one negative relation the effect of resources is unclear.

The next series of columns contain the estimated effects of the school processes. First, the three achievement press variables are considered.

Focus on higher order problem-solving skills (only in Iceland) and improvement orientation (Portugal, Spain and Switzerland) only seldomly have a positive association with adjusted reading achievement. The pattern for focus on reading is slightly more encouraging, with positive effects in Finland, Hungary, Indonesia, Sweden and the U.S.A.

For educational leadership there were four indicators, listed in the next four columns. The results indicate that only rarely is educational leadership consistently positively associated with adjusted school effects in reading (notice for instance for the U.S.A. the negative effect of principal involvement and the positive effect of evaluation of teachers by the principal). Consensus and cooperation as indicated by staff discussions shows a positive effect in two countries (Denmark and New Zealand), but a negative effect in three others (Canada, Hungary and Ireland).

Teaching hours, a variable that showed remarkably large differences among the 27 countries, only has an effect in three countries: positive in Finlandand the U.S.A., and negative in Singapore. The climate indicator "problem in schools" has a positive effect in six countries (Canada, France, Germany (West), Greece, New Zealand, and Sweden). The evaluative potential of schools, indicated by the number of evaluative procedures, has a positive sign of the effect in France and Slovenia, but a negative effect in Ireland.

Table 7.6 Within- and between-countries effects of input, context, school and classroom variables on reading literacy, in terms of regression coefficients (× 100), with significant effects ($\alpha<0.05$) in bold

Country	Constant	Age	Sex	Grade retention	Home	Teacher language	Private	Community type
Belgium	76	**−15**	**13**	−15	**19**	−8	−74	0
Canada	−76	−6	**21**	−114	**14**	46	16	**12**
Cyprus	−39	−3	10	−31	**17**	24	0	**10**
Denmark	25	−6	**23**	−39	**12**	−52	12	1
Finland	−25	**6**	**19**	**−90**	**16**	16	0	0
France	−20	**−14**	5	**−21**	**19**	−5	25	1
Germany (East)	−13	2	**18**	**−64**	**17**	26	−23	−1
Germany (West)	−11	**−11**	6	−7	**21**	12	−5	−3
Greece	12	**8**	4	**−132**	**17**	−12	−7	**12**
Hong Kong	−4	−1	**10**	−7	0	−2	1	**20**
Hungary	−5	**−12**	**8**	**−32**	22	0	1	**11**
Iceland	**−109**	**5**	**20**	−43	3	65	27	**8**
Indonesia	−12	**−7**	−3	−9	**7**	−12	16	**29**
Ireland	1	**−9**	22	−38	23	−5	−5	**13**
Italy	−25	1	5	**−47**	**17**	24	0	−3
Netherlands	−23	**−16**	5	13	16	−3	11	4
New Zealand	−50	−1	**26**	**−78**	**27**	16	20	0
Norway	−20	3	**16**	**−88**	**15**	10	1	−3
Portugal	−32	**−11**	4	**−20**	**18**	0	29	9
Singapore	−10	7	13	−73	23	10	0	0
Slovenia	12	−3	**20**	**73**	**20**	14	−35	**9**
Spain	−9	**−4**	**7**	−61	**16**	1	4	3
Sweden	−2	**5**	**13**	**−94**	**21**	−2	0	2
Swiss	−26	−3	**8**	**−39**	**28**	8	12	1
Trinidad Tobago	−5	**−5**	**18**	−14	**17**	28	14	3
U.S.A.	−46	**−7**	**11**	−47	**10**	32	11	**−6**
Venezuela	−30	**−5**	**6**	−4	**8**	−11	**36**	8
TOTAL	−14	**−3**	**11**	**−31**	**18**	0	**9**	**4**

Training	Class size	Parent	Resources	Higher order	Focus on reading	Improvement reading	Principal involv.	Eval. tea.
5	1	9	−2	−1	2	**7**	−1	−1
1	−8	7	−8	−3	3	9	−5	2
2	−3	6	0	−1	−4	−4	−1	−4
−3	1	3	−1	−2	−3	4	4	−2
−3	−2	0	**−11**	−3	**11**	−3	−9	3
1	−1	−3	2	1	−3	−1	1	2
−1	1	−1	6	5	−3	7	−8	**9**
−1	4	**11**	−4	−1	0	1	−3	4
3	−4	**10**	7	5	−6	−5	−1	1
0	**16**	6	−1	2	1	6	−1	−4
4	0	3	4	7	**10**	1	−6	−2
2	−3	1	1	**13**	1	4	1	−2
0	**10**	2	**10**	5	**12**	−2	−6	−4
3	−7	5	0	1	7	−3	**14**	7
10	−2	−1	9	12	7	1	5	1
0	8	**14**	−5	5	3	5	−10	6
−2	−2	**16**	−1	−1	−4	−5	3	−3
0	4	**5**	0	6	−3	−2	−4	−7
−3	0	**−10**	−6	−0	1	**13**	−4	1
4	14	**11**	−4	−1	−3	−2	3	2
−3	−1	4	−5	1	−1	−5	1	**−7**
−1	4	6	4	6	2	**6**	**6**	2
0	−1	**9**	1	−4	**11**	−2	−1	−6
0	**7**	5	1	5	0	**6**	1	0
6	−1	**22**	−4	−3	4	−1	4	5
6	−4	**18**	−1	0	**8**	1	**−10**	−3
−4	−2	8	−12	−1	0	9	2	3
1	**3**	**8**	0	**2**	**2**	1	0	0

(cont'd)

Table 7.6 (continued)

Country	Eval. by princ.	Leader-ship	Staff discus-sion	Teaching hours	Problems	Eval. proce-dure	Time for instruc-tion	Time for reading
Belgium	–1	3	–7	5	**9**	2	1	–5
Canada	0	2	**–17**	–4	**17**	0	7	2
Cyprus	1	4	–5	3	–2	4	–16	**–7**
Denmark	0	–4	**0**	–6	–2	0	**8**	1
Finland	**11**	5	2	**12**	4	5	–3	–9
France	–3	7	4	3	**8**	**11**	**8**	–2
Germany (East)	4	**11**	8	1	5	–4	–4	0
Germany (West)	4	–2	2	–7	**10**	–1	–1	0
Greece	2	13	–3	0	**9**	4	1	2
Hong Kong	3	–2	–8	–2	–1	–2	–1	1
Hungary	–3	–3	**–8**	2	1	–7	2	–4
Iceland	–2	**–7**	–5	3	5	–5	1	**–13**
Indonesia	3	3	8	4	–2	1	–1	20
Ireland	0	2	**–10**	6	4	**–17**	–5	–5
Italy	–5	0	–5	–2	3	9	0	**–12**
Netherlands	–2	5	0	7	1	–1	11	8
New Zealand	2	–2	8	4	**8**	3	0	6
Norway	–2	–2	4	–3	–5	4	4	–1
Portugal	–2	–4	–6	6	–1	–1	0	–5
Singapore	0	2	–3	**–24**	–1	2	**25**	–2
Slovenia	–1	–4	–6	–4	–1	**7**	–1	2
Spain	–2	1	–2	4	2	–2	3	**–8**
Sweden	–3	0	–1	–4	7	5	0	–3
Swiss	–2	0	1	3	3	0	–4	**–6**
Trinidad Tobago	8	4	–6	–5	6	–1	3	–7
U.S.A.	9	0	–6	9	4	–1	–1	–3
Venezuela	5	3	3	6	0	–8	3	2
TOTAL	1	0	–2	1	4	0	0	–2

The next series of columns contains the three direct instruction indicators. Instructional strategies shows a positive effect in one country (Sweden) only, and in Hungary its estimated effect is negative. With respect to the two needs assessment variables Ireland shows two estimated negative effects, Hong Kong one positive effect, the U.S.A. one negative effect, and Hungary one positive and one negative effect.

The last two process variables dealing with pupil monitoring show a similarly odd pattern.

Home-work	Instr. strategies	Needs asses. writ.	Needs asses. obs.	Oral assess.	writ assess.	% school var.	% var. accounted	% by educ. vars
3	2	−2	6	−5	−5	16	43	15
−5	10	8	−4	4	**15**	10	25	19
1	2	3	−4	1	−1	6	24	14
1	6	−2	0	−1	2	6	22	17
−1	1	−1	5	3	−5	6	40	37
2	−4	−5	6	**8**	−1	11	26	13
5	−10	2	−6	1	2	13	32	19
−3	−6	−1	1	2	−1	14	37	20
3	5	−3	−5	4	8	31	33	21
1	10	**10**	−5	−4	−6	31	36	33
−4	**−9**	**11**	**−9**	6	−4	21	54	24
3	−4	0	−3	6	0	8	18	15
−7	3	6	−6	−3	**−11**	38	49	33
−2	3	**−10**	**−8**	**−7**	**8**	14	46	29
2	−1	3	1	3	−2	30	18	15
−8	−7	3	−3	4	3	13	39	31
−2	−5	0	2	−5	1	15	59	26
0	−4	2	−3	3	2	4	17	10
−1	−8	6	8	4	2	28	54	18
−1	−3	−3	−2	−4	5	21	48	21
−5	−2	−2	2	−1	−3	10	35	18
−3	−5	−1	−1	1	**6**	20	46	19
2	**7**	5	−4	0	**−7**	9	44	26
−2	−6	−2	−3	2	0	10	25	12
−2	1	1	2	12	3	40	41	23
−1	−3	**−9**	0	−4	−1	21	51	33
4	5	6	−2	4	−6	34	30	19
−1	−1	1	−1	1	1	20	30	9

The last column of the table contains a fit-parameter for the educational effectiveness model. It is the amount of between-schools variation uniquely accounted for by the input, context and process variables at school and classroom level. The model has the highest explanatory power in Finland, the U.S.A., Indonesia, Hong Kong, Ireland and Sweden, where approximately one-third of the between-schools variation is uniquely accounted for by the educational effectiveness model. The model does rather poorly in Norway, France, Switzerland, Italy, Cyprus, Belgium and Iceland (only 10–15% variance accounted for).

Results Part 3: The International Effectiveness Model

The pattern presented in the previous paragraph was rather diverse. The cross-country generalizability of the educational effectiveness model appears to be problematic, at least when looking into school effects on reading achievement of 9-year-old pupils. In this section an attempt is made to estimate an across-countries educational effectiveness model. One should bear in mind that Z-scores have been used for each country, so that variables cannot explain between-country differences in reading achievement.

The results are contained in the last row of Table 7.6. Since information was combined from 27 countries, each containing more than approximately 100 schools in their samples, the statistical power to detect any effects is almost perfect, so that small estimated effects may show up as significant. Both context indicators public–private and rural–urban show a positive association with adjusted school effects in reading, showing advantages for private and urban schools. From the input indicators class size has a small and meaningless positive effect, and parental involvement has a clear positive effect (0.08).

From the school process variables two achievement pressure variables (focus on higher order problem-solving skills and focus on reading) have significant but small (0.02) positive effects. The consensus and cooperation indicator has a significant but small (–0.02) negative effect. The climate indicator shows a somewhat higher association (0.04).

The other school process variables have estimated effects that are, statistically speaking, not discernable from zero. Of all teacher/classroom process variables only one, the effect of time for reading, has a negative effect (–0.02), which was unexpected.

All in all, the model for the international data performs poorly, with only 9% of unique variation between schools accounted for by the educational effectiveness variables.

Conclusions

In this chapter we have tried to add to the existing knowledge base on educational effectiveness by applying a rigorous analytical strategy on a database containing reading achievement scores for 9-year-old pupils. Having first reviewed earlier attempts at cross-national school and teacher effectiveness studies, which only showed promising results when applying a technique using outliers (a data-analysis strategy that carries a strong risk of capitalizing on chance), the IEA Reading Literacy dataset was used for a new attempt.

First, differences between countries were investigated in the means of input, context, school process and teacher/classroom process variables. There appear to be numerous differences in all categories of indicators. Educational leadership, to name one instance, is far more prominent in some countries (including the U.S.A.) than in others. Classroom practices also differ widely among countries. Direct instruction is poorly developed in some countries (including The Netherlands), but far more developed in others (including Cyprus, Greece and Italy).

With regard to the tenability (validity) of the educational effectiveness model each country appeared to have its own unique, sometimes more and sometimes less powerful, "educational production function". Factors that work in some countries do not work in others. The model does only poorly even in its country of origin (the U.S.A.).

An attempt to fit one model for all countries was only successful in a limited sense: most effects were small and insignificant, others small but significant, and only two context variables (public–private and urban–rural), one input variable (parental involvement), and one school process variable (safe and orderly climate) survived the significance and relevance test.

It must be stressed that the generalizability has only been investigated for 9-year-old students and their reading achievement. Moreover, there were several insufficiencies in the research design:

- no pretest available;
- process indicators were measured using questionnaires;
- variables were measured using self-reports by teachers and principals;
- many variables seem to have been operationalized on an *ad hoc* basis, without apparent careful conceptual thinking;
- school and classroom variables were measured with one question: it seems more appropriate to measure climate using answers from several teachers (cf. Raudenbush *et al.*, 1991).

PART THREE: REDIRECTION

CHAPTER EIGHT

THEORIES ON SCHOOL EFFECTIVENESS

Introduction

What is a theory?

A theory is seen as an explanation of an observed relationship between phenomena (Odi, 1982, p. 55). It consists of (a) a set of units (facts, concepts, variables), (b) a system of relationships among units, and (c) interpretations about (b) that are comprehensible and predict empirical events (Snow, 1973, p. 78). As stated in Chapter Two, a model can be seen as a prerequisite for a theory in the sense that it specifies or visualizes in a simplified or reduced way phenomena that cannot easily or directly be observed. A model, thus, contains the (a) and (b) parts of the definition of theory stated above, but does not necessarily contain element (c) (Scheerens, 1992, p. 13).

What do we mean by theorizing about school effectiveness?

When applied to school effectiveness, theorizing means going beyond the statement of factors that work [see element (a) in the above definition of theory] and also beyond the models as described in Chapter Two [see element (b) in the definition of theory] to the identification of explanatory principles [element (c)]. There are two approaches to this identification: an inductive approach and a deductive approach. The inductive approach is evident in the type of model building discussed in Chapter Two. Elaborate models may point to levers that are critical in creating equilibrium or disequilibrium in complex causal models. For example, in the systems model described by Clauset and Gaynor (1982), standard setting can be seen as such a lever.

In this chapter a deductive approach is chosen. This means that more generally applicable, more or less established theory is examined in order to lay bare explanatory mechanisms which in their turn will be examined for their usefulness in explaining school effectiveness phenomena. The exercise is to be seen as a first step, in which the emphasis will be on making explicit a match between "what we know" about school effectiveness (Part Two of this book) and the theoretically embedded principles that will be identified. A more ambitious further step, deducting hypotheses for further, more theory-driven school effectiveness research will only be dealt with in a rudimentary way, by indicating the general fruitfulness of a particular theory.

Why theorize about school effectiveness research?

So far a pragmatic, empirical approach has dominated the study of school effectiveness. The inductive approach to theory building is more in line with this tradition than the deductive approach. Several researchers active in this field have recently made a plea for more theory in school effectiveness research (Slater & Teddlie, 1991; Stringfield, 1995; Mortimore, 1992). There is an obvious need to take one reflective step backwards, and try to integrate the mass of fragmented empirical research results.

An objection that might be raised with respect to theorizing about school effectiveness research is the relative weakness of the knowledge base. In Parts One and Two of this book these weaknesses, for example the lack of convergence in operationalization of effectiveness-enhancing factors, the small direct impact of school-level characteristics and the scant evidence on more complex indirect effects, have been amply documented. When one thinks of the interplay that research and theory should ideally have, this objection can be categorized as a "chicken and egg" problem. When the heuristic value of theory building is underlined, this very endeavor could be seen as one possible approach to *improve* the knowledge base.

Three more specific reasons for theoretical reconstruction are:

- to focus more on future research, *vis-à-vis* the broadness of concepts and complexity of possibly interesting relationships;
- to obtain, by means of more theory-driven research, results that are easier to interpret;
- to discover factors and critical relationships that can be used as levers for school improvement.

With respect to this latter aim the old platitude, about there being nothing more practical than a good theory, still applies.

Which Theories are Relevant?

Since the focus of this book is on school organization, the main concern is with organization and management theory. Yet, from the perspective of integrated, multilevel models on school effectiveness, it follows that principles governing the primary process of learning and instruction, including ideas about educational technology and classroom management, should also be considered. As shown later, the proposition that organizational structure should be in line with characteristics of the technology behind the organization's primary process is one of the central theses from an organizational theory known as contingency theory (see Chapter One).

Application of this principle to the modeling and design of school organizations could be seen as a bottom-up process, whereby learning processes of individual pupils are the starting point. Conditions of learning could be the basis for propositions about instructional strategies, which in turn would shape teaching behavior, educational technology and classroom management. Next, classroom practice could be analyzed with respect to required managerial and organizational conditions.

Here, instructional theory will not be pursued any further than a schematic presentation of the contrast between structured instructional models, such as mastery learning and direct instruction versus constructivist-oriented perspectives on learning and instruction and the likely implications for school management and organization (also Murphy, 1993, 1994; Scheerens, 1994).

In the diagram depicted in Fig. 8.1, cited from Scheerens (1994), constructivist principles on learning and instruction are associated with current ideas on restructuring educational administration and school management and contrasted with aspects of comprehensive educational effectiveness models.

At the macrolevel, that is the interface between school and environment, there is a lot of common ground between the two frameworks under comparison. Accountability requirements from higher administrative levels and incentives that would force schools to be more responsive to demands from external constituencies (e.g. parents) have a place in both approaches.

In current ideas about restructuring educational systems, the premium is put on market mechanisms rather than on formal requirements (such as standards), although it is hard to conceive of accountability requirements without such standards. Next, restructuring emphasizes decentralization, deregulation and enlarging school autonomy, whereas this is not a critical feature at the macrolevel of comprehensive school-effectiveness models. Both perspectives may be united on the principle of liberating educational process or throughput and exercising a form of output control, although, as mentioned before, the consistency of the multilevel constructivism/restructuring framework would be lost to the more radical constructivists once accountability requirements were introduced.

It should also be noted that this compromise solution (deregulated processes and output control) masks the inherent tension between market-based and administrative control (cf. Chubb & Moe, 1990). Technical problems and ethical considerations of the mixture of accountability requirements and market

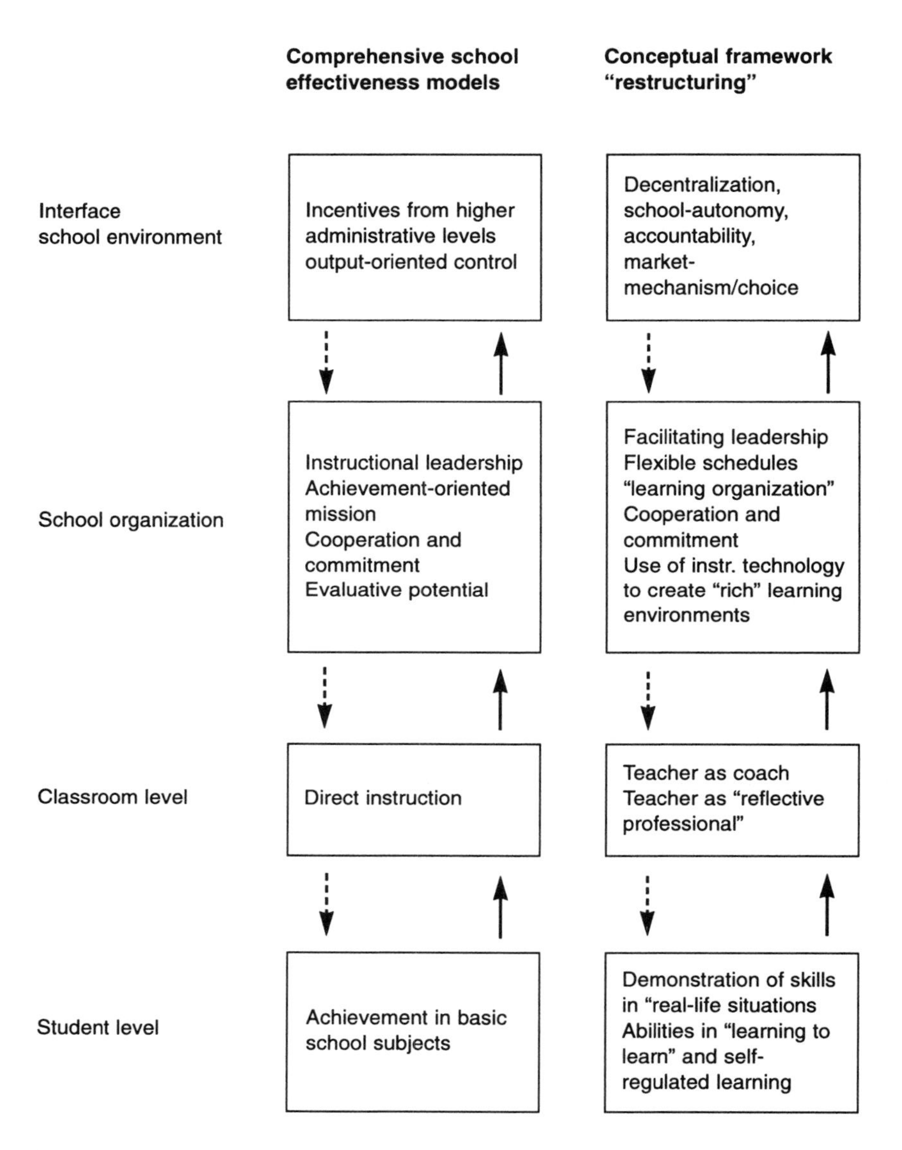

Figure 8.1 Global comparison of comprehensive school effectiveness models and conceptual framework of restructuring. Arrows in solid lines indicate the direction for design (bottom up); arrows in dotted lines indicate the actual functioning of the system where higher level processes are seen as facilitative conditions for lower levels.

mechanisms in education are left undiscussed here, although the British experience should be singled out as relevant in this respect (cf. Scheerens *et al.*, 1991).

At the mesolevel of school organizational functioning the importance of cooperation between teachers, teamwork and a commitment to the school's mission are common in both frameworks. The perspectives on leadership roles differ, however. In educational effectiveness models the construct of instructional leadership carries elements of achievement pressure and hierarchical bureaucratic control, for instance when keeping records on classroom practice and teacher assessment are emphasized. In restructuring, these bureaucratic and rational control elements are absent. Instead, the facilitative and pedagogical role of school leaders is emphasized, which also corresponds to the focus on teacher empowerment and participatory decision making in school policy matters.

Although creating conditions for adaptive instruction is an issue for both perspectives, empirical work in the area of educational effectiveness has led to a certain modesty in aspirations with respect to differentiation and individualization, recognizing that instructional time, as a crucial mediating variable, may be used more effectively when simpler organizational arrangements, such as whole-class teaching, are used. Constructivism and restructuring are far less inhibited in this respect and much more radical in recommending the redesigning of schedules and allocations.

Within the framework of school effectiveness thinking, the evaluation potential of schools (i.e. the use of evaluation mechanisms to monitor school work, within the framework of the cybernetic principle of feedback and reinforcement) is seen as an important effectiveness-enhancing vehicle (cf. Scheerens, 1992). Restructuring with constructivism at its core is likely to denounce these behavioralistic procedures. A similar note could be made with respect to the use of monetary incentives (as in merit pay) which, although being a debatable issue on both sides, would be likely to be abolished altogether from the constructivism/restructuring agenda.

Restructuring, in contrast, is more open to the scope of modern information technology for education. From this feature it is again evident that constructivism and restructuring are more future-oriented ideals when compared to the pragmatic nature of the outlook of educational effectiveness thinking on "what works" in current practice.

At the mesolevel the differences between the educational effectiveness and constructivism/restructuring outlooks appear to come down to some old antagonistic positions in organizational theory. With respect to the organization–structural dimension this is the distinction between mechanistic and organic structures (cf. Mintzberg, 1979) and for the procedural dimension, this is the distinction between rational/empirical and more interactive strategies. In the educational effectiveness outlook there is definitely room for mechanistic structure and rational–empirical strategies, although softer aspects such as a shared mission and cooperative climates are also considered important. Restructuring swings the pendulum back to the human relations approach in its emphasis on teacher autonomy, the reflective practitioner, facilitative leadership and organic structure.

Turning now to organization and management theory it should be pointed out first of all that some of the theories that will be discussed do not exclusively relate to school-level conditions, but sometimes have important implications for classroom phenomena as well.

The rationality paradigm is chosen as the encompassing framework to discuss theory-embedded principles concerning effective management and organization. This paradigm lies at the basis of theories on planning and public policy making, microeconomic theory, organizational learning theory and even contingency theory.

The basic principles of the rationality paradigm are:

- behavior is oriented toward preferred end states (as reflected in goals or individual well-being);
- in situations where there is a choice between alternative ways to attain the preferred end states, an optimal choice is made between these alternatives, which means that profit, well-being, or other preferred end states are maximized given the alternatives and constraints;
- in organizational settings the alignment of individual preferences and organizational goals is a major issue.

The rationality paradigm is applied in formal and less formal models of planning, control, design and feedback and is attached to different units: organizations as a whole, subgroups or departments and individuals. Apart from this, procedural and structural interpretations may be distinguished, the former referring to organizational processes and the latter referring to the design (division and interconnection) of units and subunits.

A further important distinction concerns the question of whether goals are considered as "given" to the social planner or designer. or whether the process of choosing particular goals is seen as part of the planning process. In the first case the approach is "instrumental", whereas the term "substantial rationality" (Morgan, 1986, p. 37) is sometimes used for the latter. Stated more popularly, the instrumental approach is inherent in the phrase "doing things right" whereas the substantial perspective asks the additional question about "doing the right things".

In general terms, the model that is implicitly used in school effectiveness research fits the economic rationality model quite well (see Chapter One of this book). Economic rationality applies the rationality paradigm to the organization's production process, and is therefore also frequently referred to as the productivity model. The basic means-to-end relationships considered in the productivity model are situated in the organization's primary process. This is also the case in economically oriented research on "education production functions" (Monk, 1992) and educational productivity schemes that largely depend on research on teaching and learning environments (Walberg, 1984).

Usually, in school effectiveness research, the instrumental interpretation of the rationality paradigm is implicitly chosen, since basic school competencies to be attained by pupils are usually considered as the given criteria to evaluate effectiveness.

Merely classifying school effectiveness research in terms of the rationality paradigm in itself does not help very much in the search for underlying principles or mechanisms that could explain why certain conditions or factors appear to work in education. It should be noted, however, that the rationality paradigm is not just an analytical tool to *describe* social reality, but also has very strong *prescriptive* connotations. Depending on a particular interpretation of the overall paradigm, specific principles are emphasized as conducive to the improvement of the effective functioning of organizations. Three of these principles will be discussed, which can be labeled by the following three imperatives:

- "plan synoptically and structure formally";
- "align individual and organizational goals by creating market conditions";
- "plan retroactively by means of proper evaluation and feedback".

Organizational images that are related to these three principles are, respectively: the bureaucracy, the autonomous or privatized school and the school as a learning organization. The theoretical backgrounds are: classical planning theory and scientific management, public choice theory and cybernetics.

In the case of contingency theory one of the two key principles is: "fit the organization structure to contingency factors, like environmental complexity and characteristics of the core technology"; the second calls for a consistent pattern of structural characteristics. Contingency theory can be subsumed under the general rationality paradigm, since "fit" is considered as instrumental to goal attainment (cf. Thompson, 1967).

Each of these four theory-embedded principles will be explained in the subsequent sections. In each case the correspondence with parts of the empirically supported knowledge base on educational effectiveness will be indicated, along with a general evaluation with respect to the heuristic value for further research.

Synoptic Planning and Bureaucratic Structuring

The pure rationality model (Dror, 1968) formally enables the calculation of the optimal choice among alternatives after a complete preference ordering of the end states of a system has been made. This ideal is approached in mathematical decision theory, as in game theory where different preference orderings of different actors can also be taken into account. For most real-life situations of organizational functioning, however, the assumptions of pure rationality are too strong. Simon's (1964) construct of bounded rationality modifies these assumptions considerably by recognizing that the information capacity of decision makers is usually limited to taking into consideration only a few possible end states and alternative means.

Cohen *et al.* (1972) and March and Olsen (1976) go even further in criticizing the descriptive reality of the pure rationality model. Cohen *et al.* (1972) describe organized anarchies as characterized by "problematic preferences", "unclear technology" and "fluid participation". With respect to problematic preferences, they state that the organization can "better be described as a loose collection of ideas than as a coherent structure; it discovers preferences through action more than it acts on the basis of preferences" (Cohen *et al.*, 1972, p. 1). Unclear technology means that the organization members do not understand the organization's production processes and that the organization operates on the basis of trial and error, "the residue of learning from the accidents of the past" and "pragmatic inventions of necessity". When there is fluid participation, participants vary in the amount of time and effort they devote to different domains of decision making (Cohen *et al.*, 1972, p. 1).

According to Cohen *et al.*, decision making in organized anarchies is more like rationalizing after the fact than rational, goal-oriented planning. "From this point of view, an organization is a collection of choices looking for problems, issues and feelings looking for decision situations in which they might be aired, solutions looking for issues to which they might be the answer, and decision makers looking for work" (Cohen *et al.*, 1972, p. 2). They see educational organizations as likely candidates for this type of decision making. In terms of coordination, organized anarchies have a fuzzy structure of authority and little capacity for standardization mechanisms.

March and Olsen (1976) describe their reservations with respect to rational decision making in terms of limitations in the complete cycle of choice (see Fig.8.2, where this cycle is depicted). The relationship between individual cognitions and preferences on the one hand and individual action on the other is limited, because of limitations in the capacity and willingness of individuals to attend to important preferences and because of discrepancies between intentions and actions: "...the capacity for beliefs, attitudes, and concerns is larger than the capacity for action" (March & Olsen, 1976, p. 14).

At the same time there may be a loose connection between individual action and organizational action, because internal individual action may be guided by principles other than producing substantive results (e.g. allocation of status, defining organizational truth and virtue) and organizational actions by external events. In the same vein they observe that actions and events in the environment sometimes have little to do with what the organization does and that it is sometimes hard to learn from environmental response.

Despite all these limitations on the descriptive reality of rational decision making and planning in organizations, even the most critical analyses leave some room for shaping reality more to the core principles. The first type of activity which could bring this about is synoptic planning.

The ideal of synoptic planning is to conceptualize a broad spectrum of long-term goals and possible means to attain these goals. Scientific knowledge about instrumental relationships is thought to play an important role in the selection of alternatives. Campbell's (1969) notion of "reforms as experiments" combines a rational planning approach to social (e.g. educational) innovation with the scientific approach of (quasi-) experimentation. The general idea of linking school effectiveness research to school improvement, where the results of school effectiveness research are seen as guidelines for school improvement projects, also fits the idea of rational, synoptic planning quite well, although many intricacies of this general idea have been pointed out (Reynolds, 1992), whereas other authors (Brown *et al.*, 1995) have even denounced it as simplistic. Other educational applications of the idea of synoptic planning include prescriptive models of instructional design, such as the famous Tyler model (Tyler, 1950), and spin-offs such as the model developed by Gage, teaching models such as the model of "direct instruction" (see Creemers, 1994) and frameworks for school development planning (Van der Werf, 1988; Hargreaves & Hopkins, 1991).

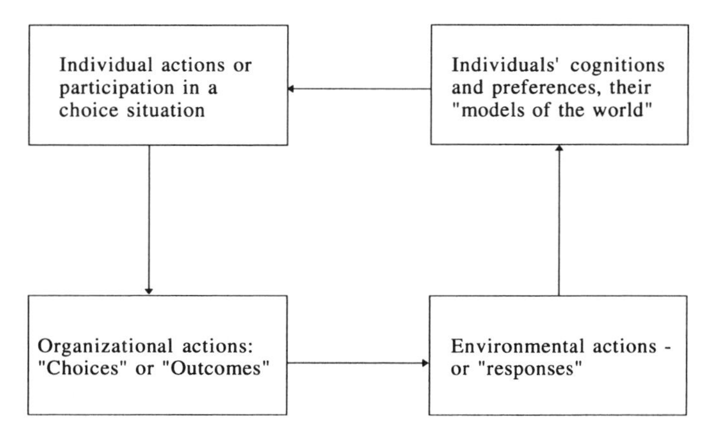

Figure 8.2 The complete cycle of choice (cited from March & Olsen, 1976).

The main characteristics of synoptic planning as a prescriptive principal conducive to effective (in the sense of productive) organizational functioning, as applied to education, are:

- "proactive" statement of goals, careful deduction of concrete goals, operational objectives and assessment instruments;
- decomposition of subject matter, creating sequences in a way that intermediate and ultimate objectives are approached systematically;
- alignment of teaching methods (design of didactical situations) to subject matter segments;
- monitoring of the learning progress of students, preferably by means of objective tests.

As stated before, given the orientation towards the primary process, inherent in economic rationality, the synoptic planning approach in education applies most of all to curriculum planning, design of textbooks, instructional design and preparation of (series of) lessons.

When the ideal of rational planning is extended to organizational structuring, related principles about "controlled arrangements" are applied to the division of work, the formation of units and the way supervision is given shape. "Mechanistic structure", "scientific management" and "machine bureaucracy" are the organizational–structural pendants of rational planning (cf. Morgan, 1986, Chapter 2). The basic ideas go back to Max Weber, who stated the principles of bureaucracy as "a form of organization that emphasizes precision, speed, clarity, regularity, reliability, and efficiency achieved through the creation of a fixed division of tasks, hierarchical supervision, and detailed rules and regulations". Educational organizations, i.e. schools and universities, are usually thought of as *not* fitting the overall image of a machine bureaucracy. Mintzberg (1979) describes a specific variant of the classical bureaucracy, namely the professional bureaucracy that is specifically inspired by educational organizations. In the professional bureaucracy formalization and standardization by rules, close hierarchical supervision and minute job specification are replaced by standardization through training and professional norms. Although it is not unusual to interpret the professional bureaucracy as the complete antithesis of classical bureaucracy, given, for instance, the considerable autonomy of teachers, it should be underlined that the basic notion of standardization and predictability of work processes (albeit within a considerable bandwidth of individual leeway) is retained. Also, in some educational systems, for example Germany, principals and teachers are part of the state's bureaucracy, like any other civil servant (meaning, for instance, that teachers and school heads are appointed by the state departments of education).

When considering the question of to what extent the principles of synoptic rational planning and bureaucracy correspond to the process indicators that have received empirical support in school effectiveness research, the list of factors presented by Sammons *et al.* (1995b) and cited in Chapter Five can be used as a basis for comparison. The striking characteristic of this list of 11 factors is the mixture of elements with respect to either bureaucratic or mechanistic principles or elements that fit a more cultural, organic and participative image of organization. "Firm and purposeful leadership", "unity of purpose", "consistency of practice", "maximization of learning time", "academic emphasis", "focus on achievement", "efficient organization", "clarity of purpose", "structured lessons", "clear and fair discipline", "feedback", "monitoring pupil performance" and "evaluating" are all factors that fit the bureaucratic and rational planning model, whereas others such as "colleagability and collaboration" and "high expectations" are more in line with organic and participative structure.

In other conceptual models on school effectiveness, e.g. Creemers' (1994), important notions such as consistency, consensus and control bear a close resemblance to the overriding principles of ordered structure inherent in the bureaucratic image. The principles of "direct instruction" as stated by Rosenshine (1987), for example ,"proceed in small steps", "give detailed and redundant instructions", provide another case in point. In Dutch studies where systematic school development and lesson planning were specifically studied for their possible effectiveness-enhancing potential, disappointing or ambiguous results were found (van der Werf, 1988; Friebel, 1994).

A fascinating piece of conceptual work and related empirical investigation is provided in Stringfield's description of "high reliability organizations" (Stringfield, 1995; Stringfield *et al.*, 1995).

The principles of high-reliability organizations (e.g. nuclear power plants and air navigation systems) are:

1. the notion that failures within the organization would be disastrous;
2. clarity regarding goals and a strong sense of the organization's primary mission held by the staff;
3. use of standard operating procedures (e.g. scripts);
4. importance of recruitment and intensive training;
5. initiatives that identify flaws (e.g. monitoring systems);
6. considerable attention to performance, evaluation, and analysis to improve the processes of the organization;
7. monitoring is seen as mutual, without counterproductive loss of overall autonomy and confidence;
8. alertness to surprises or lapses (notion that small failures could cascade into major system failures);
9. hierarchical structure, allowing for collegial decision making during times of peak loads;

10. equipment is maintained in the highest working order;
11. high-reliability organizations are invariably valued by their supervising organizations;
12. "short-term efficiency takes a back seat to high reliability" (from Stringfield, 1995, pp. 83–91).

In both the evaluation of major effectiveness-oriented improvement projects in the U.S.A. and the evaluation of a highly structured primary school program (the Calvert-Barclay school project) evidence was found that supported the validity of the high-reliability organization's image. The Calvert-Barclay project is particularly illustrative. It describes the implementation of a highly structured and traditional academically oriented private school program in an inner-city school. The success of the program in these two strongly divergent settings provides additional support to the generalizability of this structured approach.

Despite well-known criticisms of the usefulness of rational, planning and mechanistic structuring approaches in educational organizations (e.g. Lotto & Clark, 1986), these latter examples show that a plea can be made for formalized and highly structured educational programs, supported by structures that emphasize order, coordination and unity of purpose. The major challenge seems to be to combine effectively standardized procedures and partial mechanistic structuring to conditions that nevertheless are sufficiently motivating to educational professionals and likewise keep appealing to the creative insights of all the members of the organization.

Alignment of Individual and Organizational Rationality

A central assumption in the synoptic planning and bureaucracy interpretation of the rationality paradigm is that organizations act as integrated purposeful units. Individual efforts are expected to be jointly directed at the attainment of organizational goals. In the so-called political image of organizations (Morgan, 1986, Chapter 6) this assumption is rejected, emphasizing that "organizational goals may be rational for some people's interests, but not for others" (Morgan, 1986, p. 195). The fact that educational organizations consist of relatively autonomous professionals and loosely coupled subsystems is seen as a general condition stimulating political behavior of the members of the organization.

Microeconomic theory describes organizational behavior (in the case of schools: pupils, teachers and head teachers) in terms of utility functions and production functions (Correa, 1995). A basic distinction is the amount of time and energy an individual organization member is willing to invest in task-related action, as opposed to other directed activity, e.g. enjoying leisure. The amount of task-related activity (e.g. time on task) of each main type of actor within a school organization can be inserted as one of the explanatory variables in an education production function. Alternatively, the importance of effect attainment (the argument in the production function) can determine the utility of task-related effort of a particular individual. From this perspective the question of how to improve organizational effectiveness can now be stated in terms of creating conditions to the extent that organization members are stimulated and rewarded for task-related behavior.

In public choice theory the lack of effective control by democratically elected bodies over public-sector organizations marks these organizations as being particularly prone to inefficient behavior, essentially caused by the leeway that is given to managers and officers to pursue their own goals besides serving their organization's primary mission.[1]

Public choice theory provides the diagnosis of instances of organizational ineffectiveness, such as goal displacement, overproduction of services, purposefully counter-productive behavior, "making work" (i.e. officials creating work for each other), hidden agendas and time- and energy-consuming schisms between subunits. When discretional leeway of subordinate units combines with unclear technology this also adds to the overall nourishing ground for inefficient organizational functioning (see Cohen's famous garbage can model of organizational decision making; Cohen *et al.*, 1972). Not only government departments but also universities are often mentioned as examples of types of organizations where these phenomena are likely to occur.

The remedy against these sources of organizational malfunctioning would theoretically be a close alignment, and ideally even a complete union, of individual, subunit and organizational goals. The practical approach towards this offered by public choice theory is to create external conditions that will force at least part of the inefficient divergency of individual level and organizational rationality out of the system. The recommended lever is the creation of market mechanisms replacing administrative control. The competition resulting from these market conditions will be an important incentive to make public sector organizations more efficient. The essence of choice as an alternative to the bureaucratic controls that result from the way representative democracy works is that a completely different, more local type of democracy is required, in which most authority is vested directly in the schools, parents and students (Chubb & Moe, 1990, p. 218). In their "proposal

[1]A more extensive treatment of the implications of public choice theory for school effectiveness research is given elsewhere (Scheerens, 1992, Chapter 2).

for reform" these authors draw a picture of an educational system where there is plenty of liberty to found schools, a funding system that is largely dependent on the success of schools in free competition for students, freedom of choice for parents and freedom for schools to have their own admission policies.

It should be noted that the leverage point of choice differs from that of synoptic planning and bureaucracy as an alternative mechanism that might explain educational effectiveness phenomena. Whereas the latter applies to the design of the primary process and supportive managerial conditions in the areas of supervision and coordination, choice points to external, school environmental conditions. This means that, perhaps surprisingly, both mechanisms could theoretically be employed simultaneously. Although internal bureaucratic functioning (in the sense described in the previous section) will probably be seen as being embedded in the larger central or state bureaucracy there is no necessity that this is indeed the case.

Notes of criticism that have been made with respect to the propagation of choice are that parents' choices of schools are based on criteria other than performance (Riley, 1990, p. 558), that choice might stimulate inequalities in education (Hirsch, 1994) and that completely autonomous primary and secondary schools create problems in offering a common educational level for further education (Leune, 1994).

Scheerens (1992, pp. 17, 18) mentions the following three instances in which deductions from public choice theory are in line with the results of empirical school effectiveness research.

1. To the extent that public choice theory draws attention to overemphasizing general managerial and maintenance functions at the cost of investment in the organization's primary process, the results of U.S. and British studies showing that instructional leadership is associated with relatively high performance are in line with this observation.
2. The construct of "opportunity costs", which draws attention to the phenomenon that functionaries in public-sector organizations have opportunities to be active in non-task-related activities, can be seen as indicative of the general finding that more "time on task", and thus "less foregone teaching and learning", leads to better educational achievement (examples of foregone teaching and learning are lessons not taught, truancy and time required to maintain discipline).
3. Public choice theory offers a general explanation for the results of comparisons between private and public schools. Generally in developed countries, private schools appear to be more effective, even in countries where both private and public schools are financed by the state, as is the case in The Netherlands (Dijkstra, 1992).

Explanations for the alleged superiority of private schools are (a) that parents who send their children to these schools are more active educational consumers and make specific demands on the educational philosophy of schools, and (b) a greater internal democracy of private schools (the latter conclusion was reached on the basis of an empirical study by Hofman *et al.*, 1995). The evidence that schools that are more autonomous (regardless of religious denomination or private/public status) are also more effective is not very strong, however. Although Chubb and Moe (1990) claim to have shown the superiority of autonomous schools their results have been criticized on methodological grounds (Witte, 1990). At the macrolevel, there is no evidence whatsoever that national educational systems where there is more autonomy for schools perform better in the area of basic competencies (Meuret & Scheerens, 1995).

The political image of organizational functioning and public choice theory rightly challenge the assumption of synoptic rationality and bureaucracy that all units and individuals jointly pursue the organization's goal. The arguments and evidence concerning the diagnosis (inefficiency caused by a failing alignment between individual level and organizational level rationality) are more convincing than the cure (privatization, choice) as far as the effectiveness of schools is concerned. The critical factor appears to be that market forces (e.g. parents' choice of a school) may not be guided by considerations concerning school performance, so that schools may be rewarded for other than efficient goal-oriented performance.

It should be emphasized again that our perspective is the internal or instrumental interpretation of school effectiveness (see Introduction) and that the evaluation of the merits of choice may be quite different when the responsiveness of schools is seen as the central issue, particularly for higher educational levels (e.g. vocational schools, higher education) as compared to primary and general secondary schools.

Although in many industrialized countries there are tendencies towards decentralization and increased autonomy of schools, for primary and secondary education these tendencies are stronger in the domains of finance and school management than in the domain of the curriculum (Meuret & Scheerens, 1995). The U.K. is a case in point, where local management of schools is combined with a national curriculum and a national assessment program. In case studies of restructuring programs in the U.S.A. and Canada (Leithwood *et al.*, 1995) increased school autonomy is concentrated in (school-based) management and teacher empowerment whilst curriculum requirements and standards are maintained or even further articulated at an above-school level.

Stringfield (1995, p. 70) notes that several U.S. states have created new curriculum standards and new, more demanding and more performance-based tests.

What then remains as a possible fruitful direction for future school effectiveness research as a result of this analysis of the "political" image of organizational functioning? For primary and secondary education the market metaphor appears to be only useful in a limited sense for primary and secondary education, because governments will generally see the need for a certain standardization in key areas of the curriculum to provide a common base for further education. At the same time choice behavior of the consumers of education may diverge from stimulating schools to raise their performance, and undesired side-effects (more inequalities) cannot be ruled out. The critical factor appears to be that schools experience external pressures and incentives to enhance performance in key areas of the curriculum. Consumers of education, if properly informed, may well be one source for creating these conditions, but not the only source. From this perspective, contrary to the strong adherents of choice, consumerism could well be seen as compatible with accountability requirements from higher educational levels. These different external conditions that may stimulate school performance have not been the object of many empirical studies (exceptions are Kyle, 1985; Coleman & Laroque, 1990; Hofman *et al.*, 1995) and deserve to be investigated further, as well as in an international comparative context. As a second area for further research the statements about "bad" internal organizational functioning of public sector organizational deducted from public choice theory might be used as guidelines in studying unusually ineffective schools.

Retroactive Planning and the Learning Organization

A less demanding type of planning than synoptic planning is the practice of using evaluative information on organizational functioning as a basis for corrective or improvement-oriented action. In that case planning is likely to have a more step-by-step, incremental orientation, and goals or expectations are given the function of standards for interpreting evaluative information. The discrepancy between actual achievement and expectations creates the dynamics that could eventually lead to more effectiveness.

The main reason for considering this type of retroactive planning as less demanding than proactive, synoptic planning is that it enables a more pragmatic, practical and partial approach. Yet, according to March and Olsen (1976), learning from experience meets the same fundamental limitations as rational planning.

When goals are ambiguous, which these authors assume, so are norms and standards for interpreting evaluative information. Another limitation is to determine the causality of observed events. They discern four major limitations to organizational learning:

- role-constrained experiential learning: if evaluative information is contrary to established routine and role definition it may be disregarded and not used for individual action;

- superstitious experiential learning: organizational action does not evoke an environmental response (i.e. is ineffective);
- audience experiential learning: learning of individual organization members does not lead to organizational adaptation;
- experiential learning under ambiguity: it is not clear what happened or why it happened (March & Olsen, 1976, pp. 56–58).

The literature on the use of evaluation research for public policy decisions is supportive of these limitations (e.g. Weiss & Bucuvalas, 1980). Nevertheless, these limitations and constraints can also be taken as challenges for better evaluative practices (see examples in the evaluation literature such as stakeholder-based evaluations and utility-focussed evaluation).

In cybernetics the cycle of assessment, feedback and corrective action is one of the central principles. Morgan (1986, pp. 86, 87) states four key principles of cybernetics, constituting a "theory of communication and learning":

- "systems must have the capacity to sense, monitor and scan significant aspects of their environment;
- they must be able to relate this information to the operating norms that guide system behaviour;
- systems must be able to detect significant deviations from these norms;
- they must be able to initiate corrective action when discrepancies are detected".

In Morgan's statements of these key principles, evaluation–feedback–corrective action cycles have an external orientation ("scanning the environment"). This orientation is closer to the notion of organizational responsiveness to environmental constraints than to effectiveness in the sense of productivity and goal attainment. A related distinction is Argyris' notion of single and double loop learning (Argyris, 1982). "Single loop learning rests in an ability to detect and correct error in relation to a given set of operating norms" (Morgan, 1986, p. 88). Double-loop learning also questions the relevance of the operating norms.

Regardless of the distinction between single- and double-loop learning it should be noted that evaluation–feedback–corrective action and learning cycles comprise four phases:

- measurement and assessment of performance;
- evaluative interpretation based on "given" or newly created norms;
- communication or feedback of this information to units that have the capacity to take corrective action, in terms of work-related improvements or incentives or sanctions to reward or correct actors;
- actual and sustained use (learning) of this information to improve organizational performance.

In the conception of the learning organization the question of which structural arrangements are conducive to evaluation–feedback–improvement cycles is approached from the perspective of double-loop learning in particular. Some of the organizational conditions that are thought to be important to double-loop learning, however, also seem to apply to single-loop learning. Examples are: the encouragement of openness and reflectivity, recognition of the importance of exploring different viewpoints and avoiding the defensive attitudes against bureaucratic accountability procedures (Morgan, 1986, p. 90).

Other organizational features that are considered important to stimulate double-loop learning in learning organizations are flexible structure and a bottom-up or participative approach to decision making. Morgan (1986, pp. 96–109) uses the term "holographic system" to sketch the structural characteristics of a learning organization. In a hologram all of the information necessary to produce an image of the whole is contained in each of its parts. The holographic metaphor when applied to organizations signifies that a subset of parts of the organization can keep the whole system functioning when other parts malfunction or are removed. Subunits, in this way, are seen as having similar design, although they may have quite different functions. This characteristic creates what Morgan terms "rich connectivity" between parts, which in its turn makes the whole structure flexible and amenable to learning and "learning to learn" (Morgan, 1996). In organizational practice these abstract principles may be made concrete in the sense that extra functions are given to units (e.g. teachers are given certain school-management responsibilities), in other words by creating multifunctioned employees and multifunctioned teams. In order to enhance flexibility in functioning and adapting to the external environment formalization, advance specification and standardization should be minimized. Subsystems, although being "rich in interconnectivity" by design, should be relatively autonomous at the same time in order to make efficient functioning of the whole organization possible.

The image of the learning organization is developed in reference to "today's current fast-moving business world" (Rist & Joyce, 1995, p. 131). Here, survival in a rapidly changing environment sets high demands on flexibility and capacity to anticipate the future creatively. Although quite a few authors (e.g. Simons, 1989; Murphy, 1992; Southworth, 1994) find referring to schools as "learning organizations" quite appealing, the notion that this image could indeed be seen as a kind of ideal-type school-organizational structure should not be accepted uncritically. The key question as to the appropriateness of this metaphor for schools is the dynamic complexity of the environment. In this respect there are, of course, important distinctions between educational levels. In primary and secondary education, as argued in an earlier section, an important degree of standardization on desired educational attainment is necessary to provide a common basis for further education. But in the area of middle-level and higher-level vocational education, too, there is an ongoing debate about a common set of key qualifications on the one hand, and a curriculum that would be more directly adaptive to, for instance, the needs of local industry on the other hand. So even in these higher sectors of the educational system a considerable amount of standardization in output, possibly formalized in national examinations, will probably be present. Given these relatively important areas of stability in school environments the call for constant revision of norms and standards as in double-loop learning appears to be unwarranted, as would the necessity of the related structural characteristics of learning organizations.

Therefore, perhaps a more modest interpretation of the image of the learning organization is more appropriate, "modest" meaning a set of features such as concentration on single-loop learning, the optimization of evaluation–feedback–corrective action cycles, given a set of relatively stable performance standards, the creation of sufficient opportunities for staff development and work-oriented consultation between staff.

The beneficial effects of "frequent monitoring of students' progress" belongs to the common-sense knowledge about effectiveness-enhancing school processes. It also has received some support from empirical school effectiveness research, although there are also several studies in which this factor could not be shown to be positively associated with performance (Scheerens, 1992; Scheerens & Creemers, 1996). Our meta-analyses, summarized in Chapter Six, show an overall positive correlation of 0.15.

From a theoretical point of view the cybernetic principle of evaluation–feedback–action is very powerful as an explanatory mechanism of organizational effectiveness. It should be noted that evaluation and feedback also have a place in synoptic planning and bureaucratic structure, as well as in the perspective from public choice theory. In the former case evaluations are most likely to be used for control purposes, while in the latter case there would be an emphasis on positive and negative incentives associated with review and evaluations. From the organizational image of the learning organization as discussed in this section, the cognitive, adaptive and learning implications of evaluations are highlightened.

The action potential, or the potential for school improvement resulting from the comparison of actual performance and standards, is a central factor in dynamic system models like those of Clauset and Gaynor (1982) and de Vos (1989). It can be concluded that in-depth empirical study on school-based evaluations and pupil monitoring, with respect to both the evaluation procedures and the impact on school-organizational functioning deserves a high place on the agenda of theory-driven school effectiveness research.

Fitting Structure and "Contingencies"

"Contingency" is described as a "thing dependent on an uncertain event" and "contingent" as "true only under certain conditions" (*Concise Oxford Dictionary*). In organizational science "contingency theory", also referred to as the "situational approach" or contingency approach (Kieser & Kubicek, 1977), is taken as the perspective in which the optimal structure of an organization is seen as dependent on a number of "other" factors or conditions (de Leeuw, 1982, p. 172). These other factors are mostly referred to as "contingency factors" (Mintzberg, 1979). Contingency factors comprise a heterogeneous set of conditions, both internal and external to the organization. Major categories are: age and size of the organization, the complexity of the organization's environment, power, and the technology of the organization's primary process.

Some well-known general hypotheses about effective combinations of contingency factors and structural configurations are:

- "the older the organization, the more formalized its behaviour";
- "the larger the organization, the more elaborate its structure, that is, the more specialized its tasks, the more differentiated its units, and the more developed its administrative components";
- "the more sophisticated the technical system, the more elaborated the administrative structure, specifically the larger and more professional the support staff, the greater the selective decentralization (to that staff), and the greater the use of liaison devices (to coordinate the work of that staff)";
- "the more dynamic the environment, the more organic the structure" (Mintzberg, 1979, Chapter 12).

The terms in which organizational structure is described are organiza - tional dimensions such as the division of work (specialization) and authority or vertical decentralization, the use of prestructured arrangements (standardization) and the use of written regulations or formalization, the level of skills needed to carry out tasks or professionalization and the interdependence of units (coordination requirements). A distinction is frequently made between mechanistic and organic structure. A mechanistic structure is characterized by high levels of standardization and supervisory discretion and low levels of specialization, intra-unit interdependence and external communication. An organic design is characterized by low levels of standardization and supervisory discretion and high levels of specialization, interdependence and external communication and is less prone to information saturation (Gresov, 1989, p. 432). Mechanistic structure is likely to be efficient when tasks are simple and repetitive and environmental uncertainty is low. When tasks are uncertain, interdependence is high and the environment is dynamic, the organic structure would be more fitting.

The central and global thesis from contingency theory which says that organizational effectiveness results from a fit between situation and structure has been interpreted in various ways.

The congruence thesis states that effective structure requires a close fit between the contingency factors and the design parameters (Mintzberg, 1979, p. 217). Drazin and van de Ven (1985) specify congruence as theories that just hypothesize fit between contingency factors and structure, without examining whether this context–structure relationship actually affects performance. They save the term contingency hypothesis to analyses that look into the joint influence of context and structure on performance.

The configuration thesis is focussed on the internal consistency of structural characteristics, whereas the "extended configuration hypothesis" states that effective structure requires a consistency among the structural characteristics and the contingency factors. These four different hypotheses call for four different research designs that are progressively more complicated.

1. Configuration is technically the least demanding since it could look only at structural parameters and their supposed harmonious fit, for instance by computing simple correlations.
   At the same time interpretation would be circumscribed since no claims about context–structure fit or organizational effectiveness in terms of performance are involved.
2. Congruence (in Drazin & van de Ven's more limited sense) would likewise call for simple correlations, this time between structural characteristics and contingency factors. Much of the early empirical literature on contingency theory in organizational science went no further than testing congruence relationships (e.g. Kieser & Kubicek, 1977).
3. Testing "real" contingency hypotheses would involve some kind of pattern of three types of variables: contingency factors, organizational structure variables and an operational effectiveness criterion (e.g. profit, goal achievement, performance). Kickert (1979) has shown that at least three different interpretations of such patterns are possible (see Fig. 8.3). In Fig. 8.3a, structural characteristics (x) are seen as an intermediary factor in a causal chain with contingency factors as a primary cause. Path analysis would be a fitting technique. In Fig. 8.3b, structure (x) and context (z) are seen to interact in their influence on performance (y). This would call for RANCOVA design (random effects analysis of covariance) with a specific interest in the significance of interaction effects. In the third interpretation (c), structure and context would have an independent additive effect on performance (main effects RANCOVA model).
4. The extended configuration hypothesis in its simplest form would investigate configurations of at least one contingency factor, a pattern of

structural variables and performance. More realistically, there would also be more than one contingency factor involved. The problem of multiple contingencies could be that one contingency factor would pull the structure in one direction, whereas another contingency factor could exercise influence in the opposite direction. For example, in a situation where the task environment becomes simplified as less interdependent, while the external environment becomes less predictable and dynamic, there would be opposing forces towards mechanistic and organic structure, respectively. Gresov (1989) offers three ways for acting managers or organizational advisors to resolve problems of multiple and opposing contingencies. Two of these imply changing or redesigning the organization or the organizational context.

For researchers, only the third approach appears to be realistic, namely to presuppose a dominance ordering of contingency factors, e.g. if technology and environment exercise opposing demands on structure then the influence of technology is the most important. There is, however, no common understanding about a universal importance ranking of contingency factors. If a solution were sought in statements like "the predominance of one contingency factor over the other depends on certain situational characteristics", one would presuppose a "second-degree" contingency theory, which would end up with an infinite regress.

There are still more problematic aspects of contingency theory, at least from a researcher's perspective. The concept of "equifinality" recognizes the possibility that different patterns of context–structure relationship may be equally effective.

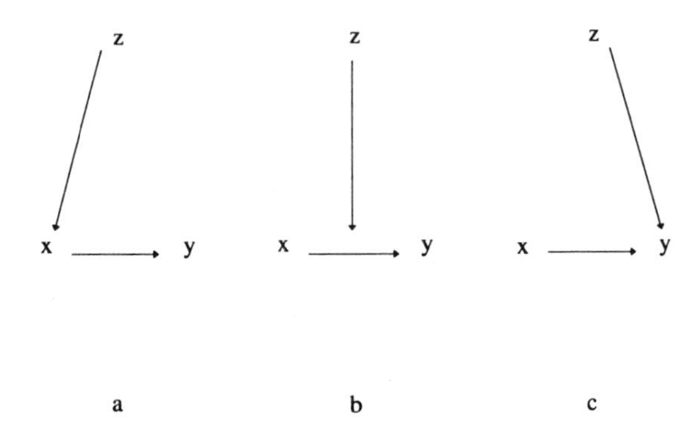

Figure 8.3 Three different interpretations of contingency hypotheses; z = contingency factors, x = structural characteristic(s), y = outcome variables (from Kickert, 1979, p. 193).

In such a situation, effectiveness researchers might propose outcomes like the identification of ranges of effective and ineffective context–structure interrelationships. Probing deeper would inevitably lead to hypothetical second-order contingency factors, namely those background factors that would interact with different context–structure patterns to produce equal performance.

Next, contingency theory is sometimes criticized for its static nature, while at the same time it is recognized that organizations (i.e. management) may not only adapt the structural characteristics of the organization but also actively shape their environment (e.g. through cooptation and marketing).

A third questionable area is the specificity of application of contingency theory. In organizational science it has frequently been used to compare broad categories of organizations, for instance industries using craft, batch, mass or continuous process production. The question could be raised of how useful contingency theory could be in comparing the effectiveness of context–structure relationship for a particular type of organization, in this case educational organizations. Here one should be more demanding than applying contingency hypotheses to explain differences between primary, secondary and tertiary (i.e. universities) educational organizations. For this example a general hypothesis could be that, given an average position of educational organization on the dimension mechanistic–organic, lower education levels would tend to the mechanistic and higher education levels to the organic pole.

For our purposes, theory development on educational effectiveness, contingency theory should be able to predict results of what has been termed as "contextual effectiveness", the idea that different performance-enhancing conditions of educational organizations work in different contexts. Drawing on Teddlie's recent review on "context in school effects research" (Teddlie, 1994), it becomes clear that contextual effectiveness is indeed a viable part of educational effectiveness research. The most frequently studied context variables are: average socioeconomic status of students, urbanicity and educational level (i.e. primary versus secondary schools). A relatively new type of contextual analysis is the comparison of performance-enhancing educational conditions across nations (Scheerens *et al.*, 1989; Postlethwaite & Ross, 1992; Creemers, 1995; and Chapter Seven of this book).
Some illustrative results from contextual effectiveness studies are:

- a shorter term orientation towards educational expectations in effective low as compared to middle socioeconomic status schools (Teddlie *et al.*, 1989);
- more pronounced external reward structure in effective low as compared to high socioeconomic status schools (Teddlie *et al.*, 1989);
- invitational versus a more careful (sometimes even buffering) attitude of middle as compared to effective low socioeconomic status schools towards parents (Teddlie *et al.*, 1989);

- controlling educational leadership with respect to instruction and task orientation in effective low socioeconomic status schools as compared to low or moderate controlling leadership in these areas in effective high socioeconomic status schools (Hallinger & Murphy, 1986, cited by Teddlie, 1994);
- primarily academic goals in effective elementary schools as compared to both personal as well as educational goals in secondary schools (various authors, cited by Teddlie, 1994);
- strong instructional leadership required for success in urban elementary schools, whereas a managerial leadership style often proved successful in suburban elementary schools and an "intermediate" between manager and initiator worked best in rural elementary schools (Teddlie & Stringfield, 1993).

It appears that the research results on contextual effectiveness, as is almost always the case with educational effectiveness research, have mostly been induced from exploratory correlational analysis, rather than being the result of the testing of hypothesis deduced from theory. The outcomes of contextual effectiveness studies make a lot of sense, and are only vaguely related to contingency hypotheses from the general organization science literature. The more controlling leadership in effective urban, low socioeconomic status schools than in suburban, middle socioeconomic status elementary and secondary schools could be connected to the general contingency hypothesis that relatively simple, rather unconnected tasks in a generally stable environment call for a more mechanistic type of organization structure.

The question should be raised as to what can be expected of contingency theory when it comes to giving a more theoretical turn to educational effectiveness research. Can substantive hypotheses be deduced from contingency theory that could further research into contextual effectiveness? It is our impression that a careful study of the general organization literature on contingency, particularly the empirical research literature, can yield interesting hypotheses. In this sense the promising areas appear to be:

- hypotheses concerning changes in the technological system situated at the level of the primary process of learning and instruction;
- hypotheses concerning increased environmental uncertainty for educational organizations.

There is sufficient variety in the technology of learning and instruction to expect concrete implications for the structuring of educational organizations. If one compares, for instance, very structured approaches such as the Calvert-Barclay-program, described by Stringfield (1995), with educational practice inspired by constructivism this variation is evident, even within a certain educational level (elementary education in this case). In the first instance, structural patterns would be expected to be in line with a mechanistic type of organization (little interdependence between units, standardization and direct control), whereas in the second instance a more organic type of organiziation would be expected (horizontal decentralization, plenty of interdependence between units, less standardization).

These overall patterns are more or less confirmed in publications describing the organizational context of the two polar types of educational technology; see Stringfield's description of "high reliability organizations" and Murphy's analysis of the managerial context of constructivist teaching (Murphy, 1992), also see Scheerens (1994). Barley's reference to role theory, already referred to in an earlier section (Barley, 1990), could be used as a guideline to study the more minute ways in which these task characteristics have an impact on organizational structures.

The second type of hypothesis could be applied to the variation in patterns of school autonomy versus dependency that arises from the current restructuring of educational governance in many countries. Policies of functional and territorial decentralization (cf. Meuret & Scheerens, 1995) as well as deregulation, paired to ideas about "school-based management" and "teacher empowerment", as well as the stimulation of market mechanisms, all make for a less stable and predictable school environment and would therefore be expected to force educational organizations to more organic structure, in order to remain effective. Since these configurations of centralization and decentralization differ strongly between countries (Meuret & Scheerens, 1995), international comparative educational effectiveness studies could be used to test this type of hypothesis.

Finally, it should be noted that the logic of the contingency approach is very close to substantive and formal methodological interpretations of multilevel school effectiveness models, which were referred to in the Introduction. The general idea of higher level facilitation of lower level conditions can be seen as an instance of the configuration thesis. Moreover, some of the intricacies of formalizing across-level relationships (e.g. interactive, additive, non-recursive) in multilevel school effectiveness models are also present in the alternative formal specifications of contingency hypotheses.

In making up the balance, in spite of all methodological problems, contingency theory appears to have more content than has currently been used in educational effectiveness research. In this way it offers certain possibilities to improve the theoretical basis on educational effectiveness so that research could be driven by specific hypotheses deduced from theory.

Discussion

Before making up the balance with regard to the various theory-embedded principles discussed in this chapter and their relevance to further school effectiveness research, two alternative theoretical perspectives need to be referred to briefly. The first is the possible challenge of chaos theory and the second pertains to the cultural view on organizations.

Chaos theory

Chaos theory is concerned with "the exploration of patterns emerging from apparently random events within a physical or social system" (Griffith *et al.*, 1991). The qualification "apparently" is crucial in this definition, because the very objective in this branch of systems theory is to discover patterns that may be very complex but can nevertheless be visualized. Chaotic equations, for example, can be visualized by means of fractal geometry, fractals being figures that are inherently identical, which means that upon enlarging a detail the same pattern appears as the whole. This can be seen in clouds and bronchia (van Lidt de Jeude & Brouwer, 1992, p. 16); also compare the holographic metaphor referred to in an earlier section.

One of the basic principles of chaos theory is that "small causes may have large effects" and that relatively minor variations in entrance conditions may have enormous consequences when interrelationships between phenomena develop over time. In some mathematical functions there may be linear growth for a particular range of values of a critical parameter, but when this range of values is exceeded chaotic patterns may appear. Examples are the constraint growth function and Kaldor's macroeconomic model (van Lidt de Jeude & Brouwer, 1992, pp. 14, 15). The point at which the function starts to meander is called the bifurcation point. Although in the chaotic region, beyond the bifurcation point the interrelationships become unpredictable, shapes or patterns may still be identified. These shapes are seen as being determined by "strange attractors". "Attractors" because states of the system will be attracted by the lines of the structure, and "strange" because values of the function close together at one point in time may be far removed as time evolves. An example is the Lorenz attractor (van Lidt de Jeude & Brouwer, 1992).

New patterns of subsystems that may emerge in a seemingly chaotic environment are sometimes interpreted in terms like synergism, self-organization or autopoiesis. Autopoiesis concerns the tendency of self-reproduction of systems and organism, whereas synergism concerns the evolvement of new macrolevel structures when microlevel subsystems interact in a complex way. It should be noted that the construct of self-organization inspired by chaos theory is more extensive than the principle of double-loop learning in Morgan's characterization of the learning organization, as discussed in an earlier section. Chaos theoretical conceptions of self-organization include positive feedback cycles, next to negative feedback (emphasized in cybernetics) and developments that do not confirm to the phenomenon of homeostasis.

It is beyond the scope of this chapter to give more than a very general impression of some of the principles of chaos theory. Nevertheless, implications for theory formation and research on school effectiveness appear to be present. These will be looked at both from a substantive and from a methodological point of view.

Environmental turbulence can be seen as forcing systems (or organizations) into disequilibrium and give rise to the type of complex developments as are the objects of chaos theory. Some authors feel that the fit paradigm of contingency theory fails to describe adequately the way organizations should cope with increasing external complexity and uncertainty.

> The environment ceases to be a stable ground on which organizations can play out their games and countergames ... the ground is in motion with autochthonous processes ... the turbulence results from the complexity and multiple character of the causal connections (Emery & Trist, 1965, cited by Broekstra, 1992, p. 115).

In a general state of disequilibrium of an organization "small causes" may indeed have "big effects" in which an individual, an idea or a new behavior can upset the global state (Priogine & Stengers, cited by Broekstra, 1992, p. 116). These "big effects" can be seen as a new order that has arisen out of chaos, induced by actions of actors within the system able to exploit inlinearities or perhaps by mere serendipity.

In a more prescriptive sense the following principles are implied for organizational functioning:

- a further departure of hierarchical control, blue-print planning and mechanistic structure;
- less reliance on attempts to fit structure to environment, since in general the situation is thought to be too unstable and dynamic for this;
- flexibility with respect to organizational closure, which could mean frequent use of semi-autonomous small subsystems, but also strategic alliances (interorganizational networks) with other organizations which are in a similarly turbulent field;
- heightened attention to endogenous developments within the organizations, an active attitude towards influencing environmental conditions;
- focus on feedback and learning functions;
- increased tolerance for intuition, political vision and ethos in management style (in this connection, Broekstra coins the term "creative organization"

which he prefers to the purely intellectual notion of the learning organization; Broekstra, 1992, p. 122).

To the extent that schools' environments can be seen as turbulent, these managerial factors could, likewise, be seen as relevant to enhancing school effectiveness and thus as relevant "process features" or effectiveness-enhancing conditions. However, the same reservations are in place here as were made in the section on the learning organization: in general, school environments may not be that turbulent.

In their analysis of alternative models of school effectiveness, Bosker & Scheerens (1994) conclude that "the actual problem might in fact be that the most likely model might be indirect, non-recursive, contextual, synergetic, and interactive at the same time". All of these characteristics imply increased complexity as opposed to simpler (direct, recursive, additive) versions. This means that state-of-the-art multilevel and structural modeling techniques based on the general linear model could fall short of dealing with the complexity inherent in conceptual school effectiveness models.

Another basis for concern about the current state of the art of school effectiveness research are doubts about the progress that is being made in building a well-established knowledge base and even, to a lesser degree, whether school effectiveness can be studied as a relatively stable phenomenon.

Doubts about the solidity of the current knowledge base on school effectiveness are expressed by Tymms (1994), who presents a first attempt to apply chaos-theoretical reasoning to school effectiveness by means of a simulation study. As progress is being made in properly adjusting outcome data for prior achievement and other relevant background variables at the student level, the proportion of the total student variation that can be attributed to attending a particular school appears to drop to values of 5–7% (see Bosker & Witziers, 1995; Hill *et al.*, 1995a). If roughly half of this proportion of variance can be accounted for by a set of effectiveness-enhancing process conditions of schooling, it is clear that it is extremely difficult to accumulate evidence on a stable set of malleable correlates of net achievement.

In the area of stability of school effects the authors of recent studies conclude that "there is a substantial amount of instability across years" (Luyten, 1994, p. 56) and "that there is a significant degree of change over time" (Thomas *et al.*, 1995, p. 15). Correlations of effects across years vary according to subjects, roughly from 0.40 to 0.90. These authors found that only at the extremes (either effective or non-effective) consistent (across subjects) and stable (across years) positioning patterns were found.

Considering these areas of unexplained variation [with respect to both process correlates of net school effects and (in)stability over time], two approaches can be considered:

- continue using the current linear analysis techniques, but preferably with larger samples, more reliable measures and controlled (quasi-experimental) research designs;
- employ the more complex systems dynamics modeling techniques that are associated with the general label of chaos theory.

Preferably, both approaches should be followed. There appears to be insufficient ground (in terms of lack of consistency and stability) to abandon the more traditional methods. It seems worthwhile, however, to explore further the possibilities of dynamic systems theory approaches to try and discover patterns in the so-far unexplained regions in our knowledge on what causes schools to be effective. In a more practical sense this approach could have the following implications for school effectiveness research:

- efforts in examining the long-term effects of differences in entrance conditions (achievement of students, contextual effects of grouping of students, matching of teachers and groups of pupils);
- efforts in creating longitudinal data files on outcomes, student background and process characteristics per school;
- perhaps a new source of inspiration to carry out qualitative studies of school-specific measures that might explain the improvement or deteriation of a school's functioning and increased effort to try and reach some type of accumulation and synthesis from numbers of qualitative studies.

A note on culture

In the search for theory-embedded principles of effective schooling, much has been said about procedures, organizational processes and structural arrangements. Another important mode of organizational reality, namely culture, has received relatively little attention (see also the section on culture as a mode of schooling in Chapter One). This may seem strange, considering the fact that several of the effectiveness-enhancing conditions referred to in the introduction are concerned with ethos and expectations (e.g. "firm and purposeful leadership", "collegiality and collaboration", "high expectations all around" and "clear and fair discipline"). The very nature of the "mission" of educational organizations, which is concerned with issues about pedagogy that are strongly normative and value-laden, could be seen as further underlining the importance of culture. Finally, the nature of schools as "professional bureaucracies", which implies certain limitations with respect to structural variation, can also be seen as a context in which shared meanings, collective norms and views on interaction and collaboration (i.e. culture and climate) are of great importance to provide the "normative glue" that holds the organization together (Morgan, 1986, p. 135). Empirical evidence on the association of "strong" versus "weak" school culture with overall academic achievement in secondary schools is provided by Cheng (1993).

In the literature on educational innovation and school improvement (e.g. Kieviet & Vandenberghe, 1993; Hargreaves & Hopkins, 1991) culture, particularly in the sense of attitudes towards change, is considered of great importance, so why not include something like "development of a strong school culture" as an alternative effectiveness-enhancing principle? The first reason is that the distinction between formal structure and culture may not be altogether clear. For example, Price (1972, cited by Cheng, 1993, p. 86) says that the essential part of organizational culture is "organizational ideology", further described as characterized by: "sharing of clear and consistent beliefs among members about the organization's role and mission in society; sharing of definite and consistent beliefs and ideas among members about the task's nature and goals and the technology used to achieve goals; and the existence of artifacts such as stories, sayings and traditions reinforcing the above sharings". In this definition it is clear that there is a fine line of distinction between several of these cultural elements and structural arrangements such as designing and reaching formal agreement on a school policy plan, the structuring of tasks by means of formally agreed teaching methods and overt coordination methods (particularly goal coordination) by the school leader. The principal additional element in the cultural image appears to be the informal and shared interpretations of structural arrangements by the members of the organization.

Second, having noted that the cultural identity of an organization depends on informal relationships and shared interpretations, the question of whether school culture should be seen as a cause, an object or an effect of school improvement should be raised. Hargreaves (1995, p. 41) contests that school culture may be all of these.

From our perspective, which is largely instrumental, we are inclined to see culture as an important mode of organizational reality, but as an effect of structural arrangements rather than as a cause. Despite the existence of direct means to change culture (see the reference made to Schein in Chapter One), it appears to be more pragmatic to use structural levers for change and improvement, while explicitly considering (and in school practice, actively managing) cultural implications.

Therefore the main reason for not treating culture as the other principles that could explain effectiveness phenomena is that it is not considered as an area that is amenable to instrumental malleability as the other principles, to some extent, are. Second, each of the other theory-embedded principles discussed in the previous sections may be seen as bearing particular consequences for organizational culture. For example, synoptic planning and bureaucracy call for unity of purpose and harmonious cooperation, the choice perspective encompasses value conflicts between individual and organizational goals and the metaphor of the learning organizations emphasizes openness to new developments and participative planning. Therefore, despite the crucial importance of issues of culture and ethos for effective organizational functioning, from the perspective chosen in this presentation it is seen more as a by-product of other effectiveness-enhancing mechanisms or as a "maintenance condition" (see Chapter One and Hargreaves, 1995) that should always receive sufficient managerial attention, and not as an additional explanatory principle.

Conclusions

From the references made to the state of the art on educational effectiveness research in the introductory section it became clear that the margins for enhancing school effectiveness in the sense of value-added productivity, by means of malleable process conditions, are relatively small. Now the issue as to what the implications of the identified mechanisms are for future school effectiveness research and for school improvement design and practice, should be further pursued. The theories and core principles are schematically summarized in Table 8.1. First, it appears that the ideal of rational synoptic planning, combined with supportive structural and cultural elements in the areas of coordination between subunits, achievement-oriented ethos and a collaborative attitude, explains the research-based effectiveness correlates cited in earlier chapters fairly well. This calls for a resurrection of proactive, science-based and structured approaches to school improvement. The success of practical applications of "high reliability organizations" (Stringfield, 1995) supports this assertion, as do experiences with a quasi-experimentally designed improvement project in The Netherlands (Houtveen & Osinga, 1995).

Table 8.1 Key mechanisms within organization theories that fit the rationality paradigm.

Theory	*Core mechanism*
rational control	proactive structuring
public choice	market mechanism
retroactive planning	cybernetic principle
contingency theory	fit

With respect to rational planning it sometimes seems as if educationalists active in the domain of school improvement have a "double morale". For instance, Hargreaves and Hopkins (1991, p. 8) warn against the simple notion in which schools are seen as rational organizations. Yet, the systematic approach to school development which constitutes the main body of their book is a perfect example of a rational planning procedure. Perhaps what could be seen as separating these authors from a purely rational approach is the predominance of bottom-up, school-based procedures, as opposed to the implementation of externally (maybe expertly) developed material and working procedures. However, the evidence from projects like the Barclay-Calvert program (Stringfield *et al.*, 1995) indicates that this aspect of synoptic planning should be reconsidered against the current trend in school improvement to denounce "fidelity" approaches and that evaluations of school improvement projects based on RDD-type (research, development and diffusion) strategies constitute one of the most promising areas for future school effectiveness research.

Second, the single most interesting explanatory mechanism for organizational effectiveness is the cybernetic principle of evaluation–feedback–(corrective) action. The analysis of alternative theoretical positions also indicates that proper evaluation and review is a critical feature inherent in diverging organization theories such as scientific management, public-choice theory and the learning organization. The orientation with respect to single- and double-loop learning is likely to vary according to educational level, although it appears that references to "turbulent environments" with respect to educational organizations to some extent reflect the current fashion of comparing educational organizations to the world of business and industry. In primary and general secondary education it would be worthwhile to concentrate further empirical investigations on the practice and use of (school-based) evaluations within the perspective of single-loop learning.

Third, contingency theory has both formal methodological and substantive implications. With respect to the formal methodological aspect it appears that conceptual and formal multilevel school effectiveness models, such as the ones developed by Scheerens (1992), Stringfield and Slavin(1992) and Creemers (1994) strongly reflect a contingency approach (also see Bosker & Scheerens, 1994). For instance, the idea of higher level facilitation of lower level processes can be classified as a particular operationalization of the configuration thesis (i.e. the internal consistency of structural characteristics). Studies aimed at "contextual effectiveness" conform to one of the formal models of contingency theory, in this case the "real" contingency hypothesis (i.e. a configuration of contingency factors, organizational structure factors and performance data). Considering the facts that educational effectiveness research employs relatively strong data on organizational output, and that sophisticated multi-leveling techniques are used to test alternative conceptual specifications, this type of research could have an important impact on contingency-oriented approaches in the broader field of organizational science.

Substantively, environmental complexity and turbulence, the nature of the primary process of teaching, and composition of the student population are considered as the most interesting contingency factors. Comparisons of didactic practice, as propagated by constructivism and more traditional approaches like direct teaching, are examples of alternative primary process models which could also be expected to require alternative organizational arrangements (Scheerens, 1994).

Fourth, the general exploration of some of the basic notions of chaos theory has yielded some new ideas on research methodological approaches (both quantitatively and qualitatively) rather than directly useable substantive ideas.

Fifth and finally, contingency theory could also be used as a meta-theoretical principle when considering school improvement options. An essential element in such a meta-theory would be the idea that the mechanisms discussed in the earlier sections could be considered as effective levers for improvement, dependent on particular contingencies. A simple example would be the contention that the choice mechanism fits middle-level vocational and higher education but not primary and secondary education. It is beyond the scope of this book to develop this idea of a situational theory any further. Elsewhere (Scheerens, 1995), other ingredients of such an encompassing theory, such as type of effectiveness orientation (e.g. productivity versus responsiveness) and mode of school functioning (technology, management, staff, staff development, finance, etc.) have been mentioned. A related effort is Cheng's dynamic model of school effectiveness, where the focus is on the usefulness of a particular organizational effectiveness orientation given the development of the school and external circumstances (Cheng, 1993).

CHAPTER NINE

REDIRECTION OF SCHOOL EFFECTIVENESS INQUIRY

Introduction

In this chapter the main conclusions reached in the preceding parts of the book will be discussed, and analyzed from a critical perspective: how useful is the current dominant approach in school effectiveness inquiry (i.e. empirical research and conceptual analysis) for scientific purposes and educational practice and which redirections are deemed necessary, given the current state of the field? After making up the balance on these issues recommendations will be made in the area of redirection in conceptual modeling and research methodology.

School Effectiveness: A Myth?

Throughout the history of educational effectiveness research critical analyses have appeared. In fact, a study that was taken as very critical at the time, the Coleman report (1966), is generally considered as the starting point of the educational effectiveness research tradition. Case studies and outlier studies on unusually effective schools were, to some extent, a reaction against the impression made by the Coleman report that schooling "did not matter". Even today, school effectiveness researchers express their opposition against this image by choosing book titles like "Schools can make a difference" and "School matters".

When considering the best of currently available recent empirical school effectiveness studies (see Chapters Five and Six) no conclusion can be reached other than admitting that Coleman was right with respect to the size of school effects in terms of the between school variance on valued-added outcomes in basic school subjects. The same applies to Coleman's conclusion about the much larger impact of student background variables as compared to the impact of malleable conditions of schooling. The question of whether interpretations of these sustaining empirical findings should be as pessimistic as they were at the time of the Coleman report should be readdressed.

At the end of the 1970s and the beginning of the 1980s several critical reviews appeared on the early "effective schools" studies (e.g. Purkey & Smith, 1983). Ralph and Fennessey (1983) were most critical in renouncing the "ideology of reform" by which these studies were colored. They also proposed more strict and demanding criteria to assess school effectiveness. Schools should only be seen as effective when relatively high performance could be demonstrated over a range of years, a range of grade levels and classrooms and a range of subjects. These latter, more demanding criteria were used as a rationale in studies on the stability, scope and consistency over subject matters of school effects (Scheerens, 1993, calls these studies "foundational studies"). In the light of the evidence of this type of study, summarized and discussed in Chapter Three, it is clear that such strict criteria of school effectiveness are rarely met, and that to some extent school effectiveness appears to be temporary rather than permanent, local rather than pervasive in all parts of the school organization and partial rather than general over subject matter areas.

A last example of a critical review of the field is provided by the empirical studies and analytical reviews on the existence of the educational production function (Hanushek, 1986; Monk, 1992). Monk's telling conclusion is that, at this stage, we cannot be certain about the existence of an educational production function. Again, further qualification and interpretation are required in the light of recent empirical findings, leading to possible redirections in the way these should be used in further empirical and analytical work.

In this book, as was the case in the studies and reviews just cited, no easy, straightforward and optimistic conclusions could be drawn. Although this is a null hypothesis that one would like to reject, the critical question, "is school effectiveness a myth?" cannot be left aside. In making up the balance of this book we shall try to shed some light on this existential question by reviewing the results of earlier chapters guided by a set of questions:

- is "school effectiveness" a useful, unambiguous concept?
- are school effects large enough to answer questions about school effectiveness?
- which factors, if any, matter?

- do we have useful theory-embedded mechanisms that could support a more theory-driven redirection of empirical school effectiveness research?
- do we know how to improve school effectiveness?

Is "school effectiveness" a useful, unambiguous concept?

School effectiveness relates to the impact of malleable conditions of schooling, referred to as modes in this book, on the outcomes of schooling. In the literature on organizational effectiveness several criteria are distinguished on which organizational effects can be assessed. To the degree that such criteria are thought to be equally valid, effectiveness is a pluralistic concept. Depending on one's perspective the choice inherent in most empirical school effectiveness studies for "productivity" as the singular effect criterion can be considered plausible or arbitrary. The solution chosen in this book, to use productivity as the key criterion and consider other criteria as possible supportive conditions, is just one of several options. In this we would like to make a distinction between the perspective of empirical research and the perspective of a school manager. We consider our choice feasible with respect to empirical research, but would admit that a contingency view (emphasizing one criterion or several criteria, depending on the situation) is more useful for practical considerations of school management.

The distinction of modes of schooling is used in order to attempt to delineate an exhaustive set of conditions that, in principle, can be actively manipulated by the school. Analysis of this set of conditions shows that the full range of possible handles to improve school effects provides a level of complexity that is never mirrored in empirical effectiveness research. A degree of reductionism is required in order to make empirical research manageable. Nevertheless, the exposition on modes has indicated some areas that should probably receive more attention in future research, for instance the selection of teachers and other aspects of human resource development for school staff.

In actual school management, differences to the degree that certain modes are under the control of school management (cf. the constructs of controllability and potential for control) will also determine the concern with a particular mode as a vehicle for the improvement of performance.

At a more operational level the relevance and "integrity" of the concept of school effectiveness might also be seen as depending on issues such as the impact of school-level versus classroom-level conditions, the stability of school effects and the scope of effectiveness with respect to subject matter areas and organizational subunits (grades and classrooms). In our opinion, this question ultimately depends on the size of school effects. Even if empirical evidence shows, as is indeed the case to a considerable extent (see Chapters Three to Five) that effects are only moderately stable, local and partial and even if there is more variance between classes within schools than between schools, and classroom-level conditions account for more variance than school-level conditions, one still cannot discard the potential relevance of the school organizational level, providing that there would be some variance between schools. The issues of the relevance and importance of the size of school effects will be addressed later.

Deliberations concerning the size of school effects also indicate a fundamental limitation of the operational definition of school effectiveness, namely its comparative and therefore relative nature. It is theoretically possible that all schools in a country would be equally successful in attaining certain fixed achievement levels. In that case the concept of school effectiveness in its comparative sense would be meaningless.

In summary, school effectiveness is to be considered as a useful concept only if inescapable reductions of the full complexity in terms of different criteria and modes are explicitly recognized.

Are school effects large enough to answer questions about school effectiveness?

The question referring to school effectiveness and the size of school effects should be distinguished from questions about the practical relevance of school effects.

There are at least two kinds of practical issues at stake when dealing with this latter question. First, one should consider the situation where information on the value added to educational outcomes by each school is used for political or practical decision making, for example, rewarding well-performing schools by providing them with extra resources, and comparing schools among each other in league tables. The purpose of these league tables is the provision of information to parents to facilitate the choice of a school for their children. Goldstein and Thomas (1996) have shown that, at least in the U.K., it is possible to discriminate significantly between schools only when extremely high and extremely low scoring schools are compared. Goldstein (1996) provides an example in which "the confidence intervals for each school cover a large part of the total range of the estimates themselves, with three quarters of the intervals overlapping the population mean ..." (p. 5), the implication being that it is generally not possible to discriminate in a precise and reliable way between schools for the practical concerns in question. Second, one should consider the consequences for individual pupils. It has been established repeatedly in the literature (Purkey & Smith, 1983; Brandsma & Doolaard, 1996) that relatively small differences between schools will nevertheless have important consequences for the educational career of individual pupils. Purkey and Smith, for instance, discerned an entire school year's difference between the average pupil in the most effective schools and the average pupil in the least effective school.

Brandsma and Doolaard (1996) demonstrate that pupils from effective schools have more chance of being advised to proceed to higher types of secondary education in the Dutch educational system. These authors distinguish between four IQ bands, and show that in highly effective schools pupils from the three lower bands score at a level that is equivalent to two curricular tracks (out of a total of 5) higher than pupils from ineffective schools. In the highest IQ band the difference is equivalent to one track.

A point made by Bosker and Witziers (1996) in this context, and stated in Chapter Three, is that school effects, small as they may be, have implications for not just one, but for all pupils of the cohort for whom the effects were measured. Questions about school effectiveness concern the supposedly causal impact of malleable educational conditions on value-added outcomes. This question can be seen as a practical question for schools which, for example, monitor their own performance by means of school self-evaluation procedures. Given the limitations cited above, it is generally impossible or at least very difficult to answer effectiveness questions in this context (Goldstein, 1996; Bosker & Scheerens, 1995).

When school effectiveness is seen as a research discipline in which, essentially, hypothetical causal models of educational performance of schools are tested, small net differences between schools are less troublesome. The law of large numbers can be applied to detect causes of small differences between schools. For example, when applying appropriate multilevel analyses, it is possible to establish which proportion of the variance between schools can be accounted for by explanatory process indicators on malleable educational conditions. Therefore, the answer to the question in the heading of this subsection is that, providing there are sufficient units of analysis (schools, classes and pupils), small net differences between schools do not preclude investigating research questions on school effectiveness.

In this context, the question of school versus classroom effectiveness should be readdressed. As shown by the illustrative studies summarized in Chapter Five, application of multilevel modeling makes it possible to distinguish school effects from the effects of classrooms within schools and to separate the impact of school-level conditions from the impact of classroom-level conditions.

Does the concept of school effectiveness become less relevant when there is considerable evidence that more variance is accounted for by the classroom level than by the school level (Hill *et al.*, 1995a; Scheerens *et al.*, 1989; Wang *et al.*, 1993; see also Chapter Three) and classroom conditions explain more variance than school-level conditions?

The answer to this question is clearly negative. Even in the hypothetical case where all the between-school variance could be accounted for by the variation between classrooms, this state of affairs would create a window of opportunity for school-level intervention. Between-classroom variation within schools implies that some classrooms, i.e. teachers, do better than others, so that measures could be taken to improve the performance of lower performing classrooms. The empirical evidence on school versus classroom effects should be taken seriously, however, in the context of conceptual modeling and the selection of school-level variables. Perhaps the selection and (teacher) recruitment function of the school should receive more emphasis next to conditions like educational leadership.

Which factors matter?

In earlier chapters several sources were tapped to assess the factors of schooling that matter with respect to enhancing school effectiveness:

- qualitative reviews;
- quantitative research syntheses;
- five exemplary empirical studies;
- international comparative analyses.

The results presented in the chapters in question are combined in Table 9.1. The most telling information from this table is the discrepancy between the results of qualitative reviews and quantitative research syntheses in the domain of school organizational factors, as well as the confirmation of a phenomenon that was also established by others, that instructional conditions matter more than school organizational conditions. The discrepancy between the results of qualitative reviews and quantitative meta-analyses in the area of school organizational factors is surprising, because the reviews in question are all based on empirical studies. The Cotton (1995) review, for example, mentions sets of individual empirical studies that support the contentions about effectiveness-enhancing conditions. Apparently, across studies, effects are inconsistent to the degree that quantitative averages are close to correlations of zero. The stronger impact of factors that are closer to the operating core of learning and instruction, as compared to more distant factors, was also highlighted by other authors (Scheerens, 1989b; Wang *et al.*, 1993; Hill *et al.*, 1995a).

How should these results be interpreted? Do they imply that the set of factors, about which there is apparently so much consensus among researchers and analysts that work in the field of school effectiveness, does not really matter?

In our opinion this conclusion is too strong, and too negative, despite the inescapable message that the impact of school organizational factors has been overrated in the past. This consideration may seem like an easy way out: one can always point to the greater research technical and methodological difficul-

Table 9.1 Review of the evidence from qualitative reviews, international studies and research syntheses

	Qualitative reviews	International analyses	Research syntheses
Resource input variables:			
Pupil–teacher ratio		–0.03	0.02
Teacher training		0.00	–0.03
Teacher experience			0.04
Teachers' salaries			–0.07a)
Expenditure per pupil			0.20b)
School organizational factors:			
Productive climate culture	+		
Achievement pressure for basic subjects	+	0.02	0.14
Educational leadership	+	0.04	0.05
Monitoring/evaluation	+	0.00	0.15
Cooperation/consensus	+	–0.02	0.03
Parental involvement	+	0.08	0.13
Staff development	+		
High expectations	+	0.20	
Orderly climate	+	0.04	0.11
Instructional conditions:			
Opportunity to learn	+	0.15	0.09
Time on task/homework	+	0.00/–0.01 (n.s.)	0.19/0.06
Structured teaching	+	–0.01 (n.s.)	0.11 (n.s.)
Aspects of structured teaching:			
—cooperative learning			0.27
—feedback			0.48
—reinforcement			0.58
Differentiation/adaptive instruction			0.22

Numbers refer to correlations, the size of which might be interpreted as: 0.10: small; 0.30: medium; 0.50: large (cf. Cohen, 1988).
+: positive influence; n.s.: statistically not significant.
[a]Having assumed a standard deviation of $5000 for teacher salary.
[b]Assuming a standard deviation of $100 for PPE.

ties in establishing the impact of more distal organizational than more proximal educational conditions.

Another explanation could be the phenomenon that in economic terms is described as the diminishing rate of return. This phenomenon, which is essentially a non-linear relationship between an antecedent condition and the effect variable, has frequently been demonstrated with respect to the influence of "time on task". After an initial steep increase the curve flattens, and above certain levels a large amount of extra input is required to attain an ever-smaller increment on the effect variable. Perhaps this phenomenon could be occurring in educational systems in which most schools do fairly well in establishing basic organizational conditions for learning and instruction. In such a situation, although there may be considerable variation among schools in the organizational conditions, the impact on the effect variable could still be very small. Evidence that this may be a realistic interpretation of the situation in Western industrialized countries is provided in a Dutch study, in which it was established that basic supportive conditions for learning and instruction are present in a large majority of schools, and that there are only a few areas, most notably those related to differentiation and adaptive instruction, where there is still room for considerable improvement (Lam, 1996).

A final consideration is that these organizational conditions apparently make so much sense to educationalists, researchers and reviewers that they are repeatedly mentioned in different settings and in different countries. This fact itself is important enough not simply to disgard these factors, and to retain them on the research agenda as hypotheses that are still expected to receive empirical support, if hard to realize, rigorous research conditions could be implemented. Another less negative interpretation could be to treat these organizational conditions as examples of "educational good practice" on which educational connoisseurs in various countries apparently agree.

Moreover, it should not be forgotten that school effectiveness research deals with relative and not with absolute achievement levels of schools. A zero correlation between a presumed effectiveness-enhancing factor and achievement can therefore never be interpreted in terms of irrelevance of this factor for educational practice: if, for example, educational leadership is of a satisfactory level in all schools then it is possible that it will not be related to achievement. That does not imply, however, that a school leader can refrain from his or her activities.

At the same time, the lack of empirical support for organizational and managerial conditions should be taken seriously in the consideration of these factors in future effectiveness-oriented research, policy and practice. The message that Wang *et al.* (1993) have for educational policy can also be repeated in the light of our results. Stated in a positive sense, this message is to direct innovational policy at conditions that are closely related to the improvement of classroom-level conditions.

From a more theoretical perspective the lack of empirical support for the impact of organizational and managerial conditions enforces the image of schools as professional bureaucracies, where central management and overt coordination measures are expected to have only a modest influence on the work of relatively autonomous professionals.

Is a theoretical redirection of school effectiveness feasible?

The basic message from the exploration of theory-embedded principles in Chapter Eight was that there are several principles stemming from established theories that can be related to the knowledge base on school effectiveness.

As stated in Chapter Eight the cybernetic principle of evaluation–feedback–corrective action has a place in several of these theories. When it comes to reflecting on the empirical support for the relevance of this principle, it is clear from Table 9.1 that this general principle is close to the characteristics of structured instructional approaches, and as such has received considerable support. As far as school level characteristics are concerned, evaluation policies and practices appear to have some impact, as evident from meta-analyses.

The interpretation of evaluation and monitoring in schools given by Sammons *et al.* (1995b), cited in Chapter Five, draws attention to the fact that high "evaluative potential" (Scheerens, 1987) more or less presupposes the availability of other hypothetical key factors. An evaluation centered approach is close to the image of "schools as output driven organizations" (Coleman *et al.*, 1993). If one knows what to evaluate this implies that one is working in a goal-oriented manner. At the same time, school-based evaluation as a means to stimulate coordinated and concerted action forms a starting point for structured planning, and almost inevitably, will have motivationally relevant implications. The use of evaluation for school-based improvement is likely to be a starting point for staff development and organizational learning and can also be used as a basis to inform and engage parents. Finally, if schools have high evaluative potential, it is hard to imagine that this could be done without the active involvement of the headteacher. In fact, the initiation of evaluation and monitoring and the use of evaluative results comprise key elements of instructional leadership.

The strength of the evaluation mechanism is that it has potential for the four main ingredients of effective schooling that can be induced from analytical and empirical work, cited in earlier chapters: direction, a basis for learning at all levels, coordination, and the application of incentives.

Other theories are helpful to elucidate smaller subsets of these four basic ingredients. Rational planning theory is relevant to both direction and coordination. The work of Slavin (*Success for All*) and Stringfield (*High Reliability Organizations*) shows that structured planning approaches have more potential for effective schooling and school improvement than has been the dominant view for several decades of conceptualizing bottom-up educational innovation. Contingency theory adds situational dimensions to both the establishment of goals (direction) and the effective alignment of subsystems (coordination). Microeconomic theory in general, and public choice theory in particular, focusses on incentives (e.g. by rewarding good performance) that stimulate the task-oriented and achievement-oriented behavior of head teachers, teachers and pupils.

Each of these core ingredients has structural, more formal and cultural, less formal aspects. It is our contention that it is more helpful and practical to consider cultural aspects as consequences of structural arrangements rather than vice versa, the pragmatic reasoning behind this being that structural factors are more directly malleable than cultural aspects.

In our opinion the theoretically embedded principles discussed in Chapter Eight have sufficient potential to be considered as a basis for a theoretical redirection of school effectiveness research. In advance of a later section in which redirection of school effectiveness research will be the subject, such a redirection calls for more in-depth process study of the mechanisms in question, as follows.

- How are evaluations planned, executed and used?
- How do staff react to structural planning and "relentless" (Slavin) monitoring of pupils' performance?
- Which arrangements enhance task-oriented incentive structures while still leaving opportunity for professional freedom?
- Which coordination mechanisms are effective, given certain situational (including cultural) arrangements?

It is now time to return to the question raised in the title of this section: is school effectiveness, given the critical points raised when summarizing the substance of this book, a myth? Clearly, several of our findings call for a modest view on the degree to which schooling in general is malleable and, more precisely, on the degree to which, in well-established educational systems, one school can do significantly better than another. In addition, our results support the view that conditions close to the primary process of learning and instruction have more impact on performance than do more distal administrative and organizational factors. Knowledge on "what works" appears to be less generalizable, more local, partial and temporal than was assumed in earlier reviews (e.g. Walberg, 1984). Finally, favorable educational conditions may be due to selection effects rather than targeted educational arrangements and actions to a still larger extent than has been assumed in the field of school effectiveness.

Despite these limitations that point away from the school organizational level as an important handle for the improvement of effectiveness, we do not think that school effectiveness is a myth. Classroom-level conditions are nested within schools and there is no reason whatsoever to deny that creating as many effective classrooms as possible is a school-level assignment for educational management. Our findings call for an approach to school management that is strongly focussed on facilitating the primary process of learning and instruction, including providing direction to this process in the light of environmental requirements.

Redirection in school effectiveness modeling

How should one conceive of school effectiveness given the evidence based on the analyses, research reviews and theoretical reflection in the previous chapters? This question needs to be addressed with respect to implications for the conceptual models that are used to guide empirical research and also with respect to general research approach and methodology. In this section several aspects of conceptual school effectiveness modeling will be discussed. The relationship between school-level organizational/managerial measures and classroom-level instruction should be reconsidered in the light of current evidence and theoretical perspective. The question about possibilities for more theory driven research will be taken up again, and the set of key factors will have to be reconsidered.

The subsidiarity principle

In political science the concept of subsidiarity means that higher administrative levels should not direct activities that can be accomplished by lower administrative levels. This is closely related to the issue of autonomy of regional offices, school management and teachers. When applied to schools it can also be extended to include pupils. In traditional organizational models of the school, the fact that subunits, particularly teachers, have a lot of autonomy is strongly underlined. Examples include the school as a professional bureaucracy (Minzberg, 1979) and the school as a loosely coupled system (Weick, 1976). In constructivist perspectives on learning and instruction, the idea of self-regulated learning and independent learning is a key element. At the organizational level, initiatives in many countries to restructure educational organizations and even national educational systems have focussed on changing patterns of centralization and decentralization. One of the more practical ideas that organizational science has adapted from chaos theory is the concept of self-organization. The subsidiarity principle is useful in coming to grips with the issue of the degree of autonomy of lower level units in a hierarchical system, because it implies not only autonomy, but also minimal control, or facilitation from higher levels.

The lack of empirical support for the direct impact of higher level, school organizational factors on achievement and none for the indirect impact on the functioning of teachers and classrooms, supports the traditional picture of professional autonomy and loose coupling. Yet the relatively sparse evidence on the effectiveness of strongly structured school improvement programs indicates that well-targeted direction has a place in the effective organization of schools.

The subidiarity principle focusses the question concerning lower level autonomy and minimal higher level control on the *type* of higher level control measures that are required. In the light of the evidence and analysis presented in the previous chapters, selection, recruitment and output monitoring should be seen as key areas of higher level control and facilitation in educational organizations. The assumption is that these mechanisms provide sufficient direction and coordination to keep the organization on the right track, while still leaving a large amount of autonomy at the level of teachers. At the instructional level critical experiments would be required to determine optimal levels of structuring versus autonomy. As stated in Chapter Two, the evidence from instructional effectiveness studies appears to be in favor of considerable structuring, while constructivist approaches propagate more independence and less structure. A final implication of the subsidiarity principle is the recognition that most of the explanatory power of antecedent conditions of the performance of certain units (teachers, pupils) will be situated at the level of these units themselves and that higher level facilitation, even if well targeted and well implemented, will be of secondary importance.

Higher level facilitation revisited

The general idea of the facilitation of lower levels by higher organizational and administrative levels is central to the conceptual school effectiveness models discussed in Chapter Two. The research evidence and the above exposition on the subsidiarity principle cast considerable doubt on the power of the idea of higher level facilitation.

The results of our meta-analyses indicate that there is hardly any evidence for direct effects of organizational key factors on student achievement. Only a few studies have investigated indirect effects of school organizational factors, mediated by classroom-level factors (e.g. Hill *et al.*, 1995a; Hofman, 1995; Bosker & Scheerens, 1994; Hofman *et al.*, 1995; van der Werf, 1988). The results of these studies generally do not support higher level facilitation, in other words consistent indirect effects have not been convincingly demonstrated.

This general picture is further qualified when considering the various interpretations given to higher level facilitation presented in Chapter Two (contextual effects, higher levels mirroring and thus enforcing lower level conditions, overt measures, incentives, material conditions and buffering). It is our impression that most of the (in itself limited) set of studies presenting results on indirect effects of school organizational factors have investigated overt measures such as educational leadership or curriculum planning. Whenever the mirroring interpretation applies, and school-level and classroom-level factors are correlated, in the statistical analysis some impact of school factors may be swamped by the classroom-level factors, when models are being built stepwise, starting from the lower levels upwards.

More specific research into particular types of higher level facilitation is required to verify whether the general negative picture shown by available studies is justified. In some cases school organizational measures are more or less implied whenever classroom level conditions can be observed. For example, certain practices of monitoring achievement presuppose the availability of a (school-level) pupil-monitoring system.

Given the negligible direct effects of school organizational factors on pupil achievement, more research effort ought to be put into studies investigating indirect causation, guided by hypotheses built on the different types of higher level facilitation that have been described.

Reconsideration of key factors in view of theoretical redirection

Returning to the summary table on the various sources of knowledge on school effectiveness at the beginning of this section (Table 9.1), the discrepancy between the positive recognition of a set of organizational key factors in qualitative reviews on the one hand and the results of meta-analyses on the other is particularly troubling. We are inclined to take seriously the generally negative results of the meta-analyses, although the support for these factors in qualitative reviews must not be simply disgarded. As stated before, one attitude could be that most of our empirical studies have not been able to surmount the considerable methodological problems, and that — if only we could do better research — it could still be shown that they enhance effectiveness. Another perspective, outside the realm of school effectiveness thinking in the productivity sense of the concept, would be to keep these factors on the agenda of empirical studies and school improvement activities as valuable in themselves and as instances of good school organizational practice.

Given this problem of deciding which potentially relevant organizational factors deserve further investigation, the solution we wish to propose is to turn to those factors that are relevant in the light of available theory. Drawing on the analysis of theory-embedded principles presented in Chapter Eight, we will examine which factors are relevant from the perspective of two conceptualizations: (a) high-reliability organizations, and (b) the evaluation centered, output-driven school. The underlying principles of high-reliability organizations are considered to be synoptic rational planning and formal structuring. The evaluation-centered school unites the cybernetic principles and the motivational aspects of aligning individual and organizational rationality.

The characteristics of high-reliability organizations were cited in Chapter Eight (source: Stringfield *et al.*, 1995), some of the most telling principles being: "the notion that failures are disastrous", "clarity regarding goals", "use of standard operating procedures", "hierarchical structure" and "considerable attention to performance, evaluation and analysis to improve the organization's primary mission".

The image of high-reliability organizations focusses the attention on factors such as curriculum and lesson planning, monitoring and diagnosis. Although these factors have been investigated frequently, with the dubious results discussed earlier, perhaps they deserve to be reconsidered. The successes described by Stringfield *et al.* (1995) and Slavin (1996) shed some light on a more promising way to do so. Characteristic features of successful programs, described by these authors, such as the Barclay-Calvert project and Success for All appear to be a very strict or even extremely serious implementation of the key factors in question. For example, Slavin uses the term "relentless" in qualifying the concern for early diagnosis and immediate remediation of learning difficulties. As stated before in somewhat different terms: the failure to demonstrate the effects of school organizational factors that intuitively make a lot of sense may be caused by "the restriction of range" in the scale in which the organizational conditions are observed. A similar phenomenon has been noted with respect to the effects of resource inputs, e.g. class size. Experimentation with large differences in class size shows clear and consistent effects (Finn & Achilles, 1990). Findings from developing countries where differences on these factors are even more extreme also show significant input–output relationships. The most likely type of research to investigate less restricted manifestations of the key factors behind the image of high-reliability organizations would be the evaluation of experimental improvement programs.

The centrality of the cybernetic principle (evaluation-feedback-corrective action) in effectiveness-oriented theory has already been explained in earlier parts of this book. According to Scheerens (1992, p. 90), evaluation is to be seen as the key mechanism of effective schooling. In a previous section of this chapter the association was discussed of an evaluation-centered approach with other hypothetical effectiveness-enhancing conditions, such as goal orientation, coordination and educational leadership.

Evaluation can be seen as a precondition for all cognitive adaptation or learning and for all conative (or motivational) stimulation. In this latter interpretation evaluation has a place in microeconomic approaches to school functioning (cf. Correa, 1995). Most of these aspects are united in the concept of "schools as output-driven organizations" (Coleman *et al.*, 1993). He defines "output-driven" as:

> An organizational form in which the rewards and punishments for performance in productive activity come from the recipient of the product. Applied to intermediate products within the organization, this means that the recipient of the intermediate product has the right to monitor the quality of that product, and thus to determine the rewards and punishments for the part of the organization from which it receives intermediate products — and in turn the obligation to satisfy the requirements that its own products must meet — as monitored by the recipients of its products (p. 17).

Elements of an output-driven school, according to Coleman, are the following.

1. Externally imposed standards as the basis for all evaluations of student performance.
2. Evaluations based on two measures: *level* of performance, and performance *gain* or *value added.*
3. Yearly rewards to teachers, students and parents for level of performance and performance gain.
4. Using the final output criteria (the externally imposed standards) as the starting point for designing evaluations at each stage of the education of a child, creating a system with short feedback loops.
5. Allocation of rights and responsibilities not only to individuals, but also to groups of teachers, groups of students and groups of parents, to encourage the development of social capital, that is, informal norms that support educational goals.
6. The use of a core of academic achievement plus an area of specialized performance (which may be academic, but need not be) as the "performance criteria" (Coleman *et al.*, 1993, pp. 24, 25).

It should be noted that the image of an output-driven school recognizes leeway and autonomy for teachers and students, while at the same time standard setting, monitoring and incentive settings are presented as a strict coupling mechanism. This implies that the concept of output-driven schools is in line with the subsidiarity principle, described in a previous subsection.

The output-driven/evaluation-centered school has also much in common with high-reliability organizations, the extra element in the latter being the acceptance and propagation of proactive detailed planning and structuring.

Given these two organizational images and the underlying theory-bound principles, structured planning, evaluation and monitoring and use of performance-oriented incentives should be considered as the most interesting hypothetical key factors of effective schooling in which, incidentally, standard setting is included.

A final observation is that, in light of the disappointing results concerning the impact of direct organizational measures to facilitate classroom-level processes, recruitment and selection of teachers also deserves more attention as a mode of schooling with effectiveness-enhancing potential.

Redirection of School Effectiveness Research: The Future of State-of-the-Art School Effectiveness Research and the Need for Foundational Studies

Given the rather disappointing results with respect to the impact of school-level organizational factors, obtained from meta-analyses, the first question to be raised is whether we should continue to invest effort in the type of empirical study on which these meta-analyses were based. Of course, this "type of study" is a rather heterogeneous conglomerate of surveys, with many differences in design, choice of variables and analysis techniques (Bosker & Witziers, 1996), but even if we think of the five exemplary studies described in Chapter Five, and if we were to focus our question on such high-quality research, it remains a valid question.

It is our contention that, ideally, future state-of-the-art school effectiveness studies (cf. Scheerens, 1993) should be preceded by more foundational work, to settle important conceptual issues and improve research methods.

The work that has already been taken up by various researchers (Bosker, 1991; Luyten, 1994; Thomas, 1995; Luyten & Snijders, 1996) in searching for one-dimensionality versus differentiation in school effects, considering grade levels, subjects and teachers, deserves to be continued. In this way full-fledged school effectiveness studies could be better targeted and start from more realistic expectations about effect sizes and the impact of effectiveness-enhancing conditions. One example is the distinction between subjects like mathematics or Latin, which depend strongly on education at school, and other skills such as language and reading. Another foundational issue that is by no means new is the question regarding the general versus curriculum-specific nature of effect measures. As shown by Madaus *et al.* (1979), effects on curriculum-bound tests are likely to be stronger than on more general cognitive achievement tests. Thought should also be given to using criterion-referenced tests as a basis for school effectiveness studies. If these tests are constructed using Item-Response Theory, then it should be feasible to position schools at given points on the underlying dimension, representing the mean ability level of its students with characteristic items belonging to that part of the scale.

Selection effects, in terms of drop-out rates and the number of pupils who repeat classes, should also be considered in estimates of school effects and the impact of effectiveness-enhancing conditions (Bosker & Scheerens, 1989).

A different type of selection effect, namely the selection bias arising from the choice of schools of parents, also deserves further attention. As shown by Grisay (1995), complex interactions between favorable process conditions and schools attracting more able students, are beyond merely controlling for intake characteristics and the common methods of obtaining value-added measures. Analytical techniques should be further explored and developed to sort out these interactions. Differential effectiveness, in terms of possibly varying effectiveness-enhancing conditions for different age and ability groups, is also a foundational issue that deserves to be kept on the research agenda. With respect to achievement in secondary schools, an interesting and innovative development is the study of cross-classification, in which the effect is assessed of a pupil having visited a particular primary school on his or her secondary school achievement (Goldstein & Sammons, 1995).

Apart from advances and redirection in the more specialized area of statistical analysis of school effects and school effectiveness, additional work in the instrumentation of process indicators is needed. The divergence in current instruments and the lack of standardized and validated instruments is an enduring setback to future school effectiveness studies. Measurement approaches that use pupil observation and evaluations on the performance of teachers, and teachers' observation and appraisal of heads (Grisay, 1996; Hill *et al.*, 1995a, b), provide an interesting methodological alternative to the more traditional self-report procedures that are usually employed. Moreover, since multiple respondents are used to rate the same aspect of, for instance, teaching behavior, the aggregated measure will be far more reliable than when using one self-report by a teacher. Process studies to investigate phenomena such as selection, recruitment, matching pupils and teachers, the way in which evaluations are carried out and used, and human resource development activities should also be seen as a most relevant type of foundational, effectiveness-oriented study.

The often-heard plea for more longitudinal research in school effectiveness can only be repeated here. Not only effects should be measured at more than one point in time, but also input and process variables.

One conclusion is that there is a lot of foundational work to be done in order to improve the relevance of future full-scale school effectiveness studies. A final point is that such studies have much more value then just an instrumental use for further school effectiveness studies. Each of these types of foundational study provides information on the nature of schooling, even though process–outcome relationships are not directly addressed.

Experimentation and the maximization of contrast

The majority of school effectiveness studies consist of surveys that depend on naturally occurring variation, in order to be able to detect differences between schools. This is a hazardous affair when one considers the unreliability in discrimination between schools (cf. Goldstein, 1996) on the basis of their value-added effects and restrictions in the variability of school process variables that are supposed to explain the differences in effects. Considering the relatively small size of school effects, samples of schools tend to be too small to detect significant effects.

There are several ways in which this small effects syndrome can be countered: the use of outlier designs, experimentation and a rigorous implementation of treatments. It should be noted that this "artificial" way of enlarging effects is in fact a normal procedure in scientific research, although to those used to field studies it may be looked upon as a violation of ecological validity. The successes of studies on classroom management and instructional procedures, for example those summarized by Slavin (1996), may be partly due to the (quasi-)experimental nature of the designs used for these studies as compared to survey research. When talking about maximization of contrast, two special target groups of pupils come to mind, namely those most likely to show progress: young children and disadvantaged and/or at-risk pupils.

Change in outliers

Studying differences between exceptionally effective and exceptionally ineffective schools has a long tradition in school effectiveness research, whereas investigating the process by which ineffective schools can become effective and effective schools can deteriorate into ineffective schools is an approach that is relatively new. Although it has been recommended by various authors (Slater & Teddlie, 1991; Reynolds *et al.*, 1994), very few examples of such studies exist.

The traditional approach of studying naturally occurring processes could be replaced by active and programmed ways to improve outlying schools on the negative pole, thus enriching the tradition of outlier studies by quasi-experimental approaches.

Experimental school improvement programs

Descriptive surveys on school effectiveness are expensive projects, particularly when large-scale assessments of outcomes and longitudinal designs are incorporated. From the position of educational practitioners it would probably be considered more attractive to have an action orientation in such studies. Such an action orientation would minimally consist of a careful and thorough feedback of research results to schools. A more involved approach would be the introduction of programs to improve educational outcomes. The gearing of school effectiveness research to school improvement in this case would comprise (a) design and implementation of a school improvement program, based on the school effectiveness knowledge base, and (b) a rigorous evaluation of the experimental improvement program. Several high-quality examples of this approach are now available (Slavin, 1996; Houtveen & Osinga, 1995). A drawback of comprehensive school improvement programs in the sense of furthering knowledge on school effectiveness is that it is usually impossible to attribute results to particular school-organizational or instructional factors.

Therefore, more targeted and narrow experimentation is also to be recommended, in which ideally only one factor or one interconnected cluster of factors (e.g. the evaluation–feedback–reinforcement cycle) should be manipulated.

An important requirement for either comprehensive or more narrow experimental improvement projects is a rigorous and closely monitored implementation of the treatment. Only in this way could the restriction of range in the independent variables of school effectiveness studies be resolved.

Research on class size can be used as an illustrative example. Most meta-analyses indicate that a reduction in class size has no effect or a very limited effect. However, an extensive and longitudinal study by Achilles *et al.* (1993) showed that substantive and enduring effects were found when class sizes of approximately 24 students per class were compared to sizes of approximately 15 students per class. A similar phenomenon might be observed when there are important differences across experimental conditions on factors such as achievement-oriented leadership and evaluating feedback or corrective action.

Other aspects of experimentation

Two aspects that might blur the evidence from (quasi-)experiments are the Hawthorne effect and the usual practice of not considering variations in the costs of improvement.

Hawthorne effects, i.e. the phenomenon that being chosen to take part in a study has influence on the participants, next to the real effect of the actual contents of the program, are hard to avoid in (quasi-)experimental studies. Within the context of relatively large-scale school improvement programs, placebos as used in experimental laboratory settings are inconceivable. The use of longitudinal effect studies, years after the experimental program was implemented, could shed some light on whether effects measured directly after program implementation are real effects or artifacts of the experimental situation.

If a methodological reorientation of school effectiveness research did, as we recommend, turn in the direction of experimentation, cost aspects of treatments would deserve more consideration. In fact, such a consideration of costs would move school effectiveness research in the direction of school *efficiency* research.

In this context, the ecological validity of experimental effectiveness studies would require that the costs of treatments—including development, implementation guidance and dissemination costs—should be taken into account to determine the feasibility among a wider range of schools.

If, for instance, a program focussing on staff development is compared to a program where head teachers receive further training in educational leadership, the costs of the two treatments should be compared in order to assess the value for money of each approach.

In summary, the main recommendation expressed in this subsection is to seek research designs in which contrast and differences between conditions are enlarged. This can be achieved either by capitalizing on differences in the outcome variables, as in outlier studies, or by ensuring that there is sufficient variation in the independent variables. In evaluation terminology the diagnosis with respect to the rather disappointing results in many school effectiveness studies could be phrased in terms of non-event evaluation, and in methodological terminology we could speak of this phenomenon as a type of restriction of range in the process conditions of schooling.

Applied, assessment-oriented research and school effectiveness

Given the expensiveness and organizational complexity of both state-of-the-art school effectiveness studies and experimental improvement programs, a third major category of research activity should be considered for its relevance to furthering our knowledge in this area. This category of research activity can be referred to in terms of "applied school effectiveness inquiry" (Scheerens, 1993).

Partly as an aspect of a growing interest in accountability, also a part of decentralization policies (van Amelsvoort *et al.*, 1995), there is growing interest among governments and regional authorities in outcome assessment and monitoring. Particularly in assessment programs with a longitudinal design, e.g. cohort studies, additional data collection on input and process conditions can yield interesting possibilities to answer effectiveness-related questions. A limitation of such studies is the fact that written questionnaires are usually the only possible method to obtain data on process indicators. Some of the methodological advances that could result from the more foundational studies, as discussed previously, might be applied to overcome these limitations as much as possible. An example could be to gather data on a particular level of school functioning, e.g. school management, at the level below; thus, for example, in the case of school management by relying on teacher descriptions and appraisals. Another example is enhanced validity and reliability of outcome measurements, by using item response modeling and correcting adjustment variables for attenuation.

A second promising applied area is the growing practice of school self-evaluation. Particularly when a pupil-monitoring system is part of overall school self-evaluation, there are interesting possibilities for analyzing school-level process–output associations (Bosker & Scheerens, 1994). There is a kind of circular dependence of school self-evaluation and answering effectiveness-oriented research questions. First, school self-evaluation instruments may be developed on the basis of the school effectiveness knowledge base, by selecting those input and process variables that are expected to work. Next, the information gathered by means of school self-evaluation instruments based on this developmental rationale could be used to further this knowledge base, as a side-product of the practical use that is of primary importance.

Statistical Advances

School effectiveness studies have advanced statistical modeling in educational research in particular but also in social, psychological, biological and medical research. The wide general class of multilevel models, also referred to as parameter-varying models, hierarchical linear models, variance component models or random coefficient models, has led to a better handling of data generated in a multilevel structure. With software widely available (e.g. MLn, HLM, VARCL, Genmod, the SAS procedure Mixed) and advances being made in the area of logistic models, survival event history models, repeated measurements, time series, multivariate models, measurement error models, etc., almost all imaginable phenomena can be handled for statistical analysis. This will lead to a better understanding of educational phenomena, since nested structures (pupils within classrooms within schools) are amenable to statistical modeling in a variety of situations. In this area school effectiveness research has advanced educational research at large. It is now time that productivity-oriented educational scientists all use these tools, for the reasons that were unfolded in Chapter Two, the argument being that working with models that do not adequately represent this hierarchical structure "is dangerous at best, and disastrous at worst" (Aitkin & Longford, 1986, p. 42). In retrospect it might even be concluded that we might as well reconsider the knowledge base of educational science at large, as we did in this book for the area of school effectiveness only.

The next wave of statistical innovations, recently completed, may help us to build structural multilevel educational effect models with corrections for unreliability in input, process and output variables. A breakthrough in this area is the work of Muthén (1994), who succeeded in partitioning within- and between-group variance–covariance matrices. With some special tricks one can then use ordinary LISREL, AMOS or EQS models to estimate these complex models. An important tool in this respect is the preprocessor software STREAMS produced by Gustafsson and Stahl (1996). Another area of statistical advancements comes from Gibbs sampling (which does not use analytical solutions, but sampling from constrained distributions to estimate parameters), which might also lead towards multilevel structural and measurement models (Spiegelhalter *et al.*, 1994). The main advantage of these new statistical models is that we are in a good position to model unreliably measured school conditions as they may have indirect effects only (see Chapter Two).

Although these techniques may be helpful in unraveling educational phenomena, they can only do so to the extent that theories are only strong enough to indicate possible relationships, the way that these are moderated, and their form. In that respect it is very helpful to formulate theoretical models more precisely in mathematical notation, so as to refine the central propositions of such a theory, and to study their consistency and consequences using computer simulation models (e.g. Clauset & Gaynor, 1982; Bosker & Guldemond, 1994; de Vos & Bosker, forthcoming). Moreover, more psychometric work needs to be done in constructing valid and reliable measures of process variables in school effectiveness research, if possible starting "one level below". With this latter recommendation, more reliable measures of, for instance, teaching behavior can be obtained by using multiple raters, e.g. students taught by that teacher.

Conclusions: The Viability of School Effectiveness Research

This book contains evidence on the conceptualization, knowledge base and theoretical interpretation of school effectiveness research results that is sometimes critical and disturbing. At one point in the presentation we even had to address the question of whether school effectiveness is supposed to be a myth.

The conceptual analysis showed that school effectiveness is a more complex phenomenon than has usually been realized and that our knowledge is of a rather partial and limited nature. Inconsistency in the results might also to some extent be attributed to the divergence in operationalization of the process variables in empirical effectiveness studies.

Another important conclusion is the discrepancy between more qualitative reviews on the one hand and meta-analyses on the other, with respect to school-level conditions that are expected to enhance effectiveness. The overall message of these analyses appears to be that classroom conditions have more impact than school organizational conditions in improving outcomes.

In the face of these difficult points our overall contention has *not* been that school effectiveness is an unfruitful approach to furthering our knowledge on schooling and helping to design school improvement. Instead, a redirection of school effectiveness research is required, in which foundational work, a more theory-driven approach and, where possible, experimental methodology are the core elements.

In this book the emphasis has been on conditions of schooling at the school organization level. An important issue in classroom-level instruction concerns the balance between structured and more open instructional arrangements. Although this crucial issue was only touched upon, it deserves more attention in effectiveness-oriented research at the instructional level. From our perspective meso-level or school-level implications of the redirection of this balance between structure and freedom form an important consideration (see Chapter Three).

The viability of school effectiveness research also depends on major developments in education, such as the growing use of modern information and communication technology and a reorientation on educational objectives, for example by giving more emphasis to non-cognitive goals and general cognitive skills as compared to subject-matter-related skills. It is our belief that the instrumental questions of schooling and instruction in a comprehensive approach to educational effectiveness lose none of their relevance when these types of reorientation take place. The question, "does it work?" and the more interesting question, "why does it work?" are likely to remain basic questions on the educational research agenda.

REFERENCES

Achilles, C. M., Nye, B. A., Zacharias, J. B., & Fulton, B. D. (1993). Creating succesful schools for all children: A proven step. *Journal of School Leadership,* **3,** 606–621.

Aitkin, M., & Longford, N. (1986). Statistical modelling issues in school effectiveness studies. *Journal of the Royal Statistical Society, Series A (General)*, **149**(Part 1), 1–43.

Aitkin, M., & Zuzovsky, R. (1994). Multilevel interaction models and their use in the analysis of large-scale school effectiveness studies. *School Effectiveness and School Improvement,* **5**, 45–74.

Aldridge, B. G. (1983). A mathematical model for mastery learning. *Journal of Research in Science Teaching*, **20,** 1–17.

Alexander, K. L., & Eckland, B. K. (1980). The "explorations in quality of opportunity" sample of 1955 high school sophomores. In A. C. Kerhoff (Ed.), *Research in sociology of education and socialization*, Vol. I, *Longitudinal perspectives on educational attainment*. Greenwich, CT: JAI Press.

Amelsvoort, H. W. C. H. van, Scheerens, J., & Branderhorst, E. M. (1995). *Decentralization in education in an international perspective*. Enschede: University of Twente, Faculty of Educational Science and Technology.

Anderson, C. S. (1982). The search for school climate: A review of the research. *Review of Educational Research*, **52**(3), 368–420.

Argyris, C. (1982). *Reasoning, learning and action*. San Francisco, CA: Jossey-Bass.

Averch, H. A., Carroll, S. J., Donaldson, T. S., Kiesling, H. J., & Pincus, J. (1974). *How effective is schooling? A critical review of research*. Englewood Cliffs, NJ: Educational Technology Publications.

Bangert, R. L., Kulik, J. A., & Kulik, C. C. (1983). Individualized systems of instruction in secondary schools. *Review of Educational Research*, **53**, 143–158.

Barley, S. R. (1990). The alignment of technology and structure through roles and networks. *Administrative Science Quarterly*, **35,** 61–103.

Batenburg, Th. A. van (1990). Variatie in schoolgemiddelden op de CITO-eindtoets basisonderwijs. *Tijdschrift voor Onderwijsresearch,* **15**, 362–369.

Bereiter, C., & Kurland, M. (1982). A constructive look at follow through results. *Interchange*, **12,** 1–22.

Berliner, D. C. (1985). Effective classroom teaching: The necessary but not sufficient condition for developing exemplary schools. In G. R. Austin & H. Garber (Eds), *Research on exemplary schools*, pp. 127–154. Orlando, FL: Academic Press.

Block, J. H., & Burns, R. B. (1970). Mastery learning. *Review of Research in Education*, **4,** 3–49.

Blok, H., & Hoeksma, J. B. (1993). De stabiliteit van het schooleffect in de tijd: Een analyse op basis van vijf jarr Eindtoets Basisonderwijs van het Cito. *Tijdschrift voor Onderwijsresearch*, **18**, 331–342.

Bloom, B. S. (1968). *Learning for mastery*. Washington, DC: ERIC.

Boekaerts, M. (1991). *Gedragsverandering en Onderwijs*. Rede Rijks Universiteit Leiden.

Boekaerts, M., & Simons, P. R. J. (1993). *Leren en instructie. Psychologie van de Leerling en het Leerproces*. Assen: Dekker & Van de Vegt.

Boorsma, P. B., & Nijzink, J. P. (1984). *Doelmatigheidsprikkels en hoger onderwijsbeleid*. Enschede: Technische Hogeschool Twente.

Borger, J. B., Ching-Lung Lo, Sung-Sam-Oh, & Walberg, H. J. (1984). Effective schools: A quantitative synthesis of constructs. *Journal of Classroom Interaction*, **20,** 12–17.

Bosker, R. J. (1990a). Theory development in school effectiveness research: In search for stability of effects. In P. van den Eeden, J. Hox & J. Hauer (Eds), *Theory and model in multilevel research: Convergence or divergence?* Amsterdam: SISWO.

Bosker, R. J. (1990b). *Extra kansen dankzij de school?* (Dissertation). Nijmegen: ITS/OoMO.

Bosker, R. J. (1991). De consistentie van schooleffecten in het basisonderwijs. *Tijdschrift voor Onderwijsresearch,* **16**, 206–218.

Bosker, R. J. (1995). *De stabiliteit van Mattheus-effecten*. Paper presented at the annual conference of the Dutch Association for Educational Research (Onderwijs-researchdagen), Groningen, June.

Bosker, R. J., & Dekkers, P. J. M. (1994a). School differences in producing gender-related subject choices. *School Effectiveness and School Improvement,* **5**, 178–195.

Bosker, R. J., & Dekkers, P. J. M. (1994b). School- en sekseverschillen in vakkenkeuzes in het VWO. *Tijdschrift voor Onderwijsresearch,* **19**, 214–226.

Bosker, R. J., & Guldemond, H. G. (1990). The interdependency of performance indicators. An empirical study in a categorical school system. In S. W. Raudenbush & J. D. Willms (Eds), *Pupils, classrooms, and schools: International studies of schooling from a multilevel perspective*. New York: Academic Press.

Bosker, R. J., & Guldemond, H. G. (1994). *A hierarchical simulation model to study educational interventions*. Enschede/Groningen: OCTO/RION.

Bosker, R. J., Guldemond, H. G., Hofman, R. H., & Hofman, W. H. A. (1989a). *Kwaliteit in het voortgezet onderwijs*. Groningen: RION.

Bosker, R. J., Guldemond, H. G., Hofman, R. H., & Hofman, W. H. A. (1989b). De stabiliteit van schoolkwaliteit. In J. Scheerens & J. C. Verhoeven (Eds), *Schoolorganisatie, beleid en onderwijskwaliteit*. Lisse: Swets & Zeitlinger.

Bosker, R.J. & Hofman, W. H. A. (1994). School effects on drop out: a multi-level logistic approach to assessing school-level correlates of drop out of ethnic minorities. *Tijdschrift voor Onderwijsresearch,* **19**, 50–64.

Bosker, R. J., Kremers, E. J. J., & Lugthart, E. (1990). School and instruction effects on mathematics achievement. *School Effectiveness and School Improvement*, **1**, 233–248.

Bosker, R. J., & Scheerens, J. (1989). Issues in the interpretation of the effects of school effectiveness research. *International Journal of Educational Research*, **13**, 741–751.

Bosker, R. J., & Scheerens, J. (1994). Alternative models of school effectiveness put to the test. In R. J. Bosker, B. P. M. Creemers & J. Scheerens (Eds), Conceptual and methodological advances in educational effectiveness research [Special issue]. *International Journal of Educational Research*, **21**, 159–180.

Bosker, R. J., & Scheerens, J. (1995). A self-evaluation procedure for schools using multilevel modelling. *Tijdschrift voor Onderwijsresearch*, **20**(2), 154–164.

Bosker, R. J., & Velden, R. K. W. van der (1989a). The effects of schools on the educational career of disadvantaged pupils. In B. P. M. Creemers & B. Reynolds (Eds), *Proceedings of the 2nd International Congress for School effectiveness*, Rotterdam.

Bosker, R. J., & Velden, R. K. W. van der, (1989b). Schooleffecten en rendementen. In J. Dronkers & J. van Damme (Eds), *Loopbanen doorheen het onderwijs*. Lisse: Swets & Zeitlinger.

Bosker, R. J., Velden, R., van der, & Loo, P. van de, (1996). *Do colleges have an effect on the labour market success of their graduates?* Paper presented at AERA annual meeting, New York.

Bosker, R. J., & Witziers, B. (1995). *School effects, problems, solutions and a meta-analysis*. Paper presented at the International Congress for School Effectiveness and School Improvement, Leeuwarden, the Netherlands, January.

Bosker, R. J., & Witziers, B. (1996). *The magnitude of school effects. Or: Does it really matter which school a student attends?* Paper presented at AERA Annual meeting, New York.

Bosker, R.J., & Witziers, B. (forthcoming). *The magnitude of school effects.*

Brandsma, H. P. (1993). *Basisschoolkenmerken en de kwaliteit van het onderwijs*. Groningen: RION.

Brandsma, H. P., & Doolaard, S. (1996). The effects of between school differences in effectiveness on advice for secondary education for individual pupils. *School Effectiveness and School Improvement* (forthcoming).

Brandsma, H. P., & Knuver, J. W. M. (1988). Organisatorische verschillen tussen basisscholen en hun effect op leerlingprestaties. *Tijdschrift voor Onderwijsresearch*, **13**, 201–212.

Broekstra, G. (1992). Chaossystemen als metafoor voor zelfvernieuwing van organisaties. In C. van Dijkum & D. de Tombe (Eds), *Gamma chaos: Onzekerheden en orde in de menswetenschappen*. Bloemendaal: Aramith.

Brookover, W. B., Beady, C., Flood, P., Schweitzer, J., & Wisenbaker, J. (1979). *School social systems and student achievement–Schools can make a difference*. New York: Praeger.

Brookover, W. B., & Lezotte, L.W. (1979). *Changes in school characteristics coincident with changes in student achievment.* East Lansing: Institute for Research on teaching, Michigan State University. (ERIC Document Reproduction No. ED 181005.)

Brophy, J. (1987). On motivating students. In D. Berliner & B. Rosenshine (Eds), *Talks to teachers*. New York: Randon House.

Brophy, J. (1996). *Classroom management as socializing students into clearly articulated roles*. Paper presented at AERA annual meeting, New York..

Brophy, J., & Good, Th. L. (1986). Teacher behaviour and student achievement. In M. C. Wittrock (Ed.), *Handbook of research on teaching*, pp. 328–375. New York: Macmillan.

Brown, B. W., & Saks, D. H. (1986). Measuring the effects of instructional time on student learning: Evidence from the beginning teacher evaluation study. *American Journal of Education,* **94,** 480–500.

Brown, S., Duffield, J., & Riddell, S. (1995). School effectiveness research: The policy makers' tool for school improvement. *EERA Bulletin*, **1**(March), 6–15.

Bruner, J. S. (1966). *Towards a theory of instruction*. Cambridge, MA: Belknap Press of Harvard University.

Bryk, A. S., & Raudenbush, S. W. (1992). *Hierarchical linear models*. New York: Sage.

California State Department of Education (1980). *Report on the special studies of selected ECE schools with increasing and decreasing reading scores.* Sacramento, CA: Office of Program Evaluation and Research.

Cameron, K. S. (1984). The effectiveness of ineffectiveness. *Research in Organizational Behavior,* **16,** 235–285.

Cameron, K. S., & Whetten, D. A. (Eds) (1983). *Organizational effectiveness: A comparison of multiple models*. New York: Academic Press.

Cameron, K. S., & Whetten, D. A. (1985). Administrative effectiveness in higher education. *Review of Higher Education*, **9,** 35–49.

Campbell, D. T. (1969). Reforms as experiments. *American Psychologist*, **24**(4).

Carroll, J. B. (1963). A model of school learning. *Teachers College Record*, **64,** 722–733.

Carroll, J. B. (1989). The Carroll Model, a 25-year retrospective and prospective view. *Educational Researcher*, **18,** 26–31.

Chandler, A. (1962). *Strategy and structure: Chapters in the history of industrial enterprise*. Cambridge, MA: MMT.

Cheng, Y. C. (1993). *Conceptualization and measurement of school effectiveness: An organizational perspective.* Paper presented at AERA annual meeting, Atlanta, GA.

Chubb, J. E., & Moe, T. M. (1990). *Politics, markets and American schools*. Washington, DC: Brookings Institute.

Clauset, K. H., & Gaynor, A. K. (1982). A systems perspective on effective schools. *Educational Leadership*, **40**(3), 54–59.

Cohen, D. K. (1988). Teaching practice ... Plus ça change ... In Ph. Jackson (Ed.), *Contributing to educational change: Perspectives on research and practice*. Berkeley, CA: McCutchan.

Cohen, J. (1969). *Statistical power analysis for the behavioral sciences* (2nd edn.). Hillsdale, NJ: Lawrence Erlbaum Associates.

Cohen, M. (1982). Effective schools: Accumulating research findings. *American Education*, January–February, 13–16.

Cohen, M. D., March, J. G., & Olsen, J. P. (1972). A garbage can model of organizational choice. *Administrative Science Quarterly*, **17,** 1–25.

Coleman, J. S. (1990). *Equality and achievement in education*. Boulder, CO: Westview Press.

Coleman, J. S., Campbell, E. Q., Hobson, C. F., McPartland, J., Mood, A. M., Weifeld, F. D., & York, R. L. (1966). *Equality of educational opportunity*. Washington, DC: U.S. Government Printing Office.

Coleman, J. S., Hoffer, T., & Kilgore, S. (1981). *Public and private schools*. Chicago, IL: National Opinion Research Center, University of Chicago.

Coleman, P., Collinge, J., & Seitert, T. (1993). Seeking the levers of change: Participant attitudes and school improvement. *School Effectiveness Improvement*, **4**, 59–83.

Coleman, P., & Laroque, L. (1990). *Struggling to be "good enough", administrative practices and school district ethos*. London: Falmer Press.

Collins, A., Brown, J. S., & Newman, S. E. (1988). Cognitive apprenticeship: Teaching the craft of reading, writing and mathematics. In L. B. Resnick (Ed.), *Cognition and instruction: Issues and agendas*. Hillsdale, NJ: Lawrence Erlbaum Associates.

Collins, A., & Stevens, A. (1982). Goals and strategies of inquiry teachers. In R. Glaser (Ed.), *Advances in instructional psychology*, Vol. II. Hillsdale, NJ: Lawrence Erlbaum Associates.

Cooper, H., & Hedges, L. V. (Eds) (1994). *The handbook of research synthesis*. New York: Russell Sage Foundation.

Corcoran, Th. B. (1985). Effective secondary schools. In A. M. J. Kyle (Ed.), *Reaching for excellence: An effective schools sourcebook*. Washington, DC: U.S. Government Printing Office.

Correa, H. (1995). *The microeconomic theory of education* [Special issue]. *International Journal of Educational Research*, **23**(5).

Corte, E. De, & Lowyck, J. (1983). Heroriëntatie in het onderzoek van het onderwijzen. *Tijdschrift voor Onderwijsresearch*, **8**, 242–261.

Cotton, K. (1995). *Effective schooling practices: A research synthesis*. 1995 Update. School Improvement Research Series. Northwest Regional Educational Laboratory.

Creemers, B. P. M. (1991). *Effectieve instructie. een empirische bijdrage aan de verbetering van het onderwijs in de klas*. Den Haag: SVO Balansreeks.

Creemers, B. P. M. (1994). *The effective classroom*. London: Cassell.

Creemers, B. P. M. (1995). Process indicators on school functioning and the generalisability of school factor models across countries. In *Measuring the quality of schools*, pp. 105–119. Paris: OECD.

Creemers, B. P. M. (1996). *Common elements in theories about effective instruction*. Groningen: unpublished manuscript.

Cunningham, D. J. (1991). In defense of extremism. *Educational Technology*, **31**(9), 26–27.

Cuttance, P. (1987). *Modelling variation in the effectiveness of schooling*. Edinburgh: CES.

Cyert, R. M., & March, J. G. (1963). *A behavioral theory of the firm*. Englewood Cliffs, NJ: Prentice-Hall.

Davies, J. K. (1972). Style and effectiveness in education and training: A model for organizing, teaching and learning. *Instructional Science*, **1**, 45–88.

Dijkstra, A. B. (1992). *De religieuze factor. Onderwijskansen en godsdienst: Een vergelijkend onderzoek naar gereformeerd-vrijgemaakte scholen*. Nijmegen: ITS.

Doolaard, S. (1996). *Changes in characteristics and effects of schoolleadership over time*. Paper presented at ECER, Sevilla.

Dougherty, K. (1981). After the fall: Research on school effects since the Coleman Report. *Harvard Educational Review*, **51**, 301–308.

Doyle, W. (1985). Effective secondary classroom practices. In M. J. Kyle (Ed.), *Reaching for excellence: An effective schools sourcebook*. Washington, DC: U.S. Government Printing Office.

Drazin, R., & Ven, A. H. van de (1985). Alternative forms of fit in contingency theory. *Administrative Science Quarterly*, **30**, 514–539.

Dronkers, J. (1978). Manipuleerbare variabelen in de schoolloopbaan: Een toepassing van het Wisconsin-model op het Nederlandse primaire en secundaire onderwijs. In J. L. Peschar, & W. Ultee (Eds), *Sociale stratificatie. Boeknummer mens en maatschappij*. Deventer: Van Loghum Slaterus.

Dror, Y. (1968). *Public policy-making reexamined*. San Francisco, CA: Chandler.

Duffy, Th. M., & Jonassen, D. H. (1992). *Constructivism and the technology of instruction: A conversation*. Hillsdale, NJ: Lawrence Erlbaum Associates.

Edmonds, R. R. (1979). *A discusssion of the literature and issues related to effective schooling*. Cambridge, MA: Center for Urban Studies, Harvard Graduate School of Education.

Elberts, R. W., & Stone, J. A. (1988). Student achievement in public schools: Do principals make a difference? *Economical Education Review,* **7,** 291–299.

Emery, F. E., & Trist, E. L. (1965). The causal texture of organizational environments. *Human Relations*, **18,** 21–32.

Erbring, L., & Young, A. A. (1979). Individual and social structure: Contextual effects as endogeneous feedback. *Sociological Methods and Research*, **7**, 396–430.

Essink, L. J. B., & Visscher, A. J. (1987). The design and impact of management information systems in educational organizations. *Journal of Information Resources Management*, **1,** 23–51.

Etzioni, A. (1964). *Modern organizations*. Englewood Cliffs, NJ: Prentice-Hall.

Faerman, S. R., & Quinn, R. E. (1985). Effectiveness: The perspective from organization theory. *Review of Higher Education*, **9,** 83–100.

Finn, J. D., & Achilles, C. M. (1990). Answers and questions about class size: A statewide experiment. *American Educational Research Journal,* **27**(3), 557–577.

Fitz-Gibbon, C. T. (1992). School effects at a level: Genesis of an information system? In D. Reynolds & P. Cuttance (Eds), *School effectiveness: Research, policy and practice*. London: Cassell.

Fraser, B. J., Walberg, H. J., Welch, W. W., & Hattie, J. A. (1987). *Syntheses of educational productivity research* [Special issue]. *International Journal of Educational Research*, **11**(2).

Friebel, A. J. J. M. (1994). *Planning van Onderwijs en het Gebruik van Planningsdocumenten: Doet dat ertoe?* Oldenzaal: Dinkeldruk.

Fuller, B., Wood, K., Rapoport, T., & Dornbusch, S. M. (1982). The organizational context of individual efficacy. *Review of Educational Research*, **52**, 7–30.

Gage, N. (1965). Desirable behaviors of teachers. *Urban Education*, **1,** 85–95.

Gamoran, A. (1991). Schooling and achievement: Additive versus interactive models. In S. W. Raudenbush & J. D. Willms (Eds), *Schools, classrooms and pupils*. San Diego, CA: Academic Press.

Geurts, J. L. A. (1983). Sociale planning, systeemdenken en simulatie. *Simulatie en Sociale Systemen*, **3**(2/3), 31–54.

Glasman, N. S., & Biniaminov, J. (1981). Input–output analyses of schools. *Review of Educational Research*, **51**, 509–539.

Glass, G. V., Cahen, L. S., Smith, M. L., & Filby N. N. (1982). *School class size*. London: Sage.

Glass, G. V., & Stanley, J. C. (1970). *Statistical methods in education and psychology*. Englewood Cliffs, NJ: Prentice-Hall.

Glenn, B. C. (1981). *What works? An examination of effective schools for poor black children.* Cambridge, MA.: Center for Law and Education, Harvard University.

Goldstein, H. (1987). *Multilevel models in educational and social research*. London: Charles Griffin & Co.

Goldstein, H. (1995). *Multilevel statistical models*. London: Edward Arnold.

Goldstein, H. (1996). *Methodological aspects of school effectiveness research*. Unpublished manuscript. London: Institute of Education.

Goldstein, H., & Sammons, P. (1995). *The influence of secondary and junior schools on sixteen year examination performance: A cross-classified multilevel analysis*. London: Institute of Education.

Goldstein, H., & Thomas, S. M. (1996). *School effectiveness and "value added" analysis*. Manuscript. London: Institute of Education.

Good, T. L., & Brophy, J. E. (1986). School effects. In M. C. Wittrock (Ed.), *Handbook of research on teaching*, pp. 328–375. New York: Macmillan.

Gooren, W. A. J. (1989). Kwetsbare en weerbare scholen en het welbevinden van de leraar. In J. Scheerens & J. C. Verhoeven (Eds), *Schoolorganisatie, beleid en onderwijskwaliteit*. Lisse: Swets & Zeitlinger.

Gray, J., Jesson, D., Goldstein, H., Hedger, K., & Rasbash, J. (1995). A multi-level analysis of school improvement: Changes in schools' performance over time. *School Effectiveness and School Improvement*, **6**, 97–114.

Gray, J., Jesson, D., & Jones, B. (1986). The research for a fairer way of comparing schools examination results. *Research Papers in Education*, **1**(2), 91–122.

Gresov, C. (1989). Exploring fit and misfit with multiple contingencies. *Administrative Science Quarterly*, **34,** 431–453.

Griffith, D. E., Hart, A. W., & Blair, B. G. (1991). Still another approach to administrate: Chaos theory. *Educational Administration Quarterly*, **17,** 430–451.

Grift, W. van de (1987). *De rol van de schoolleider bij onderwijsvernieuwingen*. 's-Gravenhage: VUGA Uitgeverij B.V.

Grisay, A. (1996). *Evolution des acquis cognitifs et socio-affectifs des eleves au cours des annees de college*. Liège: Université de Liège.

Gustafsson, J. E., & Stahl, P. A. (1996). *Streams user's guide. Version 1.6 for Windows*. Mölndal: Göteborg University.

Hallinger, P., & Murphy, J. (1986). The social context of effective schools. *American Journal of Education*, **94,** 328–355.

Hanushek, E. A. (1979). Conceptual and empirical issues in the estimation of educational production functions. *Journal of Human Resources*, **14,** 351–388.

Hanushek, E. A. (1986). The economics of schooling: Production and efficiency in public schools. *Journal of Economic Literature*, **24,** 1141–1177.

Hargreaves, D. H. (1995). School culture, school effectiveness and school improvement. *School Effectiveness and School Improvement*, **6**, 23–46.

Hargreaves, D. H., & Hopkins, D. (1991). *The empowered school*. London: Cassell.

Hauser, R. M. (1974). Contextual analysis revisited. *Sociological Methods and Research*, **2,** 365–375.

Hauser, R. M., Sewell, W. H., & Alwin, D. F. (1976). High school effects on achievement. In W. H. Sewell, R. M. Hauser, & D. L. Featherman (Eds), *Schooling and achievement in American society*. New York: Academic Press.

Haywood, H. C. (1982). Compensatory education. *Peabody Journal of Education*, **59,** 272–301.

Hedges, L. V., Laine, R. D., & Greenwald, R. (1994). Does money matter? A meta-analysis of studies of the effects of differential school inputs on student outcomes. *Educational Researcher*, **23**(3), 5–14.

Hendriks, M. A., & Scheerens, J. (1996). *Zelfevaluatie in het basisonderwijs. Constructie van een instrumentarium "School- en klaskenmerken"*. Enschede: OCTO.

Hill, P. W., & Rowe, K. J. (1996) Multilevel modelling in school effectiveness research. *School Effectiveness and School Improvement*, **7**, 1–34.

Hill, P. W., Rowe, K. J., & Holmes-Smith, P. (1995a). *Factors affecting students' educational progress: Multilevel modelling of educational effectiveness*. Paper presented at the 8th International Congress for School Effectiveness and Improvement, Leeuwarden, The Netherlands, January.

Hill, P. W., Rowe, K. J., & Jones, T. (1995b). *SIIS: School Improvement Information Service. Version 1.1, October 1995.* Melbourne: University of Melbourne, Centre for Applied Educational Research.

Hirsch, D. (1994). *School: A matter of choice*. Paris: OECD/CERI.

Hoeven-van Doornum, A. A. van der (1990). *Effecten van leerlingbeelden en streefniveaus op schoolloopbanen*. Nijmegen: ITS/OoMO.

Hofman, A. W. H. (1995). Cross-level relationships within effective schools. *School Effectiveness and School Improvement,* **6,** 146–174.

Hofman, R., Hoeben, W., & Guldemond, H. (1995). Denominatie en effectiviteit van schoolbesturen. *Tijdschrift voor Onderwijsresearch*, **20**(1), 63–78.

Houtveen, A. A. M., & Osinga, N. (1995a). *A case of school effectiveness: The Dutch national improvement project*. Paper presented at the ICSEI Conference, Leeuwarden, the Netherlands.

Houtveen, A. A. M., & Osinga, N. (1995b). *The Dutch national school improvement project. A case of school effectiveness*. Paper presented at ICSEI, Leeuwarden, The Netherlands.

Hoy, W. K., & Ferguson, J. (1985). A theoretical framework and exploration of organizational effectiveness of schools. *Educational Administration Quarterly*, **21,** 117–134.

Irwin, C. C. (1986). What research tells the principal about educational leadership. *Scientica Paedagogica Experimentalis*, **23,** 124–137.

Jencks, C., Smith, M. S., Ackland, H., Bane, M. J., Cohen, D., Grintlis, H., Heynes, B., & Michelson, S. (1972). *Inequality: A reassessment of the effect of family and schooling in America.* New York: Basic Books.

Jencks, C., Bartlett, S., Corcoran, M., Crouse, J., Eaglessield, D., Jackson, G., McLelland, K., Mueser, P., Olneck, M., Schwartz, J., Ward, S., & Williams, G. (1979). *Who gets ahead? The determinants of economic success in America*. New York: Basic Books.

Johnston, K. L., & Aldridge, B. G. (1985). Examining a mathematical model of mastery learning in a classroom setting. *Journal of Research in Science Teaching*, **22**(6), 543–554.

Jonassen, D. H. (1992). Evaluating constructivist learning. In Th. M. Duffy & D. H. Jonassen (Eds), *Constructivism and the technology of instruction: A conversation*, (pp. 138–148). Hillsdale, NJ: Lawrence Erlbaum Ass.

Joyce, B., & Showers, B. (1988). *Student achievement through staff development.* New York: Longman.

Karweit, N., & Slavin, R. E. (1982). Time on task: Issues of timing, sampling and definition. *Journal of Educational Psychology*, **74,** 844–851.

Keefe, J. (1994). CASE/IMS (Computer program) (Nederlanse licentie Roders, R., Van der Wolf, J. C., Amsterdam: Seneca). U.S.A.:NASSP.

Kerr, S. (1977). Substitutes for leadership: Some implications for organizational design. *Organizational and Administrative Sciences*, **8**, 135–146.

Kesteren, J. H. M. van (1996). *Doorlichten en herontwerpen van organisatiecomplexen*. (Thesis), Groningen: University of Groningen.

Kickert, W. J. M. (1979). *Organization of decision-making: A systems-theoretical approach*. Amsterdam: North-Holland.

Kieser, A., & Kubicek, H. (1977). *Organisation*. Berlin: De Gruyter Lehrbuch.

Kieviet, F. K., & Vandenberghe, R. (1993). *School culture, school improvement and teacher development*. Leiden: DSWO Press.

Klerk, L. F. W. de (1985). ATI-onderzoek en differentiatie: Een reactie. *Pedagogische Studiën*, **62**, 372–375.

Knuver, J. W. M. (1989). Schoolkenmerken en leerlingfunctioneren; Een replicatie-onderzoek. *Tijdschrift voor Onderwijsresearch*, **14**, 329–337.

Knuver, J. W. M., & Doolaard, S. (1996). *Rekenen/wiskunde en natuuronderwijs op de basisschool*. Enschede: OCTO.

Kreft, G. G. (1985). Enige aantekeningen bij een empirisch onderzoek in het lager onderwijs te Amsterdam. *Tijdschrift voor Onderwijsresearch*, **10**, 189–194.

Kreft, G. G. (1987). *Models and methods for the measurement of schooleffects*. Amsterdam: Proefschrift Universiteit van Amsterdam.

Kreft, G. G. (1992). *The analysis of small group data: A reanalysis of Webb 1982 with a random coefficient model*. Paper presented at AERA annual meeting, San Francisco, CA.

Kreft, I. G. G., & Aschbacher, P. R. (1994). Measurement and evaluation issues in education: The value of multivariate techniques in evaluating an innovative high school reform program. *International Journal of Educational Research*, **21**, 181–196.

Kreft, G. G., & Leeuw, J. de, (1991). Model based ranking of schools. *International Journal of Educational Research*, **15**, 45–60.

Kulik, C. L. C., & Kulik, J. A. (1982). Effects of ability grouping on secondary school students: A meta-analysis of research findings. *American Educational Research Journal*, **19**, 415–428.

Kyle, M. J. (Ed.) (1985). *Reaching for excellence: An effective schools sourcebook*. Washington, DC: U.S. Government Printing Office.

Laarhoven, P. van, & Vries, A. M. de (1987). Effecten van de interklassikale groeperingsvorm in het voortgezet onderwijs: Resultaten van een literatuurstudie. In J. Scheerens & W. G. R. Stoel (Eds), *Effectiviteit van onderwijsorganisaties* [Effectiveness of school organizations]. Lisse: Swets & Zeitlinger.

Lam, J. F. (1996). *Tijd en kwaliteit in het basisonderwijs*. Enschede: Universiteit Twente.

Lambert, P. C., & Abrams, K. R. (1995). Meta-analysis using multilevel models. *Multilevel Modelling Newsletter*, **7**(2), 17–19.

Leeuw, A. C. J. de (1986). *Organisaties: management, analyse, ontwerp en verandering. Een systeemvisie,* 2nd edn. Assen: Van Gorcum.

Leithwood, K., Jantzi, D., & Steinbach, R. (1995). *Centrally initiated school restructuring in Canada*. Paper presented at the ICSEI Conference, Leeuwarden, the Netherlands.

Leithwood, K. A., & Montgomery, D. J. (1982). The role of the elementary school principal in program improvement. *Review of Educational Research*, **52**, 309–399.

Leune, J. M. G. (1994). Onderwijskwaliteit en de autonomie van scholen. In B.P.M. Creemers (Ed.), *Deregulering en de kwaliteit van het onderwijs*. Groningen: RION.

Levine D. U. (1992). An interpretive review of US research and practice dealing with unusually effective schools. In D. Reynolds & P. Cuttance (Eds), *School effectiveness: Research, policy and practice*. London: Cassell.

Levine, D. U., & Lezotte, L. W. (1990). *Unusually effective schools: A review and analysis of research and practice*. Madison, WI: National Center for Effective Schools Research and Development.

Lidt de Jeude, J. van, & Brouwer, T. (1992). Kleine oorzaken, grote gevolgen: Een inleiding op de chaostheorie. In C. van Dijkum & D. de Tombe (Eds), *Gamma chaos: Onzekerheden en orde in de menswetenschappen*. Bloemendaal: Aramith.

Lockheed, M. E. (1988). *The measurement of educational efficiency and effectiveness*. Paper presented at AERA annual meeting, New Orleans.

Lockheed, M. E. (1990). A multilevel model of school effectiveness in a developing country. In S. W. Raudenbush & J. D. Willms (Eds), *Schools, classrooms, and pupils: International studies of schooling from a multilevel perspective*. Edinburgh: Academic Press.

Longford, N. L. (1994). *Random coefficient models*. Oxford: Clarendon Press.

Lortie, D. C. (1973). Observations on teaching as work. In R. M. W. Travers (Ed.), *Second handbook of research on teaching*. Chicago, IL: Rand McNally.

Lotto, L. S., & Clark, D. L. (1986). Understanding planning in educational organizations. *Planning and Change*, **19**, 9–18.

Lugthart, E., Roeders, P. J. B., Bosker, R. J., & Bos, K. T. (1989). *Effectieve schoolkenmerken in het voortgezet onderwijs: Een literatuuroverzicht*. Groningen: RION.

Luyten, J. W. (1994). *School effects: Stability and malleability*. Enschede: University of Twente.

Luyten, H. (1996). *School effectiveness and student achievement, consistent across subjects? Evidence from Dutch primary and secondary education.* Paper presented at the annual conference of the Dutch Association for Educational Research (Onderwijsresearchdagen), Tilburg, June.

Luyten, J. W., & Snijders, T. A. B. (1996). School effects and teacher effects in Dutch elementary education. *Educational Research and Evaluation,* **2**, 1–24.

Madaus, G. F., Kellaghan, T., Rakow, E. A., & King, D. (1979). The sensitivity of measures of school effectiveness. *Harvard Educational Review*, **49**, 207–230.

Mandeville, G. K. (1988). School effectiveness indices revisited: Cross-year stability. *Journal of Educational Measurement*, **25**, 349–356.

Mandeville, G. K., & Anderson, L. W. (1987). The stability of school effectiveness indices across grade levels and subject areas. *Journal of Educational Measurement*, **24**, 203–216.

March, J. T., & Olsen, J. P. (1976). *Ambiguity and choice in organizations*. Bergen: Universitetsforlaget.

Maslowski, R. (1995). *Organisatiecultuur systematisch benaderd*. Enschede: Universiteit Twente.

McBeath, J. (1994). *Making schools more effective: A role for parents in school self-evaluation and development*. Paper presented at AERA annual meeting, New Orleans.
McDonald, R. P. (1994). The bilevel reticular action model for path analysis with latent variables. *Sociological Methods and Research,* **22**, 399–413.
Medley, D., & Mitzel, H. (1963). Measuring classroom behavior by systematic observation. In N. L. Gage (Ed.), *Handbook of research on teaching*. Chicago, IL: Rand McNally.
Merrill, M. D. (1991). Constructivism and instruction design. *Educational Technology,* **31,** 45–53.
Meuret, D., & Scheerens, J. (1995). *An international comparison of functional and territorial decentralization of public educational systems*. Paper presented at AERA annual meeting, San Francisco.
Mintzberg, H. (1979). *The structuring of organizations*. Englewood Cliffs, NJ: Prentice-Hall.
Mitchell, D. E., & Tucker, Sh. (1992). Leadership as a way of thinking. *Educational Leadership*, **49**(5), 30–35.
Monk, D. H. (1989) The education production function: Its evolving role in policy analysis. *Educational Evaluation and Policy Analysis,* **11**(1), 31–45.
Monk, D. H. (1992). *Microeconomics of school productions*. Paper for the Economics of Education Section of the International Encyclopedia of Education.
Morgan, G. (1986). *Images of organization*. London: Sage.
Mortimore, P. (1992). Issues in school effectiveness. In D. Reynolds & P. Cuttance (Eds), *School effectiveness research, policy and practice*. London: Cassell.
Mortimore, P. (1993). School effectiveness and the management of effective learning and teaching. *School Effectiveness on School Improvement,* **4,** 290–310.
Mortimore, P., Sammons, P., Stoll, L., Lewis, D., & Ecob, R. (1988). *School matters: The junior years*. Somerset: Open Books.
Mosteller, F., & Moynihan, D. D. (Eds) (1972). *On equality of educational opportunity*. New York: Random House.
Murmane, R. J. (1981). Interpreting the evidence on school effectiveness. *Teachers College Record*, **83,** 19–35.
Murphy, J. (1989). Principal instructional leadership. In P. W. Thurson & L. S. Lotto (Eds), *Advances in educational leadership*. Greenwich, CT: JAI Press.
Murphy, J. (1992). School effectiveness and school restructuring: Contributions to educational improvement. *School Effectiveness and School Improvement*, **3**, 90–109.
Murphy, J. (1993). Restructuring schooling: The equity infrastructure. *School Effectiveness and School Improvement*, **4**, 111–130.
Murphy, J. (1994). The changing role of the superintendency in restructuring districts in Kentucky. *School Effectiveness and School Improvement,* **4,** 349–375.
Muthén, B. O. (1994). Multivariance covariance structure analysis. *Sociological Methods and Research,* **22**, 376–398.
Neufeld, E., Farrar, E., & Miles, M. B. (1983). *A review of effective schools research: The message for secondary schools*. Huron Institute, Cambridge, MA: National Commission on Excellence in Education.
Niskanen, W. A. (1971). *Bureaucracy and representative government*. Chicago, IL: Aldine-Atherton.

Nuttall, D. L., Goldstein, H., Prosser, R., & Rasbash, J. (1989). Differential school effectiveness. *International Journal of Educational Research*, **13**, 769–776.

Oakes, J. (1987). *Conceptual and measurement problems in the construction of school quality*. Paper presented at AERA annual meeting, Washington..

Odi, A. (1982). The process of theory development. *Journal of Research and Development in Education*, **15**(2), 53–58.

Pfeffer, J., & Salancik, G. R. (1978). *The external control of organizations: A resource dependence perspective*. New York: Harper & Row.

Postlethwaite, T. N., & Ross, K. N. (1992). *Effective schools in reading. Implications for educational planners*. The Hague: IEA.

Price, J. L. (1972). The study of organizational effectiveness. *Sociological Quarterly*, **13,** 3–15.

Purkey, S. C., & Smith, M. S. (1983). Effective schools: A review. *Elementary School Journal*, **83,** 427–452.

Ralph, J. H., & Fennessey, J. (1983). Science or reform: Some questions about the effective schools model. *Phi Delta Kappan*, **64,** 689–694.

Rasbash, J., & Woodhouse, G. (1995). *MLn command reference*. London: University of London, Institute of Education, Multilevel Models Project.

Raudenbush, S. W. (1989). The analysis of longitudinal, multilevel data. *International Journal of Educational Research*, **13**, 721–740.

Raudenbush, S. W. (1994). The random effects models. In H. Cooper & L. V. Hedges (Eds), *The handbook of research synthesis*. New York: Russell Sage Foundation.

Raudenbush, S. W., & Bryk, A. W. (1985). Empirical Bayes meta-analysis. *Journal of Educational Statistics*, **10,** 75–98.

Raudenbush, S. W., & Bryk, A. S. (1986). A hierarchical model for studying school effects. *Sociology of Education*, **59,** 1–17.

Raudenbush, S.W., Rowan, B., & Kang, S.J. (1991). A multilevel, multivariate model for studying school climate with estimation via the EM algorithm and application to U.S. High-school data. *Journal of Education Statistics*, **16**, 295–330.

Raudenbush, S. W., & Willms, J. D. (1996). The estimation of school effects. *Journal of Educational and Behavioral Statistics*, **20**, 307–335.

Reezigt, G. J. (1993). *Effecten van differentiatie op de basisschool*. Groningen: RION.

Reezigt, G. J., Guldemond, H., & Creemers, B. P. M. (1994). *Empirical validity for a comprehensive model on educational effectiveness*. Groningen: GION.

Resnick, L. B. (1987). *Education and learning to think*. Washington, DC: National Academic Press.

Reynolds, A. J., & Walberg, H. J. (1990). *A structural model of educational productivity*. Unpublished manuscript, Northern Illinois University

Reynolds, D. (1992). School effectiveness and school improvement: An updated review of the Britisch literature. In D. Reynolds & P. Cuttance (Eds), *School effectiveness: Research, policy and practice*. London: Cassell.

Reynolds, D., Creemers, B. P. M., Bird, J., & Farrel, S. (1994). School effectiveness—the need for an international perspective. In D. Reynolds, B. P. M. Creemers, P. S. Nesselrodt, E. C. Schaffer, S. Stringfield & C.Teddlie (Eds), *Advances in school effectiveness research and practice*. Oxford: Pergamon Press.

Riley, D. D. (1990). Should market forces control educational decision making? *American Political Science Review*, **84,** 554–558.

Rist, R. C., & Joyce, M. K. (1995). Qualitative research and implemenation evaluation: A path to organizational learning. In T. E. Barone (Ed.), *The uses of educational research* [Special issue]. *International Journal of Educational Research*, **23**(2), pp. 127–136.

Rosenshine, B. V. (1983). Teaching functions in instructional programs. *Elementary School Journal*, **83,** 335–351.

Rosenshine, B. (1987). Direct instruction. In M. J. Dunkin (Ed.), *The international encyclopedia of teaching and teacher education*, pp. 257–263. Oxford: Pergamon Press.

Rosenshine, B. V., & Furst, N. (1973). The use of direct observations to study teaching. In R. M. W. Travers (Ed.), *Second handbook of research on teaching*. Chicago, IL: Rand McNally.

Rosenthal, R. (1994). Parametric measures of effect size. In H. Cooper & L. V. Hedges (Eds), *The handbook of research synthesis.* New York: Russell Sage Foundation.

Rosenthal, R., & Rubin, D. B. (1982). A simple, general purpose display of magnitude of experimental effect. *Journal of Educational Psychology*, **74,** 166–169.

Rowan, B., Bossart, S. T., & Dwyer, D. C. (1983). Research on effective schools. A cautionary note. *Educational Researcher*, **12**(4), 24–31.

Rowe, K. J., & Hill, P. W. (1994). *Multilevel modelling in school effectiveness research: How many levels?* Paper presented at the International Congress for School Effectiveness and Improvement, Melbourne.

Rowe, K. & Hill, P. (1996). Assessing, recording and reporting students' educational progress: The case for 'Subject Profiles'. *Assessment in Education*, **3**, 309–352.

Rowe, K. J., Hill, P. W., & Holmes-Smith, P. (1994). *The Victorian quality schools project: A report on the first stage of a longitudinal study of school and teacher effectiveness.* Symposium paper presented at the 7th International Congress for School Effectiveness and Improvement, Melbourne, January.

Rutter, M. (1983). School effects on pupil progress: Research findings and policy implications. *Child Development*, **54**(1), 1–29.

Rutter, M., Maughan, B., Mortimore, P., & Ouston, J. (1979). *Fifteen thousand hours. Secondary schools and their effects on children*. Somerset: Open Books.

Sammons, P., Hillman, J., & Mortimore, P. (1995a). *Key characteristics of effective schools: A review of school effectiveness research*. London: OFSTED.

Sammons, P., Nuttall, D. & Cuttance, P. (1993). Differential school effectiveness: Results from a reanalysis of the Inner London Education Authority's Junior School Project data. *British Educational Research Journal,* **19**, 381–405.

Sammons, P., Thomas, S. M., & Mortimore, P. (1995b). *Accounting for variations in academic effectiveness between schools and departments*. Paper presented at ECER, Bath.

Scheerens, J. (1987). *Enhancing educational opportunities for disadvantaged learners*. Amsterdam: North-Holland.

Scheerens, J. (1989a). *Wat maakt scholen effectief?* Den Haag: SVO.

Scheerens, J. (1989b). Process indicators of school functioning. *School Effectiveness and School Improvement*, **1**, 61–80.

Scheerens, J. (1990). School effectiveness and the development of process indicators of school functioning. In *School effectiveness and school improvement*, pp. 61–80. Lisse: Swets & Zeitlinger.

Scheerens, J. (1992). *Effective schooling. Research, theory and practice*. London: Cassell.

Scheerens, J. (1993). Basic school effectiveness research: Items for a research agenda. *School Effectiveness and School Improvement*, **4**, 17–36.

Scheerens, J. (1994). The school-level context of instructional effectiveness: A comparison between school effectiveness and restructuring models. *Tijdschrift voor Onderwijsresearch*, **19**, 26–38.

Scheerens, J. (1995). *School effectiveness as a research discipline*. Paper presented at the ICSEI Congress, Leeuwarden, the Netherlands, January.

Scheerens, J., & Brummelhuis, A. C. A. ten (1996). *Process indicators on the functioning of schools: Results from an international survey*. Paper presented at AERA annual meeting, New York.

Scheerens, J., & Creemers, B. P. M. (1989). Conceptualizing school effectiveness. In *Developments in school effectiveness research* [Special issue]. *International Journal of Educational Research*, **13**, 691–706.

Scheerens, J., & Creemers, B. P. M. (1996a). School effectiveness in the Netherlands; the modest influence of a research programme. *School Effectiveness and School Improvement*, **7**, 181–195.

Scheerens, J., Korevaar, G. J., & Rijcke, F. J. M. de (1991). Onderwijsindicatoren in Engeland, oftewel de vele valkuilen van een evaluatie-georiënteerde beleidsvoering geïllustreerd. *Tijdschrift voor Onderwijswetenschappen*, **21**(3/4), 221–226.

Scheerens, J., & Stoel, W. G. R. (1988). *Theory development on school effectiveness*. Paper presented at AERA annual meeting, New Orleans.

Scheerens, J., Vermeulen, C. J. A. J., & Pelgrum, W. J. (1989). Generalizability of instructional and school effectiveness indicators across nations. *International Journal of Educational Research*, **13**, 789–800.

Schein, E. H. (1985). *Organizational culture and leadership: A dynamic view*. San Francisco, CA: Jossey Bass.

Simon, H. A. (1964). *Administrative behavior*. New York: Macmillan.

Simons, P. R. J. (1989). Leren leren: Naar een nieuwe didactische aanpak. In P. R. J. Simons & J. G. G. Zuylen (Eds), *Handboek huiswerkdidactiek en geïntegreerd studievaardigheidsonderwijs*. Heerlen: Meso Consult.

Slater, R. O., & Teddlie, C. (1991). Toward a theory of school effectiveness and leadership. *School Effectiveness and School Improvement*, **3**, 247–257.

Slavin, R. E. (1987). Ability grouping and student achievement in elementary schools: A best evidence synthesis. *Review of Educational Research*, **57**, 293–336.

Slavin, R. E. (1996). *Success for all*. Lisse: Swets & Zeitlinger.

Slavin, R., Madden, N. A., Karweit, N. L., Dolan, L., & Wasik, B. H. (1990). *Success for all*. Baltimore: John Hopkins University.

Snijders, T. A. B., & Bosker, R. J. (forthcoming). *Introduction to multi-level analysis*.

Snow, R. E. (1973). Theory construction for research on teaching. In R. M. W. Travers (Ed.), *Handbook of research on teaching*. Chicago, IL: Rand McNally.

Sociaal Cultureel Planbureau (1982). Hoe verder? Advies over de nota Verder na de Basisschool. *S.C.P.—Cahier no. 31*. Rijswijk: SCP.

Southworth, G. (1994). The learning school. In P. Ribbens & E. Burridge (Eds), *Improving education: Promoting quality in schools*. London: Cassell.

Spiegelhalter, D. J., Thomas, A., Best, N. G., & Gilks, W. R. (1994). *BUGS: Bayesian inference using Gibbs sampling, Version 0.30*. Cambridge: MRC Biostatistics Unit.

Spiro, R. J., Feltowich, P. J, Jacobson, M. J., & Caulson, R. L. (1992). Cognitive flexibility, constructivism and hypertext: Random access instruction for advanced knowledge acquisition in ill-structured domains. In Th. M. Duffy, & D. H. Jonassen (Eds), *Constructivism and the technology of instruction: A conversation*. Hillsdale, NJ: Lawrence Erlbaum Associates.

Stallings, J. (1985). Effective elementary classroom practices. In M. J. Kyle (Ed.), *Reaching for excellence: An effective schools sourcebook*. Washington, DC: Government Printing Office.

Stallings, J., & Mohlman, G. (1981). *School policy, leadership, style, teacher change and student behavior in eight schools*. Washington, DC: Final report to the National Institute of Education.

Stebbins, L. B., St. Pierre, R. G., Proper, E. C., Anderson, R. R., & Cerva, T. R. (1977). *Education as experimentation: A planned variation model*, Vol. IV-A, *An evaluation of follow through*. Cambridge, MA: Abt Associates.

Stedman, L. (1987). It's time we changed effective schools formula. *Phi Delta Kappan*, **69**, 215–244.

Stringfield, S. C. (1995). Attempting to enhance students' learning through innovative programs: The case for school evolving into high reliability organizations. *School Effectiveness and School Improvement*, **6**, 67–96.

Stringfield, S. C., Bedinger, S., & Herman, R. (1995). *Implementing a private school program in an inner-city public school: Processes, effects, and implications from a four year evaluation*. Paper presented at the ICSEI Conference, Leeuwarden, the Netherlands.

Stringfield, S. C., & Slavin, R. E. (1992). A hierarchical longitudinal model for elementary school effects. In B. P. M. Creemers & G. J. Reezigt (Eds), *Evaluation of effectiveness* (ICO Publication 2). Groningen: ICO

Stringfield, S., & Teddlie, C. (1990). School improvement efforts: Qualitative and quantitative data from four naturally occurring experiments in phases III and IV of the Louisiana School Effectiveness Study. *School Effectiveness and School Improvement*, **1**, 139–166.

Sweeney, J. (1982). Research synthesis on effective school leadership. *Educational Leadership*, **39**, 346–352.

Teddlie, C. (1994). The study of context in school effects research: History, methods, results and theoretical implications. In D. Reynolds, B. P. M. Creemers, P. S. Nesselrodt, E. C. Schaffer, S. C. Stringfield & C. Teddlie (Eds), *Advances in school effectiveness research and practice*. Oxford: Pergamon.

Teddlie, C., & Stringfield, S. (1993). *Schools make a difference: Lessons learned from a 10 year study of school effects*. New York: Teachers College Press.

Teddlie, C., Stringfield, S. C., Wimpelberg, R., & Kirby, P. (1989). Contextual differences in models for effective schooling in the USA. In B. P. M. Creemers, T. Peters & D. Reynolds (Eds), *School effectiveness and school improvement*. Lisse: Swets & Zeitlinger.

Thomas, D. (1979). *Naturalism and social science. A post empiricist philosophy of social science*. Cambridge: Cambridge University Press.

Thomas, S. M. (1995). *Optimal multilevel models of school effectiveness comparative analyses across regions*. ESRC proposal. Institute of Education, University of London.

Thomas, S. M., Sammons, P., Mortimore, P., & Smees, R. (1995). *Stability and consistency in secondary schools' effects on students' GCSE outcomes over three years*. Paper presented at the ICSEI Conference, Leeuwarden, The Netherlands.

Thomas, S., & Mortimore, P. (1996). Comparison of value-added models for secondary-school effectiveness. *Research Papers in Education*, **11**, 5–33.

Thompson, J. D. (1967). *Organizations in action*. New York: McGraw-Hill.

Thorndike, R. L. (1973). *Reading comprehension education in fifteen countries*. Stockholm: Almqvist & Wiksell.

Tizard, W., Bladgeford, P., Burke, J., Swarquhar, C., & Plewis, I. (1988). *Young children at school in the inner city*. Hove: Lawrence Erlbaum.

Tobias, S. (1991). An eclectic examination of some issues in the constructivist-ISP controversy. *Educational Technology*, **31**(9), 41–43.

Tyler, R. (1950). *Basic principles of curriculum and instruction*. Chicago, IL: University of Chicago Press.

Tymms, P. (1994). *Theories, models and simulation: School effectiveness at an impasse*. Paper presented to the ESRC School Effectiveness and School Improvement Seminar, London.

Universiteit Twente, OCTO (1995a). *Derde internationale onderzoek rekenen/wiskunde en natuuronderwijs. Vragenlijst voor de schoolleider, populatie 1*. Enschede: Universiteit Twente: OCTO.

Universiteit Twente, OCTO (1995b). *Derde internationale onderzoek rekenen/wiskunde en natuuronderwijs. Vragenlijst voor de leerkracht, populatie 1*. Enschede: Universiteit Twente: OCTO.

Vedder, P. H. (1985). *Cooperative learning: A study on processes and effects of cooperation between primary school children*. The Hague: SVO.

Velden, L. F. J. van der (1996). *Context, visie, aanpak en effectiviteit*. Groningen: GION.

Venezky, R. L., & Winfield, L. F. (1979). *Schools that succeed beyond expectations in reading*. (studies on Educational Technical Report No. 1). Newark: University of Delaware. (ERIC document Reproduction Service No. ED 177484).

Visscher, A. J. (1993). *Design and evaluation of a computer-assisted management information system for secondary schools*. Enschede: University of Twente.

Vos, H. de (1989). A rational-choice explanation of composition effects in educational research. *Rationality and Society*, **1**, 220–239.

Vos, H. de, & Bosker, R. J. (forthcoming). A multi-level simulation model to study school effects.

Walberg, H. J. (1984). Improving the productivity of American schools. *Educational Leadership*, **41,** 19–27.

Wang, M. C., Haertel, G. D & Walberg, H. J. (1993). Toward a knowledge base for school learning. *Review of Educational Research*, **63**, 249–294.

Weber, G. (1971). *Inner-city children can be taught to read: Four successful schools*. Washington, DC: Council for Basic Education.

Weeda, W. C. (1986). Effectiviteitsonderzoek van scholen. In J. C. van der Wolf, & J. J. Hox (Eds), *Kwaliteit van het Onderwijs in het Geding*. Publicaties van het Amsterdams Pedologische Centrum, no. 2. Lisse: Swets & Zeitlinger.

Weick, K. (1976). Educational organizations as loosely coupled systems. *Administrative Science Quarterly*, **21,** 1–19.

Weiss, C. H., & Bucuvalas, M. J. (1980). Truth tests and utility tests: Decision-makers' frames of reference for social science research. *American Sociological Review*, **45,** 303–313.

Werf, M. P. C. van der (1988). *Het Schoolwerkplan in het basisonderwijs*. Lisse: Swets & Zeitlinger.

Werf, G. van der & Driessen, G. (1993). *Het functioneren van het voortgezet onderwijs. Kenmerken van scholen en docenten in het eerste leerjaar.* Groningen/Nijmegen: RION/ITS.

Werf, G. van der. & Guldemond, H. (1995). *Omvang, stabiliteit en consistentie van schooleffecten in het basisonderwijs*. Paper presented at the annual conference of the Dutch Association for Educational Research (Onderwijsresearchdagen), Groningen, June.

White, H. D. (1994). Scientific communication and literature retrieval. In H. Cooper & L. V. Hedges (Eds), *The handbook of research synthesis*. New York: Russell Sage Foundation.

Willms, J. D., & Raudenbush S. W. (1989). A longitudinal hierarchical linear model for estimating school effects and their stability. *Journal of Educational Measurement*, **26**(3), 209–232.

Windham, D. M. (1988). Effectiveness indicators in the economic analysis of educational activities [Special issue]. *International Journal of Educational Research*, **12**(6).

Witte, J. F. (1990). *Understanding high school achievement: After a decade of research, do we have any confident policy recommendation?* Paper presented at AERA annual meeting, San Francisco.

Witziers, B. (1992). *Coördinatie binnen scholen voor voortgezet onderwijs*. Enschede: Universiteit Twente.

Witziers, B., & Rosker, R. (1997). *A meta-analysis on the effects of presumed school effectiveness enhancing factors*. Paper presented at ICSEI, Memphis, U. S. A., January.

Woodward, J. (1965). *Industrial organization: Theory and practice.* London: Oxford University Press.

Author Index

Abrams, K. R. 75
Achilles, C. M. 312, 317
Aitkin, M. 62, 72, 80, 143, 185, 320
Aldridge, B. G. 40
Alexander, K. L. 142
Amelsvoort, H. W. C. H. van 26, 318
Anderson, C. S. 152
Anderson, L. W. 83, 84
Argyris, C. 281
Aschbacher, P. R. 64
Averch, H. A. 144

Bangert, R. L. 151
Barley, S. R. 289
Batenburg, van 83, 88–9, 91
Bereiter, C. 145
Berliner, D. C. 40, 59
Biniaminov, J. 144
Block, J. H. 40
Blok, H. 83
Bloom, B. S. 39, 149
Boekaerts, M. 44
Boorsma, P. B. 4
Borger, J. B. 152, 214
Brandsma, H. P. 146, 182–7, 302–3
Broekstra, G. 291–2
Brookover, W. B. 70, 145, 154
Brophy, J. 45, 47, 80, 151, 152, 165, 219
Brouwer, T. 290
Brown, B. W. 37
Brown, S. 273
Brummelhuis, A. C. A. 10, 99, 138
Bruner, J. S. 40
Bryk, A. S. 63, 64, 74, 92, 225
Bucuvalas, M. J. 281
Burns, R. B. 40

Cameron, K. S. 6, 11, 28, 31, 50
Campbell, D. T. 273
Carroll, J. B. 38–40, 149
Chandler, A. 24
Cheng, Y. C. 4–5, 10, 11–12, 32, 33, 215, 294, 297
Chubb, J. E. 268, 277, 279
Clauset, K. H. 265, 283, 320
Cohen, D. K. 42, 238
Cohen, J. 74
Cohen, M. D. 272, 277
Coleman, J. S. 69, 92–3, 141–5, 155, 299–300
Coleman, P. 173, 280, 307, 313
Collins, A. 42, 150
Cooper, H. 214, 227
Corcoran, Th. B. 3
Correa, H. 38, 276, 313
Corte, E. De 40
Cotton, K. 121, 155–6, 158–60, 163–4, 166–8, 171–2, 174, 177–81, 304
Creemers, B. P. M. 35, 36, 41–2, 44–5, 49, 50, 211, 213, 214, 239, 273, 275, 283, 287, 296
Cunningham, D. J. 42
Cuttance, P. 88, 90, 91
Cyert, R. M. 272–3

Davies, J. K. 148
Dekkers, H. P. J. M. 94
Dijkstra, A. B. 278
Doolaard, S. 84, 99, 100, 302–3
Dougherty, K. 152
Doyle, W. 150–1
Drazin, R. 285
Driessen, G. 100
Dronkers, J. 143
Dror, Y. 271
Duffy, Th. M. 42

Eckland, B. K. 142
Edmonds, R. R. 70, 152, 176, 239
Elberts, R. W. 37
Emery, F. E. 291
Erbring, L. 58

Essink, L. J. B. 30
Etzioni, A. 10, 11

Faerman, S. R. 6, 9, 11, 33
Fennessey, J. 28, 31, 80, 152, 300
Ferguson, J. 31
Finn, J. D. 312
Fitz-Gibbon, C. T. 72
Fraser, B. J. 218–19, 223, 224
Friebel, A. J. J. M. 275
Furst, N. 147

Gage, N. 147, 273
Gamoran, A. 61–2
Gaynor, A. K. 265, 283, 320
Glasman, N. S. 144
Glass, G. V. 242
Glenn, B. C. 154
Goldstein, H. 27, 86, 302–3, 315, 316
Good, T. L. 80, 151, 152, 219
Gooren, W. A. J. 10
Gray, J. 27, 86
Gresov, C. 284, 286
Griffith, D. E. 290
Grift, W. van de 152, 155
Grisay, A. 87, 199–206, 308, 315
Guldemond, H. G. 66, 86, 91, 320
Gustafsson, J. E. 64, 320

Hallinger, P. 288
Hanushek, E. A. 37, 144, 216–17, 300
Hargreaves, D. H. 273, 294–5
Hauser, R. M. 142
Haywood, H. C. 145
Hedger, K. 86
Hedges, L. V. 144, 214, 216–17, 227, 237
Hendriks, M. A. 100
Hill, P. W. 53, 64, 79, 80, 84, 88–9, 99, 187–91, 205, 209, 292, 303, 304, 310, 315
Hirsch, D. 278
Hoeksma, J. B. 83
Hoeven–van Doornum, A. A. van der 150
Hofman, A. W. H. 52, 310
Hofman, R. 279–80, 310
Hopkins, D. 273, 294–5
Houtveen, A. A. M. 295, 317
Hoy, W. K. 31

Irwin, C. C. 107

Jencks, C. 70, 141, 142, 145
Jesson, D. 86
Johnston, K. L. 40
Jonassen, D. H. 42
Joyce, B. 176
Joyce, M. K. 282

Kang, S. J. 261
Karweit, N. 40
Keefe, J. 100
Kerr, S. 108
Kesteren, J. H. M. van 14, 23, 34
Kickert, W. J. M. 285–6
Kieser, A. 284–5
Kieviet, F. K. 294
Klerk, L. F. W. de 149
Knuver, J. W. M. 80, 99, 146
Kreft, G. G. 64, 80
Kubicek, H. 284–5
Kulik, C. L. C. 149
Kulik, J. A. 149
Kurland, M. 145
Kyle, M. J. 152, 280

Laarhoven, P. van 149
Lam, J. F. 306
Lambert, P. C. 75
Laroque, L. 280
Leeuw, A. C. J. de 13, 23, 24, 284
Leithwood, K. 279
Leithwood, K. A. 107
Leune, J. M. G. 278
Levine, D. U. 70, 152, 155–8, 161–70, 173–6, 178–82
Lezotte, L. W. 70, 152, 154, 155–8, 161–70, 173–6, 178–82
Lidt de Jeude, J. van 290
Lockheed, M. E. 5, 80
Longford, N. L. 72, 75, 79, 80, 143, 185, 320
Lortie, D. C. 148
Lotto, L. S. 276
Lowyck, J. 40, 148

Lugthart, E. 80, 152
Luyten, J. W. 27, 85, 88–9, 90–1, 292, 314

McBeath, J. 173
McDonald, R. P. 64
Madaus, G. F. 52, 314
Mandeville, G. K. 81, 83, 84, 88–9
March, J. G. 272–3
March, J. T. 272, 280–1
Maslowski, R. 17
Medley, D. 147
Merrill, M. D. 44
Meuret, D. 279, 289
Mintzberg, H. 8, 13, 25, 65, 269, 274, 284–5, 309
Mitchell, D. E. 107
Mitzel, H. 147
Moe, T. M. 268, 277, 279
Mohlman, G. 149
Monk, D. H. 14, 36, 37–8, 271, 300
Montgomery, D. J. 107
Morgan, G. 270, 274, 276, 281–2, 290, 294
Mortimore, P. 84, 90–1, 99, 108, 145, 146, 158, 167, 266
Mosteller, F. 144
Moynihan, D. D. 144
Murmane, R. J. 152
Murphy , J. 169–70, 268, 282, 288–9
Muthén, B. O. 64, 320

Neufeld, E. 152
Nijzink, J. P. 4
Niskanen, W. A. 7
Nuttall, D. L. 88, 93

Odi, A. 265
Olsen, J. P. 272, 280–1
Osinga, N. 295, 317

Pfeffer, J. 8
Postlethwaite, T. N. 243–5, 255, 287
Price, J. L. 294
Priogine 291
Purkey, S. C. 70, 80, 152–5, 206, 300, 302

Quinn, R. E. 6, 9, 11, 33

Ralph, J. H. 28, 31, 80, 152, 300
Rasbash, J. 75, 86
Raudenbush, S. W. 55, 63, 64, 74, 78, 83, 85–6, 90–1, 92, 225, 261
Reezigt, G. J. 65, 149
Resnick, L. B. 42
Reynolds, A. J. 41
Reynolds, D. 70, 273, 316
Riley, D. D. 278
Rist, R. C. 282
Rosenshine, B. V. 39, 147, 275
Rosenthal, R. 227
Ross, K. N. 243–5, 255, 287
Rowan, B. 261
Rowe, K. J. 53, 64, 79, 80, 84, 88–9, 187
Rutter, M. 52, 70, 80, 152, 154

Saks, D. H. 37
Salancik, G. R. 8
Sammons, P. 27, 32, 88–9, 93, 99, 106, 155–6, 158–9, 162–3, 165–8, 170–1, 173, 176–7, 179–81, 194–99, 275, 307, 315
Schein, E. H. 17–18, 294
Showers, B. 176
Simon, H. A. 272
Simons, P. R. J. 44, 282
Slater, R. O. 266, 316
Slavin, R. E. 19, 38, 40, 45, 47–50, 149, 191–94, 221–3, 296, 307–8, 312, 316–7
Smith, M. S. 70, 80, 152–5, 206, 300, 302
Snijders, T. A. B. 56, 85, 314
Snow, R. E. 265
Southworth, G. 282
Spiegelhalter, D. J. 64, 320
Spiro, R.J. 42
Stahl, P. A. 64, 320
Stallings, J. 149–150
Stanley, J. C. 56
Stebbins, L. B. 145
Stedman, L. 173
Stengers 291

Stevens, A. 150
Stoel, W. G. R. 152
Stone, J. A. 37
Stringfield, S. C. 32, 45, 47–50, 121, 194, 266, 275–6, 279, 288–9, 295, 296, 307, 312
Sweeney, J. 152

Teddlie, C. 32, 121, 266, 287–8, 316
Thomas, S. M. 27, 90–1, 292, 302, 314
Thompson, J. D. 19, 271
Thorndike, R. L. 142
Tizard, W. 165
Tobias, S. 42
Trist, E. L. 291
Tucker, Sh. 107
Tyler, R. 6, 273
Tymms, P. 292

Vandenberghe, R. 294
Vedder, P. H. 151
Velden, R. K. W. van der 51, 209
Ven, A. H. van de 285
Venezky, R. L. 154
Visscher, A. J. 30
Vos, H. de 66, 283, 320
Vries, A. M. de 149

Walberg, H. J. 38, 39, 41, 80, 218–19, 223, 271, 308
Wang, M. C. 16, 194, 215, 219–20, 224, 303, 304, 306
Weber, G. 154
Weber, M. 274
Weeda, W. C. 148
Weick, K. 309
Weiss, C. H. 281
Werf, M. P. C. van der 91, 100, 273, 275, 310
Whetten, D. A. 6, 28, 31, 50
White, H. D. 71
Willms, J. D. 55, 78, 83, 85–6, 90–1
Windham, D. M. 4
Winfield, L. F. 154
Witte, J. F. 279
Witziers, B. 27, 215, 225, 227, 283, 292, 303, 314
Woodhouse, G. 75
Woodward, J. 24

Young, A. A. 58

Zuzovsky, R. 62

Subject Index

Ablauf structure 15–16, 21
achievement measures 51–2
achievement orientation 100, 102
additive models 60–2
aptitude–treatment–interaction studies 148, 149
assessment, constructivist 42–4
attainment levels, variance 27
attainment measures 51–2
Aufbau structure 14, 21
Australia, cross-grade stability 84

Carroll model 38–40
classroom climate 123, 124–5, 136
climate
 classroom 123, 124–5, 136
 school 112–17, 135
cohorts, stability 83–4
compensatory programs, evaluation 144
complexity, primary process 20
conceptual models 37–50
configuration plus thesis 25
consistency
 across subjects 88–91
 school effects 80–3, 89–96
constructivism 42–4
contextual effects model 62–3
contingency theory 24–6
control fuzz 24
control theory 22–4
correlational studies 28
Creemers model 49, 50
cross-level facilitation models 58–67
culture, organizational 16–18
curriculum, quality 108, 110–11, 135

data collection, methods 71–2
differential effectiveness 92–5
 secondary school project 194–99
differentiation 134, 136
direct instruction 39–40, 129

education production function 143
 models 37–8
education production process, factor analysis 4
education working factors 99–138
educational effectiveness, models 35–67, 244
educational leadership 101, 103–6
educational productivity models 38–45
effective learning time 125, 126–7, 136
effective schools
 factors 159
 practices 160
 research 145
effectiveness
 definitions 3–4
 economic 4–6
 enhancing conditions 139–209, 156, 161
 enhancing factors 135–6
 models 35–67
 modes of schooling 13–22
 organization-theoretical views 6–12
 orientation 114
 research 140
 teachers 146
efficiency, definition 4–5
England
 see also UK
 differential effectiveness 93
 school effects stability 86–7
environment, school 18
equal educational opportunity 141
evaluation potential 118–21
expectations, high 100, 102

feedback 134, 136
elements 133
foundational effectiveness studies 27
France, school effects stability 87

genuine effects model 62–3
grades, stability 84–5
gross school effects, meta-analysis 76–7

head teacher 103–6, 115

IEA see International Association for the Evaluation of Educational Achievement
independent learning 131–3, 136
indirect models 63–4
instruction effectiveness 218
instructional effectiveness models 38–45
instructional leadership 104
integrated models 45–50
interactive models 60–2
interdependence, primary process 19–20
International Association for the Evaluation of Educational Achievement (IEA) 142, 227, 240, 242–3, 252, 260
international effectiveness model 260

learning
causal influences 39
opportunity 108, 110–11, 135

management
classroom 127
organization 15–16
mastery learning 39–40
Matthew effect 93–4
meta-analysis
multilevel model 73–5, 225
results 75–9
studies selection 71–3
models
educational effectiveness 35–67
random effects 74
modes of schooling 13–22
cultural dimension 16–18, 21
environment 18, 21
goals 13–14, 21
position structure (Aufbau) 14
primary process 18–20, 21
procedure structure (Ablauf) 15–16
summary 20–2
multilevel model 74–5

net school effects, meta-analysis 77–8
Netherlands
compensatory programs 144
differential effectiveness 93–4
school effects
meta-analysis 76–8, 79
stability 83, 86, 87
subject consistency 88–9, 90–1
North America
see also USA
school effects, meta-analysis 76–8

opportunity to learn 108, 110–11, 135
organizational effectiveness
bureaucracy 8, 9
checklist 28–32
competitive values 11
dynamic application 11–12
economic rationality 6–7, 9
human relations approach 7–8, 9
means-goal relationship 10–11
models 11–12
multiple criteria 9–12
organic system model 7, 9
political model 8–9

parental involvement 121, 122–3, 136, 234–5
per pupil expenditure (PPE) 217, 237, 305
PPE see per pupil expenditure
primary education
quality 182–87
stability 83–5
subject consistency 88–9
production process, school economics 4–6
publication bias 71

QAIT/MACRO model 47–50

quality of instruction 40, 41
  model 41–3
quantitative research syntheses 211–38

random effects model 74
reading literacy data 239–61
Reading Literacy Study (RLS) 243–4
recursive model 66–7
reinforcement 134, 136
  elements 133
retroactive planning 280
RLS see Reading Literacy Study

Scheerens model 46
school characteristics 154, 155
school climate 112–17, 135
school consensus 109
school effectiveness
  checklist 28–32
  criterion definition 50–3
  enhancing factors 224
  equal educational opportunity 141
  evaluation models 11–12
  implicit definitions 53–5
  model 26–8
  redirection 299–321
  research 152, 153
  statistical models 55–8
  theories 265–97
school effects
  consistency 80–3, 89–96
  differential 92–5
  magnitude 69–79
  stability 80–8, 91–6
  subject-specific 88–91
  unidimensionality 80–96
school improvement programs 155
  evaluation 145
school leader 103–6, 107
school outputs, definition 5–6
school-level organization 224
Scotland, school effects stability 85–6
secondary education
  school effects 85–7
  subject consistency 89–91
Slavin/Stringfield model 47–50
stability, school effects 80–8, 91–6
staff, cohesion and consensus 108, 135
statistical models 55–8
statistical procedures, unsound 72
structured instruction 125, 128–31, 136
synergetic model 65

teachers, effectiveness 84, 146
teaching methods, effectiveness 146
Third World, school effects, meta-analysis 76–8

UK
  see also England, Scotland
  school effects, meta-analysis 76–8
  school effects stability 84
  subject consistency 88–9, 90–1
uncertainty, primary process 19
USA
  compensatory programs 144
  differential effectiveness 92–3
  school effects stability 83
  subject consistency 88–9

Victorian quality of schools 187–91

Walberg model 39, 41

Lewis & Clark College - Watzek Library
3 5209 00643 5107